普通高等教育“十三五”规划教材
新工科建设之路·计算机类规划教材

Android 应用开发实践教程

（第 2 版）

韩 冬 编著

電子工業出版社
Publishing House of Electronics Industry
北京·BEIJING

内 容 简 介

本书基于 Android Studio，结合 Android 应用开发的一些具体案例，由浅入深、循序渐进地阐述 Android 应用开发的基础知识、常用技巧和关键技术。对目前很流行的网络请求框架 OkHttp，以及 Retrofit 进行了较多篇幅的探讨和说明，给出了下载文件和上传文件的完整代码（包括服务器端），有较强的实用性。对服务器端接口编程、Android 异步处理技术、Fragment 与 Fragment 之间的数据传递、Fragment 与 Activity 的交互等市场急需的重要开发技能也进行了较详尽的讲述，以达到快速提升读者编程水平和实际项目开发能力的目的。

全书分基础篇、提高篇、开发篇和拓展篇四部分，共 15 章。

基础篇（1～6 章）包含开发环境搭建和 Android 应用、用户界面设计、自定义 View、Activity 和 Intent、后台服务与广播消息、数据存储与访问。

提高篇（7～10 章）包含 Java 并发编程、Fragment、Android 的一些异步处理技术、Android 网络应用。

开发篇（11～13 章）包含手机客户端解析 JSON、Maven、服务器端接口编程。

拓展篇（14～15 章）包含 MVP 模式、Java 设计模式。

本书内容翔实，实用性强，既可作为高校计算机专业学生学习 Android 应用开发的教材，也可作为软件培训机构和编程人员的参考书。

图书在版编目（CIP）数据

Android 应用开发实践教程 / 韩冬编著. — 2 版. — 北京：电子工业出版社，2020.2

ISBN 978-7-121-37818-8

Ⅰ. ①A… Ⅱ. ①韩… Ⅲ. ①移动终端－应用程序－程序设计－高等学校－教材 Ⅳ. ①TN929.53

中国版本图书馆 CIP 数据核字（2019）第 248351 号

责任编辑：戴晨辰　　文字编辑：底　波

印　　刷：北京盛通商印快线网络科技有限公司

装　　订：北京盛通商印快线网络科技有限公司

出版发行：电子工业出版社

　　　　　北京市海淀区万寿路 173 信箱　　邮编：100036

开　　本：787×1 092　1/16　印张：23　字数：589 千字

版　　次：2016 年 1 月第 1 版

　　　　　2020 年 2 月第 2 版

印　　次：2020 年 2 月第 1 次印刷

定　　价：69.00 元

凡所购买电子工业出版社图书有缺损问题，请向购买书店调换。若书店售缺，请与本社发行部联系，联系及邮购电话：（010）88254888，88258888。

质量投诉请发邮件至 zlts@phei.com.cn，盗版侵权举报请发邮件至 dbqq@phei.com.cn。

本书咨询联系方式：dcc@phei.com.cn。

前　言

本书基于 Android Studio，结合 Android 应用开发的一些具体案例，由浅入深、循序渐进地阐述 Android 应用开发的基础知识、常用技巧和关键技术。全书分基础篇、提高篇、开发篇和拓展篇四部分，共 15 章，其中各个章节的主要内容及教学注意点介绍如下。

第 1 章介绍开发环境搭建和 Android 应用，具体内容包括 Android 体系结构、搭建 Android 应用开发环境、SDK Manager 和 AVD Manager 的使用，创建并运行第一个 Android 应用、Android 项目结构、Android 应用的基本组件等。强调的教学知识点有：①对 Gradle Wrapper 的理解；②项目全局 build.gradle 和模块 build.gradle；③Activity 的生命周期；④Android 日志工具 Log 的使用。

第 2 章介绍用户界面设计。强调的教学知识点有：①Handler 消息传递机制（在教学过程中可引入对观察者模式的介绍）；②循环对象 Looper 的作用；③在子线程中创建 Handler 为何会抛出异常；④RecyclerView 的使用；⑤RecyclerView 和 ListView 的区别。

第 3 章介绍自定义 View，属选学内容，可根据学生情况酌情处理，但在相关面试中常常会问到这方面的问题。

第 4 章介绍 Activity 和 Intent。强调的教学知识点有：①Activity 的四种启动模式；②使用 Bundle 在 Activity 之间交换数据；③调用另一个 Activity 并返回结果；④显式 Intent 和隐式 Intent。

第 5 章介绍后台服务与广播消息。强调的教学知识点有：①Service 和 Thread 的区别是什么？为什么使用 Service？②Service 和 BroadcastReceiver 的使用。

第 6 章介绍数据存储与访问。强调的教学知识点有：①SQLite 数据库的使用（包括代码建库和手动建库）；②ContentProvider 的使用（包括创建数据提供者和使用数据提供者）；③对于内部存储和外部存储的理解；④Android 外部存储的公有存储空间和私有存储空间；⑤Android 的权限机制（如危险权限和普通权限的区别、在程序运行时申请权限等问题）；⑥SharedPreferences 存储。

第 7 章介绍 Java 并发编程，属教学重点章节，是成为 Android 开发者的重要基础，涉及 Android 如何异步访问网络数据，以及对所使用网络请求框架的理解。强调的教学知识点有：①Java 并发访问框架（JDK 中 java.util.concurrent 包的类使用）；②使用 LruCache 缓存图片。

对学有余力且学习态度端正的同学，可引导学生深入学习，如网络请求框架的设计与实现、ImageLoader 的设计与实现等。

第 8 章介绍 Fragment，属选学内容，可根据学生情况酌情处理。不过这部分内容对于 Android 开发者来说也是十分重要的。强调的教学知识点有：①Fragment 的生命周期；②Fragment 与 Activity 的交互；③Fragment 间的数据传递。

第 9 章介绍 Android 的一些异步处理技术，属选学内容，可根据学生情况酌情处理。具体内容包括 HandlerThread 的使用及源码分析、IntentService 的使用及源码分析、AsyncTask 的使用及工作原理。源码分析部分还是比较难的，只有认真研究相关问题的源码，才能理解得更深入透彻。教师可引导学生逐步养成遇到问题去查阅并钻研 Android 源码的习惯，这样有助于学生学习能力的提升。

第 10 章介绍 Android 网络应用，属教学重点章节。强调的教学知识点有：①使用 OkHttp 访问网络；②使用 Retrofit 访问网络；③使用 HttpURLConnection 访问网络。

第 11 章介绍如何在手机客户端解析 JSON，这是编程实践部分。强调的教学知识点有：①在手机客户端解析 JSON；②服务器端生成 JSON 数据。

第 12 章介绍 Maven，具体内容有 Maven 的安装和配置、Maven 的使用、坐标和依赖等，属选学内容，可根据学生情况及教学课时情况酌情处理。

第 13 章介绍服务器端接口编程，这是编程实践部分。强调的教学知识点有：①如何在服务器端开发 JSON 接口；②JNDI 数据源配置。

第 14 章介绍 MVP 模式，具体内容有 MVP 模式的基本概念、MVP 模式与 MVC 模式的区别、MVP 模式的优点和缺点、MVP 模式的使用示例等，属选学内容，可根据学生情况及教学课时情况酌情处理。

第 15 章介绍 Java 设计模式，具体内容有设计模式的分类和设计原则、创建模式、结构模式、行为模式等，属选学内容，可根据学生情况及教学课时情况酌情处理。

本书的编写思路基于以下教学理念：

（1）突出编程基础；

（2）突出能力本位；

（3）突出学以致用；

（4）突出内外兼修；

（5）突出编程实践。

需要说明的是，本书的定位是卓越工程师培养创新，既可面向在校学生，也可面向广大 Android 程序员、技术人员和培训机构等。

给授课教师的教学建议如下表所示，打星号的章节为可选章节。给高校学生实施教学可安排 48～72 课时，应根据具体情况灵活安排。

具体章节	建议课时	是否为可选章节	可选章节建议课时
第 1 章 开发环境搭建和 Android 应用	2		
第 2 章 用户界面设计	8		
*第 3 章 自定义 View		可选	4
第 4 章 Activity 和 Intent	4		
第 5 章 后台服务与广播消息	6		
第 6 章 数据存储与访问	8		
第 7 章 Java 并发编程	6		
*第 8 章 Fragment		可选	6
*第 9 章 Android 的一些异步处理技术		可选	4
第 10 章 Android 网络应用	10		
第 11 章 手机客户端解析 JSON	4		
*第 12 章 Maven		可选	2
第 13 章 服务器端接口编程	4		
*第 14 章 MVP 模式		可选	2
*第 15 章 Java 设计模式		可选	2
合计	52		20

本书由韩冬（苏州大学）负责编写提纲并进行主要撰写，参加本书编写的还有张建、曹国平、肖

广娣等。

本书在编写过程中，参考、借鉴了很多 IT 技术专家、学者的相关著作，对于引用的段落或文字尽可能一一列出，谨向各位IT技术专家、学者表示感谢！

本书提供的教学资源有：所有章节的程序源码、PPT 课件和模拟试卷（试题）。读者可在电子工业出版社华信教育资源网进行下载，网址为 http://www.hxedu.com.cn。

鉴于编著者水平有限，书中难免存在不足和错误之处，敬请专家和读者提出宝贵意见和建议，以便再版时改进。

韩 冬

目　录

第一部分　基础篇

第二部分 提高篇

第三部分 开发篇

第四部分 拓展篇

第一部分

基　础　篇

第 1 章　开发环境搭建和 Android 应用

本章导读

Android 是 Google 公司于 2007 年 11 月 5 日发布的基于 Linux 内核的移动平台，该平台早期由 Google 公司开发，后由开放手持设备联盟（Open Handset Alliance）开发。本章作为全书导引，主要知识点有：（1）Android 体系结构介绍和如何搭建 Android 开发环境（包括安装 Android Studio）；（2）创建并运行第一个 Android 应用（包括 SDK Manager 和 AVD Manager 的介绍）；（3）Android 项目结构分析（包括 Project 工具窗口、工程目录说明、app 目录结构、项目全局 build.gradle 和模块 build.gradle 等内容）；（4）Android 应用的基本组件介绍；（5）Activity 的生命周期；（6）Android 日志工具 Log。

1.1　Android 简介

1.1.1　Android 传奇

如今，Android 及其绿色小机器人标志和 iPhone 的苹果标志一样风靡世界，掀起了移动领域最具影响力的风暴。创造这个奇迹的人叫 Andy Rubin（安迪 • 鲁宾），他是 Google 公司的工程副总裁，Android 开发的领头人。

2002 年年初，Andy Rubin 曾在斯坦福大学的工程课上做了一次讲座，此时他还在 Danger 公司工作，听众中有 Google 公司的两位创始人 Larry Page 和 Sergey Brin，互联网手机的理念深深打动了 Larry Page，尤其是他注意到 Danger 公司的产品上默认的搜索引擎就是 Google。

离开 Danger 公司后，Andy Rubin 曾再次隐居开曼群岛，想开发一款数码相机，但是没有找到支持者。他很快又回到熟悉的领域，创办了 Android，开始启动下一代智能手机的开发。这次开发的宗旨是设计一款对所有软件开发者开放的移动平台。2005 年，Andy Rubin 靠自己的积蓄和朋友的支持，艰难地完成了这个项目。在与一家风投公司洽谈的同时，Andy Rubin 突然想到了 Larry Page，于是给他发了一封邮件。仅仅几周时间，Google 就完成了对 Android 的收购。

Andy Rubin 是 Geek（极客）文化的代表。在硅谷他的半山别墅里，从视网膜扫描门到世界上最贵的门铃，创意无所不在。他是那种既喜欢电焊枪，也着迷编写程序，并擅长业务战略的奇才。

Andy Rubin 创立了两个手机操作系统的公司：Danger 和 Android。Danger 以 5 亿美元卖给了微软公司，并于 2013 年成为 Kin。Android 以 4000 万美元卖给了 Google 公司，Google 公司在 2005 年 8 月收购了这个仅成立 22 个月的高科技企业，从此 Android 系统也开始由 Google 公司接手研发。

2007 年 11 月，开放手持设备联盟成立，由 Google 公司发起。

2008 年 9 月，Google 公司正式发布了 Android 1.0 系统，这也是 Android 最早的版本。

2010 年 10 月，Google 公司宣布 Android 系统达到了第一个里程碑，即在电子市场上获得官方数字认证的 Android 应用数量已经达到 10 万个。Android 系统的应用增长非常迅速，在 2010 年 12 月，Google 公司正式发布了 Android 2.3 操作系统 Gingerbread（姜饼）。

2011 年 7 月，Android 系统设备的用户数量达到 1.35 亿，Android 系统成为智能手机领域占有率最高的系统。

2011 年 8 月 2 日，Android 手机已占据全球智能手机市场 48%的份额，并在亚太地区市场占据统治地位，终结了 Symbian（塞班）系统的霸主地位，跃居全球第一。

1.1.2 Android 体系结构

Android 系统采用了分层架构的思想，如图 1-1 所示。从顶层到底层分别是应用层、应用框架层、系统运行库层及 Linux 内核层。

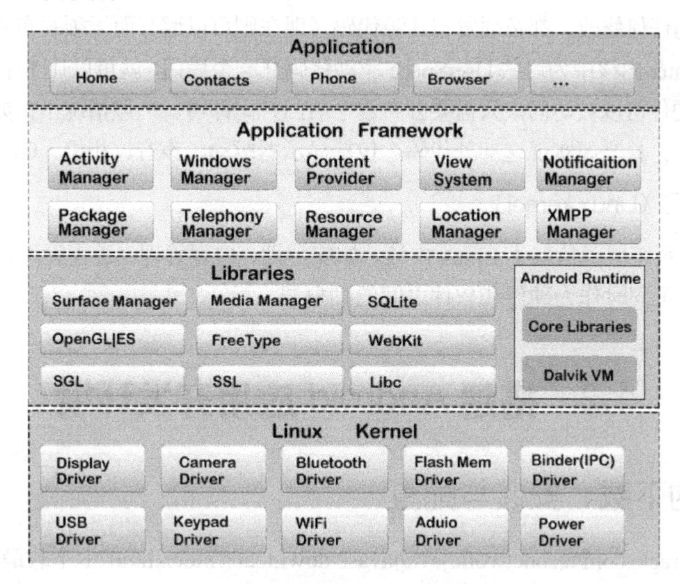

图 1-1　Android 体系结构

1．应用层（Application）

Android 平台不仅是操作系统，也包含许多应用程序，如 SMS 短信客户端程序、电话拨号程序、图片浏览器、Web 浏览器等应用。这些应用程序都是用 Java 语言编写的，并且可以被开发人员开发的其他应用程序所替换。

2．应用框架层（Application Framework）

该层是 Android 应用开发的基础。应用框架层包括活动管理器、窗口管理器、内容提供者、视图系统、包管理器、电话管理器、资源管理器、位置管理器、通知管理器和 XMPP 服务 10 个部分。在 Android 平台上，开发人员可以完全访问核心应用程序所使用的 API 框架，并且任何一个应用程序都可以发布自身的功能模块，而其他应用程序可以使用这些已发布的功能模块。基于这样的重用机制，用户可以方便地替换平台本身的各种应用程序组件。

3．系统运行库层（Libraries）

该层可以分成两部分，即系统库和 Android 运行时。

系统库是应用程序框架的支撑，是连接应用框架层与 Linux 内核层的重要纽带。系统库包括 9 个子系统，分别是图层管理、媒体库、SQLite、OpenGLES、FreeType、WebKit、SGL、SSL 和 Libc。

Android 运行时包括核心库和 Dalvik 虚拟机，前者既兼容了大多数 Java 语言所需要调用的功能函数，又包括了 Android 的核心库，如 android.os、android.net、android.media 等。Dalvik 虚拟机是一种基于寄存器的 Java 虚拟机，它主要完成对生命周期、堆栈、线程、安全和异常的管理及垃圾回收等重要功能。

每个 Android 应用程序都在自己的进程中运行，且都拥有一个独立的 Dalvik 虚拟机实例。Dalvik 虚拟机被设计成一个设备，可以同时高效地运行多个虚拟系统。它执行.dex 的 Dalvik 可执行文件，该格式文件针对小内存使用进行了优化，同时虚拟机是基于寄存器的，所有的类都经由 Java 编译器编译，然后通过 SDK 中的 dx 工具转化成.dex 格式，再由虚拟机执行。Dalvik 虚拟机依赖于 Linux 内核的一些功能，如线程机制和底层内存管理机制。

还有一个硬件抽象层在图 1-1 中没有显示出来。Android 并非将所有的设备驱动都放在 Linux 内核里，而是实现在 UserSpace 空间中，这么做的主要原因是 GPL 协议，Linux 是遵循该协议来发布的，也就意味着对 Linux 内核的任何修改，都必须发布其源代码（现在则可以避开而无须发布其源代码，毕竟它是用来赚钱的）。在 Linux 内核中为这些 UserSpace 驱动代码开一个后门，就可以让本来由 UserSpace 驱动，不可以直接控制的硬件可以被访问。只需要公布这个后门代码即可。一般情况下，如果要将 Android 移植到其他硬件去运行，只需要实现这部分代码，包括显示器驱动、声音、相机、GPS、GSM 等。

4. Linux 内核层（Linux kernel）

Android 的核心系统服务依赖于 Linux 内核，如安全性、内存管理、进程管理、网络协议栈和驱动模型。Linux 内核也可同时作为硬件和软件之间的抽象层。

1.2 搭建 Android 应用开发环境

1.2.1 JDK 的下载、安装与配置

在 http://www.oracle.com/technetwork/java/javase/downloads/index.html 中下载最新版本的 JDK，这里使用的是 JDK8（jdk-8u92-windows-x64.exe），按默认安装路径进行 JDK 安装即可。

设置环境变量如下：

```
JAVA_HOME= C:\Program Files\Java\jdk1.8.0_92
CLASSPATH=.;%JAVA_HOME%\lib\dt.jar;%JAVA_HOME%\lib\tools.jar
PATH=%PATH%;%JAVA_HOME%\bin
```

注意：%PATH%为原来的环境变量值，添加“;”和后面的内容到原来值的后面。

验证是否配置成功，可点击“开始”→“运行”，输入 cmd 打开命令行窗口，并输入 java -version，显示版本为 1.8.0_92，如图 1-2 所示。说明 JDK 安装及环境变量配置成功。

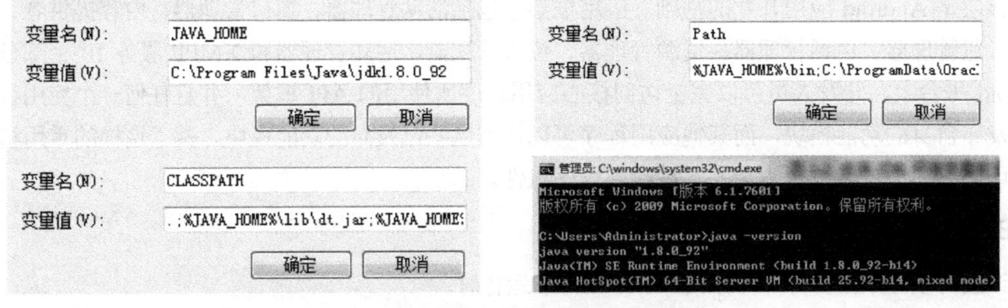

图 1-2 JDK 安装及环境变量配置和验证

1.2.2 安装 Android Studio

在相应的网站下载 Android Studio 版本，如网站 http://www.android-studio.org/或 https://developer.android.google.cn/studio/。

以 Android Studio 3.x 为例，其安装过程如图 1-3 所示。

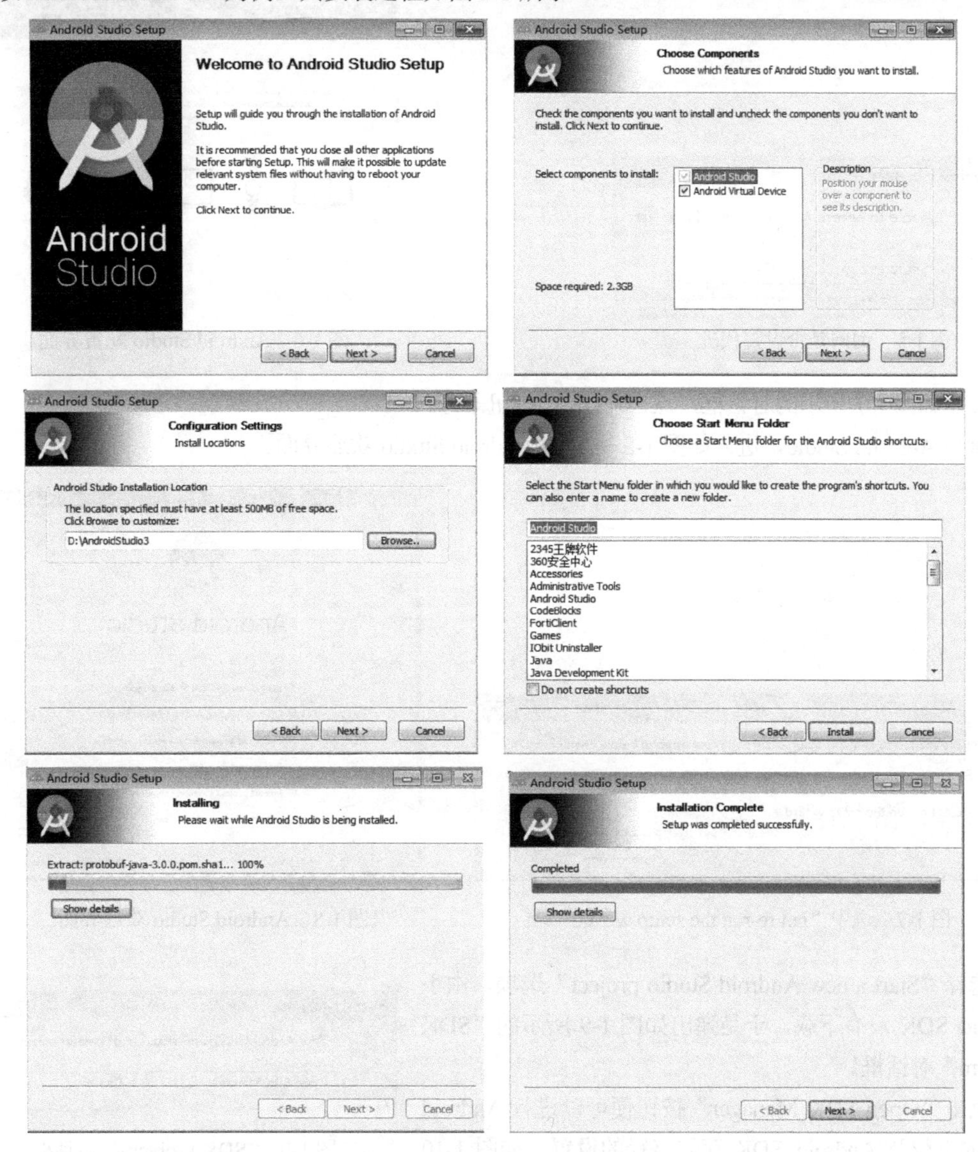

图 1-3　安装 Android Studio

完成安装后，启动“Android Studio”图标，弹出如图 1-4 所示的对话框，询问是否“导入 settings”，勾选“Do not import settings”项。

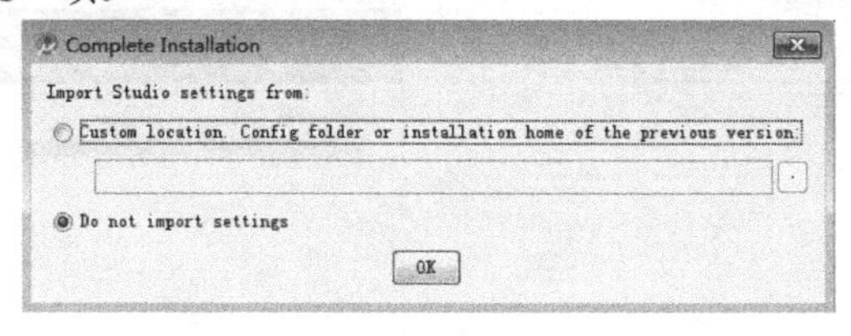

图 1-4　“Complete Installation”对话框

弹出如图 1-5 所示的对话框，询问是否设置代理，点击“Cancel”按钮。这时进入如图 1-6 所示的 Android Studio 配置界面，点击“Next”按钮进行具体配置，有 Standard 和 Custom 两个选项：Standard 表示使用默认配置，Custom 表示自定义。这里点击“Cancel”按钮，目的是以后专门设置 Android SDK 的存储路径，如果用默认 Android SDK 的存储路径，则不易查找，也不方便 Android SDK 的管理。

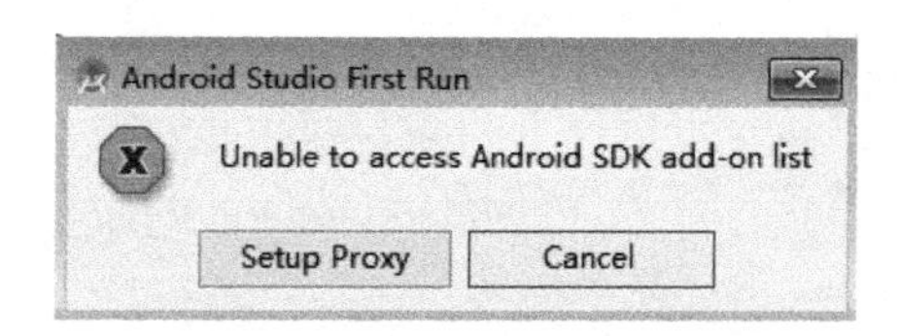

图 1-5　询问是否设置代理

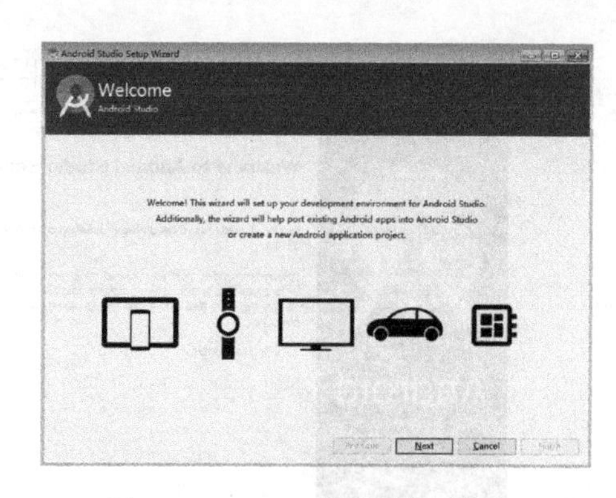

图 1-6　Android Studio 配置界面

弹出如图 1-7 所示的对话框，点击“OK”按钮即可。

重启 Android Studio，进入如图 1-8 所示的 Android Studio 欢迎界面。

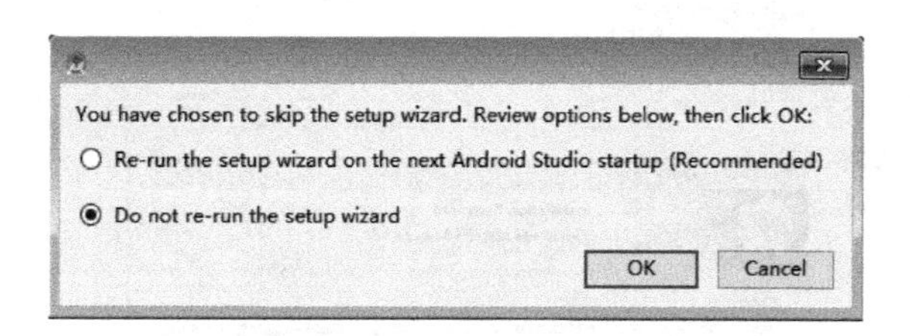

图 1-7　选中“not re-run the setup wizard”项

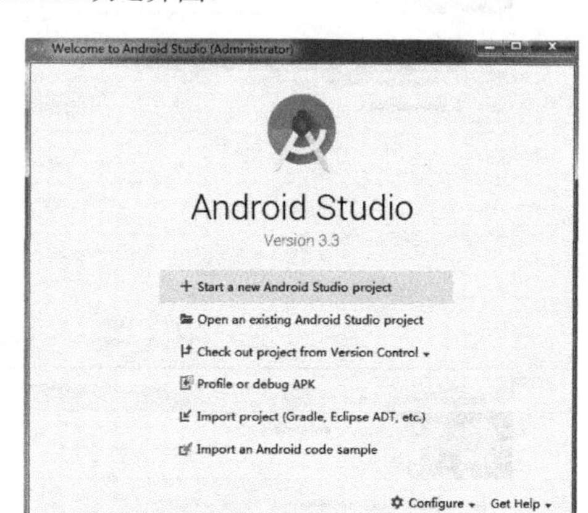

图 1-8　Android Studio 欢迎界面

选择“Start a new Android Studio project”选项，由于 Android SDK 没有下载，于是弹出如图 1-9 所示的“SDK Problem”对话框。

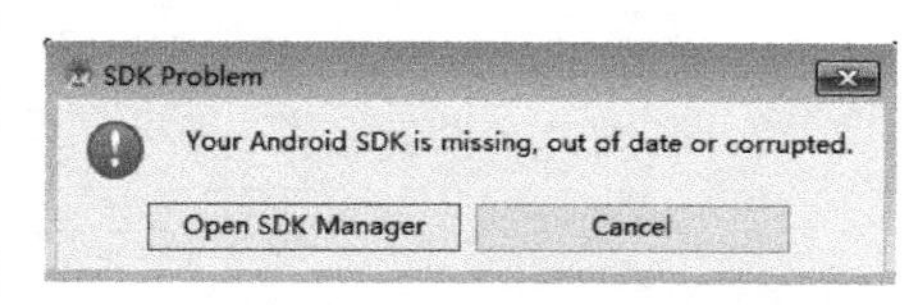

图 1-9　“SDK Problem”对话框

点击“Open SDK Manager”按钮便可以进行 Android SDK 的下载及 Android SDK 存储路径的设置，如图 1-10 所示。Android SDK 的存储路径是“D:\AndroidSDK”。

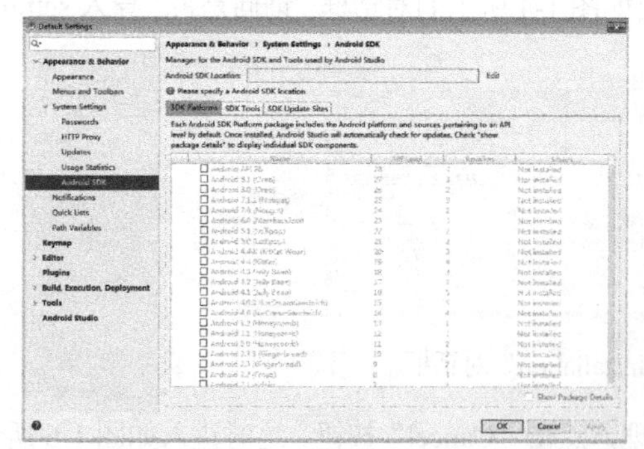
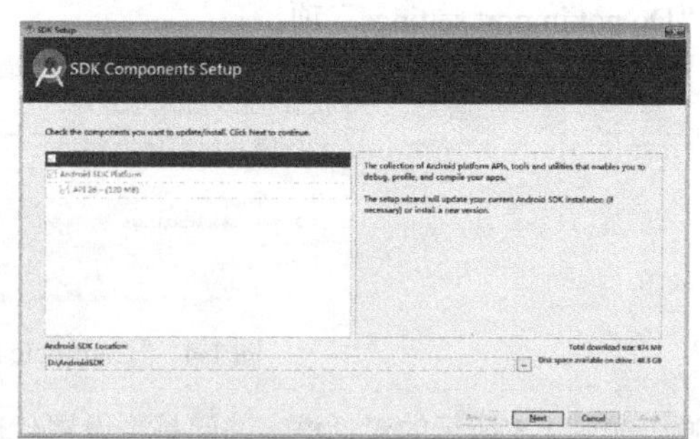

图 1-10　Android SDK 的下载和存储路径的设置

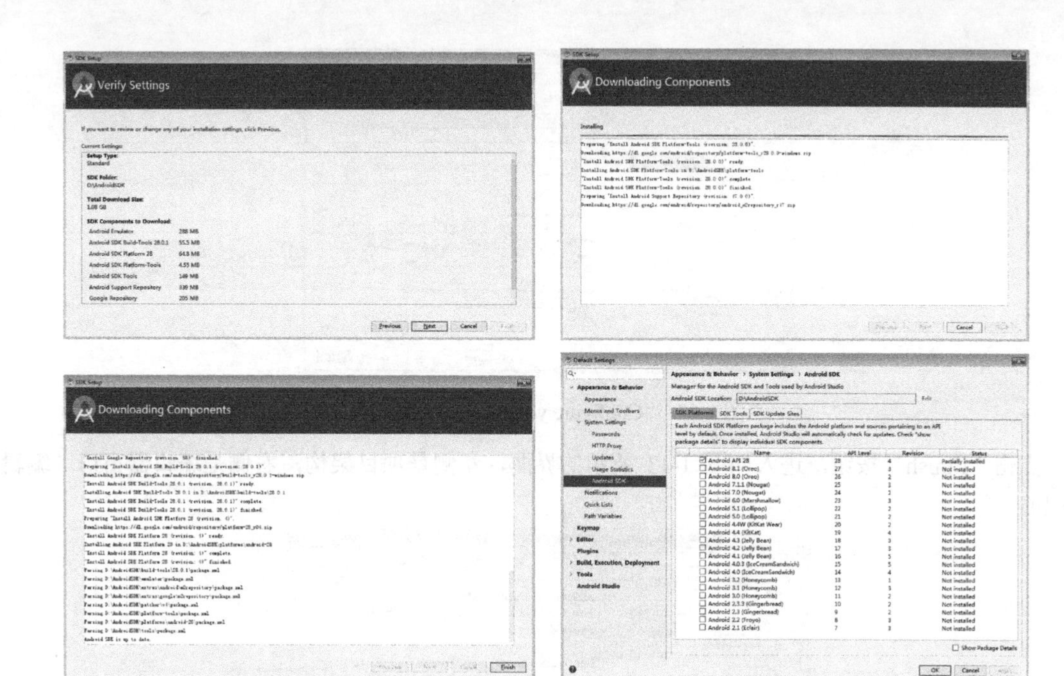

图 1-10　Android SDK 的下载和存储路径的设置（续）

1.3　第一个 Android 应用

1.3.1　创建 MyHello 项目

启动 Android Studio，出现如图 1-8 所示的 Android Studio 欢迎界面，选择“Start a new Android Studio project”选项，进入如图 1-11 所示的“Choose your project”对话框，项目类型选择“Phone and Tablet”，Activity 类型选择“Empty Activity”。

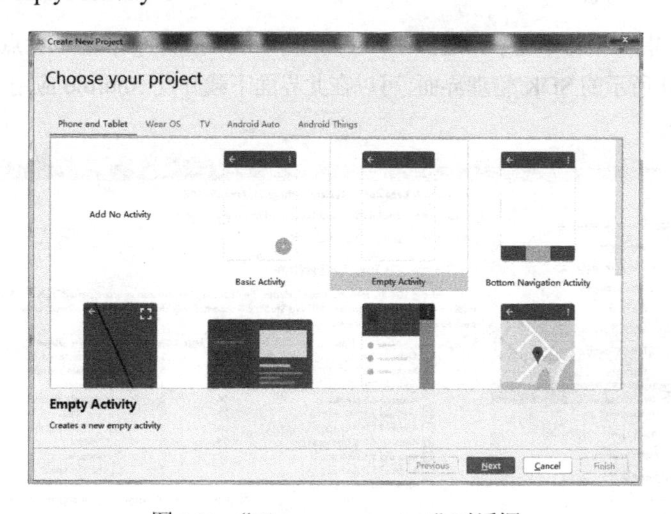

图 1-11　“Choose your project”对话框

点击“Next”按钮，进入“Configure your project”对话框，如图 1-12 所示。在 Name 中输入“MyHello”，Package name 中输入“com.mialab.myhello”，Save location 中输入“工程文件夹\MyHello”。

这里工程文件夹的路径是 D:\Android_Prj_book，用来存储 Android Project。

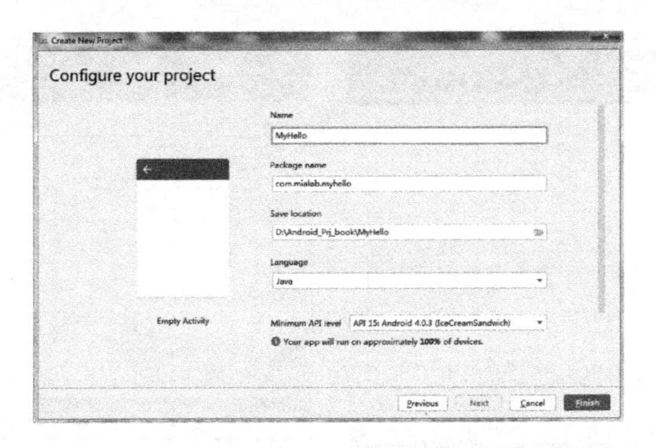

图 1-12　“Configure your project”对话框

点击“Finish”按钮，进入如图 1-13 所示的界面，左侧是项目架构，右侧是编辑器窗口（编辑代码）。

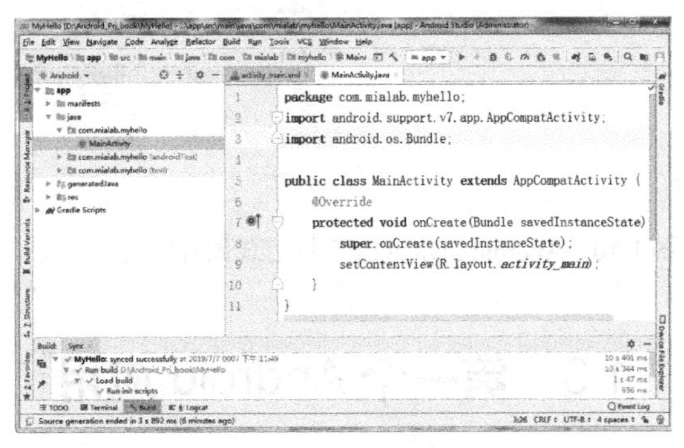

图 1-13　MyHello 项目创建成功

在图 1-13（Android Studio 集成开发环境）的右上方是一些工具栏按钮，其中，经常使用的有运行 App 按钮▶、AVD Manager 按钮和 SDK Manager 按钮。

1.3.2　SDK Manager

在 Android Studio 集成开发环境中，选择“Tools”→“SDK Manager”，或者点击 SDK Manager 按钮，弹出如图 1-14 所示的 SDK 管理界面，可以在此界面下载开发 Android 应用所需的 Android SDK 和相应的 SDK Tools。

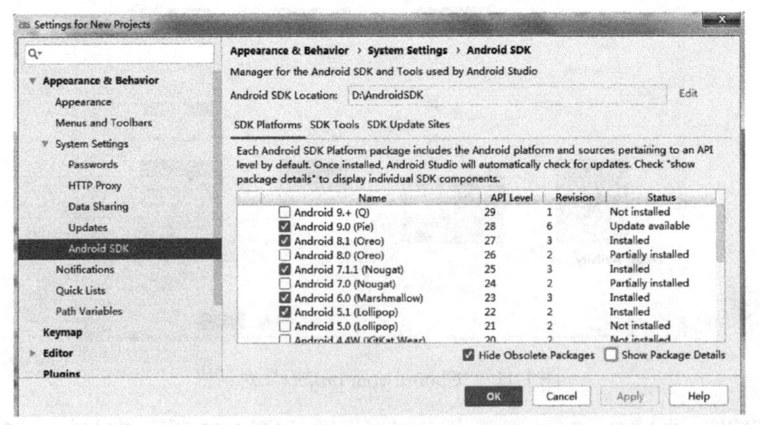

图 1-14　SDK 管理界面

勾选“Show Package Details”项，如图 1-15 所示。根据开发需要，选择下载相应的 Android SDK，并点击“Apply”按钮开始下载。

为了运行 Android 虚拟设备，应下载 SDK 包中相应的 System Image。在图 1-15 中勾选 Android 9.0（Pie）包中的“Google APIs Intel x86 Atom System Image”项进行下载，这是创建运行 Android 9.0 虚拟设备（如手机模拟器）所必备的。

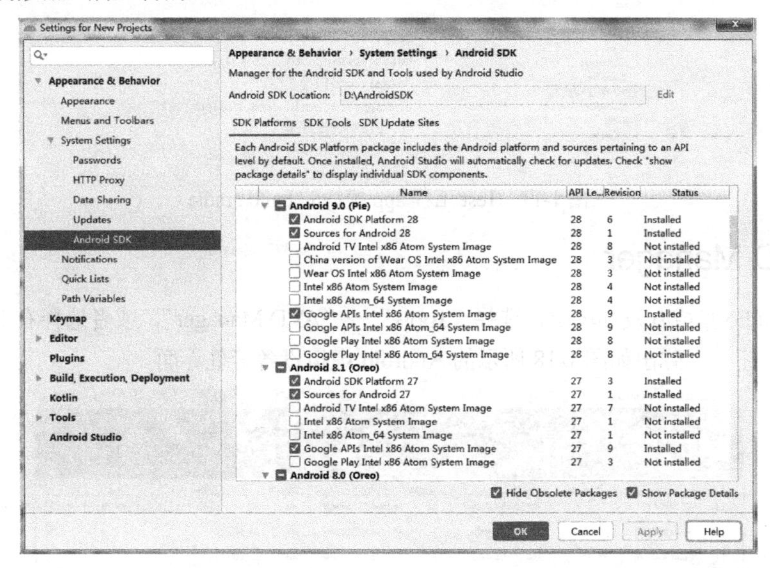

图 1-15　勾选“Show Package Details”项

选择“SDK Tools”选项，弹出“SDK Tools 标签页”界面，如图 1-16 所示，根据开发需要，可选择相应 buildToolsVersion 的 Android SDK Build-Tools 29 下载，如在新升级版本的 Android Studio 中打开原先低版本的 Android Studio 项目。

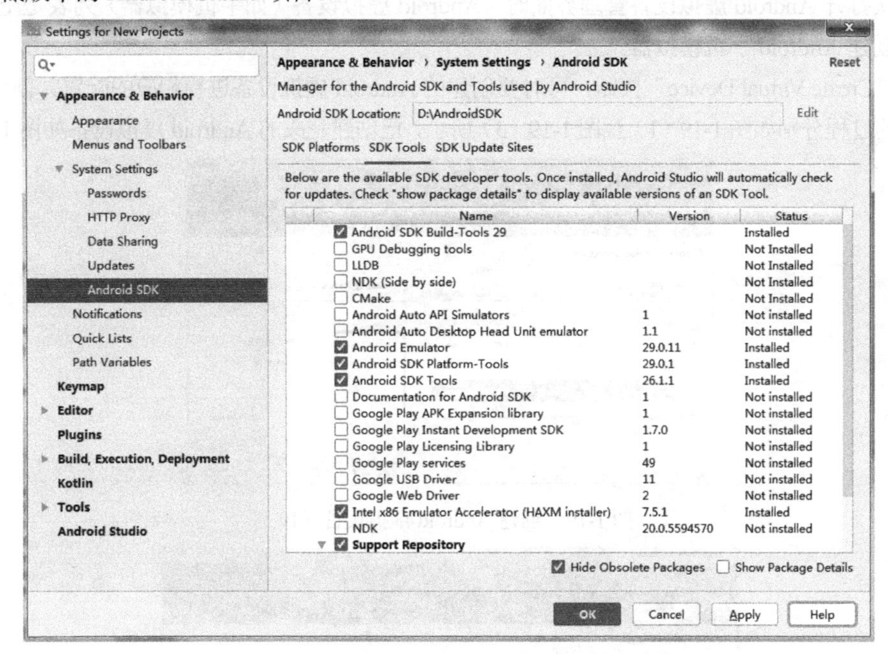

图 1-16　“SDK Tools 标签页”界面

在 Android Studio 项目（工程）的模块（如工程中的 app 模块）build.gradle 文件中，会有 buildToolsVersion 的声明，如图 1-17 所示。

为了运行 Intel x86 模拟器（如手机模拟器），还应下载“Intel x86 Emulator Accelerator(HAXM installer)”，所以需要勾选“Intel x86 Emulator Accelerator(HAXM installer)”项。

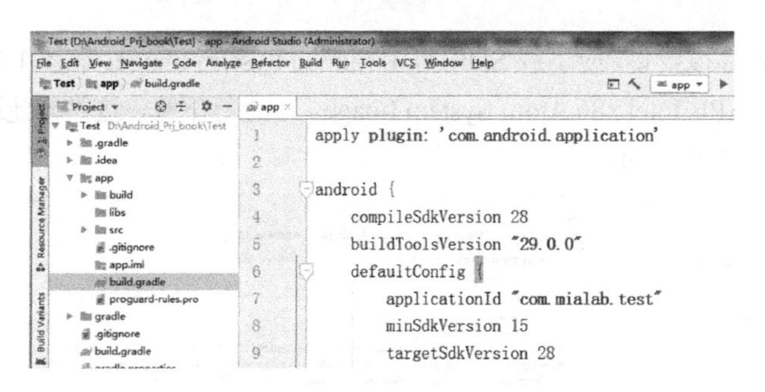

图 1-17　Test 工程 app 模块的 build.gradle

1.3.3　AVD Manager

在 Android Studio 集成开发环境中，选择“Tools”→“AVD Manager”，或者选择右上角工具栏中的 AVD Manager 按钮，弹出如图 1-18 所示的 Android 虚拟设备管理界面。

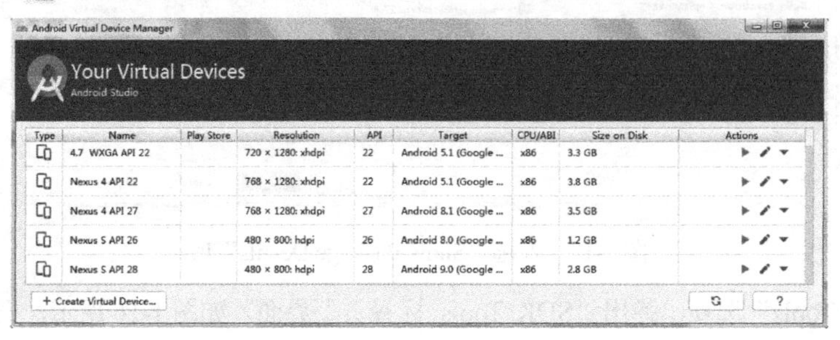

图 1-18　Android 虚拟设备管理界面

第一次打开 Android 虚拟设备管理界面时，Android 虚拟设备（如手机模拟器）列表是空的，可以在此界面创建 Android 手机模拟器。

点击“Create Virtual Device”按钮，可对要创建的 Android 虚拟设备进行相应的配置。创建 Android 虚拟设备的过程分别如图 1-19（1）至图 1-19（3）所示，已创建完成的 Android 虚拟设备如图 1-20 所示。

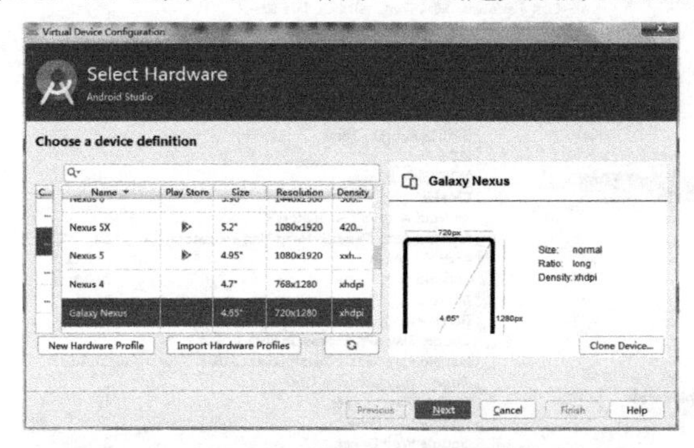

图 1-19　创建 Android 虚拟设备（1）

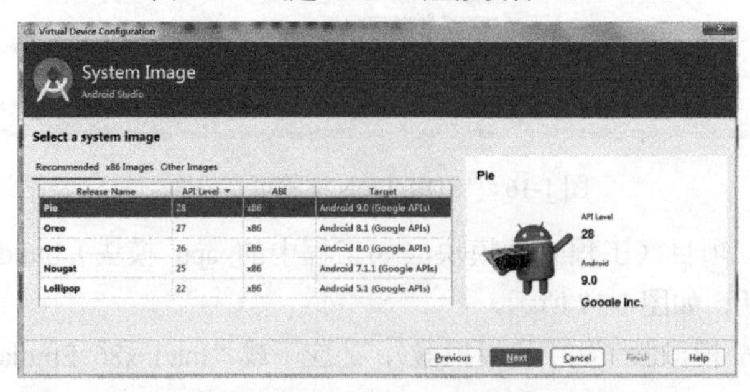

图 1-19　创建 Android 虚拟设备（2）

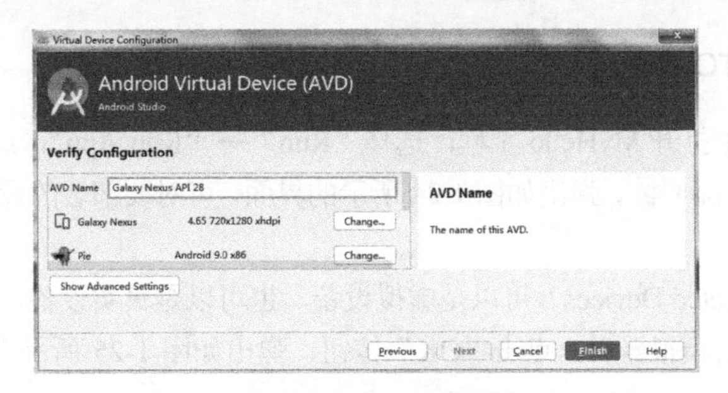

图 1-19　创建 Android 虚拟设备（3）

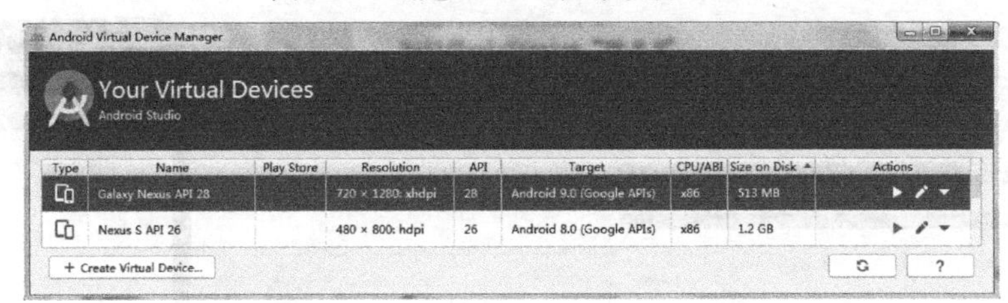

图 1-20　已创建完成的 Android 虚拟设备

点击 Actions 列下的运行按钮（Launch this AVD in the emulator），如图 1-21 所示。

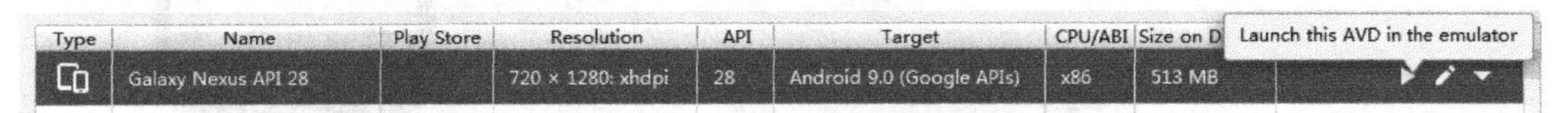

图 1-21　点击“Launch this AVD in the emulator”按钮

运行的 Android 虚拟设备（手机模拟器）如图 1-22 所示。

图 1-22　运行的 Android 虚拟设备（手机模拟器）

创建手机模拟器的时候，设置的屏幕分辨率不要太高，以免因内存问题而使手机模拟器无法运行或运行不畅。本示例测试所用的手机模拟器的屏幕分辨率为 720×1280（dpi）或 480×800（dpi）。作者在笔记本电脑上创建的手机模拟器列表如图 1-23 所示。

Android Virtual Device Manager

Your Virtual Devices

Android Studio

Type	Name	Play Store	Resolution	API	Target	CPU/ABI	Size on Disk	Actions
	4.7 WXGA API 22		720 × 1280: xhdpi	22	Android 5.1 (Google APIs)	x86	1.6 GB	
	Galaxy Nexus API 28		720 × 1280: xhdpi	28	Android 9.0 (Google APIs)	x86	3.1 GB	
	Nexus 4 API 22		768 × 1280: xhdpi	22	Android 5.1 (Google APIs)	x86	3.8 GB	
	Nexus 4 API 27		768 × 1280: xhdpi	27	Android 8.1 (Google APIs)	x86	3.5 GB	
	Nexus S API 26		480 × 800: hdpi	26	Android 8.0 (Google APIs)	x86	3.6 GB	
	Nexus S API 28		480 × 800: hdpi	28	Android 9.0 (Google APIs)	x86	2.8 GB	

图 1-23　作者创建的手机模拟器列表

1.3.4 运行 Android 应用

在 Android Studio 中打开 MyHello 工程，选择“Run”→“Run App”，或者在右上方工具栏上选择运行 App 按钮（Run App）▶，弹出如图 1-24 所示的界面，可对要部署的设备（这里指手机模拟器）进行选择。

这里的设备（Connected Devices）可以是虚拟设备，也可以是真实设备。

选择好要部署的手机模拟器后，点击“OK”按钮，弹出如图 1-25 所示界面，第一个 Android 应用运行成功。

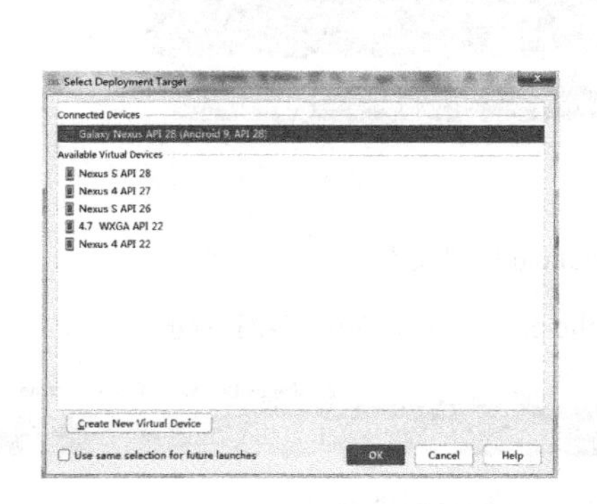

图 1-24 “Select Deployment Target”界面

图 1-25 运行第一个 Android 应用

1.4 Android 项目结构

Android Studio 默认的 IDE 布局是在左侧面板中展示 Project 工具窗口，也可通过菜单选择“View”→“Tool Windows”→“Project”，切换至 Project 工具窗口。

Android Studio 中用于导航的工具窗口有 Project、Structure、Favorites、TODO 等。

Project 工具窗口：允许浏览工程中的文件和资源。

Structure 工具窗口：显示当前文件中对象或元素的树形结构。

Favorites 工具窗口：显示收藏、书签和断点。

TODO 工具窗口：显示项目中所有有效的 TODO 列表。

1.4.1 Project 工具窗口

Project 工具窗口用途广泛且相对容易使用，是导航工具窗口中最有用的一个。要领会 Project 工具窗口的功能和范畴，需要将窗口的模式设置为 Project。Project 工具窗口的模式有 3 种：Project、Packages 和 Android。在 Android Studio 中，任何一个新建的项目都会默认使用 Android 模式的项目结构。

Android 和 Project 是比较有用的两种模式。Android 模式的项目结构简洁明了，适合进行快速开发，但它并不是项目真实的目录结构。只有将项目结构模式切换成 Project 模式，才会显示项目真实的目录结构。

Project 工具窗口提供了文件和嵌套目录的简单树形界面，可以切换显示。它为用户展示了项目中所有包、目录和文件的概况。如果在 Project 工具窗口中右击某个文件，则会出现一个上下文菜单。在这个上下文菜单中有 3 个重要的菜单项：Copy Path、File Path 和 Show in Explorer。选择“Copy Path”

项会将此文件在操作系统中的绝对路径复制到剪贴板；选择“File Path”项会以目录栈的形式显示路径，并以栈顶的文件结束，而点击任意这些目录都会在操作系统中打开；选择“Show in Explorer”项会在一个新的操作系统窗口中显示该文件。

1.4.2 工程目录说明

在 Android Studio 中打开 MyHello 工程，左侧 Project 导航工具窗口默认模式是 Android 模式，如图 1-26 所示。可将其项目结构模式切换成 Project 模式，如图 1-27 所示。

图 1-26 MyHello 工程的 Android 模式

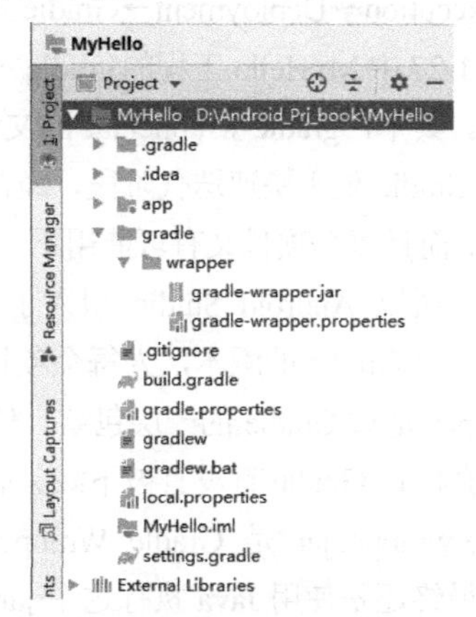

图 1-27 MyHello 工程的 Project 模式

在 Android Studio 中一个项目（Project）可以有多个模块（Module），这里 MyHello 是工程，app 是模块，模块可对应 Android 应用。

Android Studio 是基于 IntelliJ IDEA 的 IDE 工具，Android Studio 中的项目可视为对应 Eclipse ADT 的工作空间（Workspace），一个工作空间可以包含多个模块。Android Studio 中的项目可以包含多个 Android 应用（模块）。

小贴士：IDEA 全称 IntelliJ IDEA，是 Java 编程语言开发的集成环境。IntelliJ 在业界被公认为是最好的 Java 开发工具之一，尤其在智能代码助手、代码自动提示、重构、Java EE 支持、各类版本工具（git、svn 等）、JUnit、CVS 整合、代码分析、创新的 GUI 设计等方面的功能可以说是超常的。IDEA 是 JetBrains 公司的产品，这家公司总部位于捷克共和国的首都布拉格，开发人员以严谨著称的东欧程序员为主。它的旗舰版本可支持 HTML、CSS、PHP、MySQL、Python 等。免费版只支持 Python 等少数语言。

Gradle 是一个基于 Apache Ant 和 Apache Maven 概念的项目自动化构建开源工具。它使用一种基于 Groovy 的特定领域语言（DSL）来声明项目设置，目前增加了基于 Kotlin 语言的 kotlin-based DSL 功能，抛弃了基于 XML 的各种烦琐配置。Android Studio 是使用 Gradle 来构建项目的。

1．.gradle

.gradle 是 Gradle 运行时生成的文件存放目录。Android Studio 自动生成这些文件，不需要程序员改动。在这个文件夹下保存 Gradle 的临时输出、缓存和其他支持性的元数据。

2．.idea

idea 是 IntelliJ IDEA 运行时生成的文件存放目录。Android Studio 自动生成这些文件，不需要程序员改动。

3．app（module）

app 用来存放项目代码和各种资源文件、运行配置文件 AndroidManifest.xml 等，这是本书学习的重点。Android Studio 项目中的 module 相当于 Eclipse 中的 Project。

4. gradle

gradle 是项目 Gradle 的配置文件，它可以配置 gradle-wrapper.jar 的网络路径，会自动根据本地缓存来决定是否需要下载 Gradle。gradle-wrapper.jar 也称为 Gradle 包装器，包含与当前项目兼容的一个 Gradle 运行时的版本。

Android Studio 默认没有启动 Gradle wrapper 的方式，可通过选择"File"→"Setting"→"Build"，并选择 Execution→Deployment→Gradle 进行配置。

在图 1-27 中 MyHello 工程的 gradle/wrapper 目录下有两个文件：gradle-wrapper.jar 和 gradle-wrapper.properties。其中，gradle-wrapper.jar 的文件类型为 Executable Jar File。

因为 Gradle 处于快速迭代阶段，经常发布新版本，如果用项目直接去引用，那么更改版本就会变得很麻烦，而且每个项目又有可能用不一样的 Gradle 版本，用手动配置每一个项目对应的 Gradle 版本也不轻松，所以 Android Studio 引入了 gradle-wrapper.jar 文件，通过读取配置文件（gradle-wrapper.properties）中 Gradle 的版本，为每个项目自动下载和配置。

Wrapper 是对 Gradle 的一层包装，便于在团队开发过程中统一构建 Gradle 的版本。Wrapper 启动 Gradle 时会检查 Gradle 有没有被下载关联，若没有就会从配置的地址下载并运行构建。

gradle-wrapper.jar 是 Gradle Wrapper 的主体功能包。它是具体业务逻辑实现的 jar 包，gradlewrapper.jar 最终还是使用 Java 执行这个 jar 包来进行相关的 Gradle 操作，而 gradle-wrapper.properties 文件是配置文件，用于配置 Gradle 信息，主要用于指定该项目需要什么版本的 Gradle、从哪里下载该版本的 Gradle，以及放到哪里。

gradle-wrapper.properties 文件的内容如下：

```
distributionBase=GRADLE_USER_HOME
distributionPath=wrapper/dists
zipStoreBase=GRADLE_USER_HOME
zipStorePath=wrapper/dists
distributionUrl=https\://services.gradle.org/distributions/gradle-5.1.1-all.zip
```

gradle-wrapper.properties 文件的配置字段说明。

- distributionBase：下载 Gradle 压缩包解压后存储的主目录。
- distributionPath：相对于 distributionBase 解压后的 Gradle 路径。
- zipStoreBase：类似于 distributionBase，但能存放压缩包。
- zipStorePath：类似于 distributionPath，但能存放压缩包。
- distributionUrl：发行版压缩包的下载地址。

在 Android Studio 安装时自带 gradle-wrapper.jar 文件和 gradle-wrapper.properties 文件，以及 gradlew 文件和 gradlew.bat 文件，分别如图 1-28 和图 1-29 所示。Android Studio 新建一个项目，就会将 AndroidStudio 安装路径\plugins\android\lib\templates \gradle\wrapper\gradle\wrapper 目录下的 gradle-wrapper.jar 和 gradle-wrapper.properties 两个文件复制到新建项目的 Gradle 文件夹中。

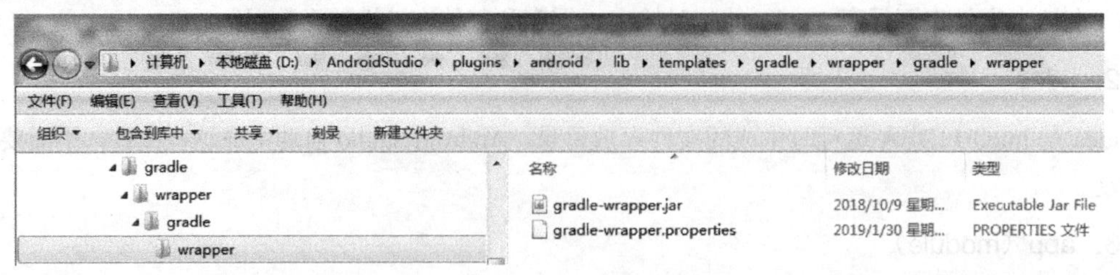

图 1-28 Android Studio 安装时自带 gradle-wrapper.jar 文件和 gradle-wrapper.properties 文件

图 1-29　Android Studio 安装时自带 gradlew 文件和 gradlew.bat 文件

gradle-wrapper 会自动在 C:\Users\<user_name>\.gradle\wrapper\dists 目录下查找对应版本的 Gradle 是否存在，如果没有，则根据 distributionUrl 的值下载相应版本的 Gradle。

gradle-wrapper.properties 文件中其他 4 个属性的作用就是指定下载 Gradle 存放的路径，其各项属性的整体含义如下：

（1）在 https://services.gradle.org/distributions/gradle-5.1.1-all.zip 中下载 Gradle 5.1.1 版本。gradle-xx-all.zip 是完整版，包含各种二进制文件、源代码文件和离线的文档。

（2）下载的 gradle-5.1.1-all.zip 存放到 C:\Users\<user_name>\.gradle\wrapper\dists 目录中。具体有 2 级目录，即全路径为 C:\Users\<user_name>\.gradle\wrapper\dists\gradle-5.1.1-all\<url-hash>，gradle-5.1.1-all 目录是根据下载 gradle 的文件名来定的，<url-hash>目录名是根据 distribution url 路径字符串计算 md5 值得来的，如图 1-30 所示。

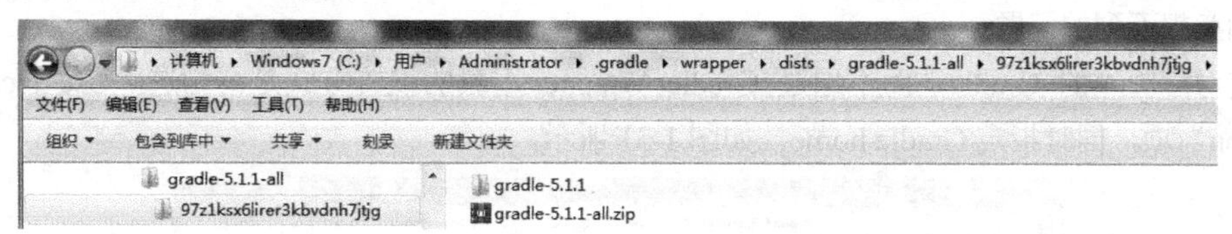

图 1-30　C:\Users\<user_name>\.gradle\wrapper\dists\gradle-5.1.1-all\<url-hash>\路径

（3）解压 gradle-5.1.1-all.zip，将解压后的文件存放到 C:\Users\<user_name>\.gradle\wrapper\dists 中。

小贴士： Android Studio 打开一个工程时，首先读取 gradle-wrapper.properties 文件，从而知道这个工程需要哪个版本的 Gradle，然后在文件夹 GRADLE_USER_HOME 中查找是否存在这个版本的 Gradle，不存在则去 distributionUrl 下载。

在打开低版本的 Android Studio 工程时，旧工程的 gradle-wrapper.properties 文件中对应的 Gradle 版本往往需要下载，但由于网络原因很可能出现卡顿。

解决方法：可以手动下载 Gradle，然后配置上去。

① 修改 gradle-wrapper.properties 文件中 distributionUrl 的值，将旧工程的 Gradle 版本改成想要的版本。

② 打开项目，Android Studio 将自动下载 Gradle，并创建好目录。这时可直接关闭 Android Studio 并退出。这一步是为了得到 Android Studio 自动创建的下载目录，如果是 Windows 系统，可定位到目录 C:\Users\<user_name>\.gradle\wrapper\dists（注意：Gradle 是共用的，一般不会把 Gradle 放在项目文件中）。

③ 接着进入对应 Gradle 版本的文件夹下，会发现有一个命名为乱码的文件夹，把手动下载对应版本的 Gradle（如 gradle-xx-all.zip）复制到命名为乱码的文件夹下，注意不要解压。重新开启 Android Studio，打开工程即可。

注意： Gradle 插件和 Gradle 是各自独立的，Gradle 插件版本是由项目最外层的 build.gradle 文件决定的。

下面仍需要搞清楚 Gradle、Gradle Wrapper 与 Android Plugin for Gradle 的区别和联系。

● Gradle 是个构建系统，能够简化项目的编译、打包、测试过程。Gradle 和 Maven 一样都是项目构建工具。

- Gradle Wrapper 的作用是简化 Gradle 本身的安装和部署。由于不同版本的项目可能需要不同版本的 Gradle，所以使用手工部署较麻烦，而且可能产生冲突，Gradle Wrapper 可有效解决这些问题，它是 Gradle 项目的一部分。
- Android Plugin for Gradle 是一系列适合 Android 开发的 Gradle 插件集合，主要由 Google 公司的 Android 团队开发，虽然 Gradle 不是 Android 的专属构建系统，但是有了 Android Plugin for Gradle 插件，使用 Gradle 构建 Android 项目变得更简单。
- Gradle、Gradle Wrapper 与 Android Plugin for Gradle 不一定要和 Android Studio 一起使用，它们可以完全脱离 Android Studio，使三者独立进行 Android 项目的构建。

下面是 Gradle、Gradle Wrapper、Android Plugin for Gradle 三者的官方指导文档，由它们的 URL 可以看出 Gradle Wrapper 是 Gradle 项目的一部分。

Gradle：https://docs.gradle.org/current/userguide/userguide_single.html。

Gradle Wrapper：https://docs.gradle.org/current/userguide/gradle_wrapper.html。

Android Plugin for Gradle：https://developer.android.com/studio/build/index.html。

是不是每个项目的 Gradle 都要通过 Gradle Wrapper 下载，所有的项目能不能共用一个 Gradle？理论上是可以的，但是由于 Gradle 本身不一定保持完全的兼容性，所以新老项目共用一个 Gradle 有时可能会遇到意想不到的问题。

指定对应版本的 Gradle，而不通过 Gradle Wrapper 下载的设置方式是：勾选“Use local gradle distribution”项，同时指定 Gradle home，如图 1-31 所示。

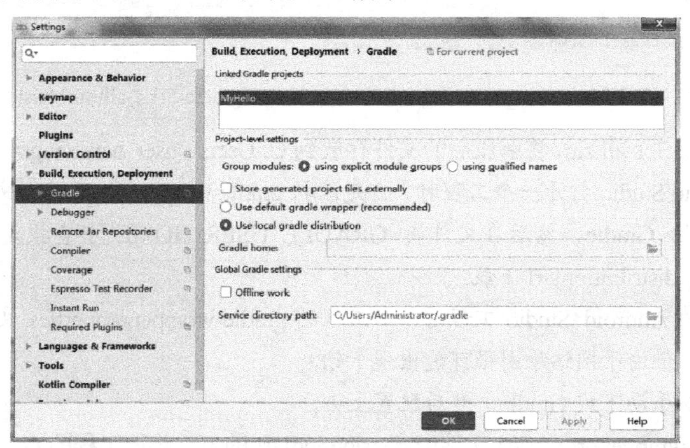

图 1-31　配置 Gradle

Gradle 对应版本下载完成后，Gradle Wrapper 的使命基本完成，Gradle 会读取顶层 build.gradle（工程级别的 build.gradle）文件，该文件中指定了该项目需要的 Android Plugin for Gradle 版本是什么，以及从哪里下载该版本。

小贴士：由顶层 build.gradle 文件知道，Android Plugin for Gradle 可从 google 和 jcenter 处下载，那么下载到本地的哪里呢？答案是它会下载到~\.gradle\caches\modules-2\files-2.1\com.android.tools.build 中。但有时候因网络原因，而选择（勾选）图 1-31 中的“Offline work”项时，可能出现“No cached version of com.android.tools.build:gradle:xxx available for offline mode”的提示，又该怎么办呢？此时只要将对应版本的 Android Plugin for Gradle 下载到本地的 C:\Program Files\Android\Android Studio\gradle\m2repository\com\android\tools\build 中即可。

5．.gitignore

.gitignore 是对项目文件管理的配置文件，用来记录排除在版本控制之外的特定目录或文件，或者说是不希望发布到 git 管理的文件。

6. gradlew 和 gradlew.bat

gradlew 和 gradlew.bat 用来在命令行中执行 gradle 命令，其中，gradlew 在 Linux 或 Mac 中使用；gradlew.bat 在 Windows 中使用。gradlew 和 gradlew.bat 用于通过包装器执行 Gradle，如果系统上没有安装 Gradle，或者没有构建兼容的版本，那么可使用 gradlew 或 gradlew.bat 来调用 Gradle。

7. build.gradle（项目全局）

build.gradle（项目全局）是项目自动编译环境配置文件，全局的 Gradle 构建脚本是对 Project 的 Gardle 进行配置，这里是所有模块通用的配置信息。每个 Module 中都有一个 build.gradle 文件对 Module 的 Gradle 进行配置。build.gradle 用来说明整体项目的构建逻辑，并负责引入所需的全部子项目及触发的构建。

8. gradle.properties

gradle.properties 是 Gradle 运行环境的配置文件，如配置 Gradle 的运行模式、运行时 jvm 虚拟机的大小。

9. local.properties

local.properties 是 Andorid NDK、SDK 环境的路径配置文件。

10. MyHello.iml

MyHello.iml 是在编译项目时自动生成的文件，命名为“项目名.iml”，用于标识一个 Intellij IDEA 项目。

11. settings.gradle

settings.gradle 用于指定项目中所有引入的模块。Android 项目是多项目 Gradle 构建的，settings.gradle 文件显示了每个子目录所对应的子项目。

12. External Libraries

External Libraries 表示外部库。

Android Studio 监控着 Gradle 的构建文件，并提供对新改动项目内容的自动同步。

1.4.3 app 目录结构

app 是 MyHello 工程的模块，这里指一个 Android 应用。app 目录结构如图 1-32 所示。

```
app
  build
  libs
  src
    androidTest
    main
      java
        com.mialab.myhello
      res
        drawable
        drawable-v24
        layout
        mipmap-anydpi-v26
        mipmap-hdpi
        mipmap-mdpi
        mipmap-xhdpi
        mipmap-xxhdpi
        mipmap-xxxhdpi
        values
      AndroidManifest.xml
    test
  .gitignore
  app.iml
  build.gradle
  proguard-rules.pro
```

图 1-32　app 目录结构

1. build

build 指构建目录，是系统自动生成的编译目录。

2. libs

libs 指使用第三方时，jar 包会自动被添加进构建的路径里。

3. src/androidTest

src/androidTest 用来编写 androidTest 测试用例，可以对项目进行一些自动化测试。

4. src/main/java

src/main/java 用来存放 Java 源代码。

5. src/main/res

res 目录用来存放 app 模块（Android 项目/Android 应用）中用到的各种资源。drawable 子目录用来存放图片及 XML 文件定义的 Drawable 资源；layout 子目录用来存放界面布局文件；values 子目录用于存放各种 XML 格式的资源文件，如字符串资源文件 strings.xml、颜色资源文件 colors.xml、尺寸资源文件 dimens.xml 等；mipmap 子目录用于保存应用程序启动图标及系统保留的 Drawable 资源，如 mipmap-mdpi、mipmap-hdpi、mipmap-xhdpi、mipmap-xxhdpi、mipmap-xxxhdpi 等子目录分别用于存放分辨率、高分辨率、超高分辨率、超超高分辨率、超超超高分辨率 5 种图片文件。

6. main/AndroidManifest.xml

main/AndroidManifest.xml 指整个 Android 项目的配置文件，在程序中定义的四大组件都需要在这个文件里注册，另外，它还可以给应用程序添加权限声明。

7. test

test 用来编写 Unit Test 的测试用例，其是对项目进行自动化测试的另一种方式。

8. .gitignore

.gitignore 用于将 app 模块内指定的目录或文件排除在版本控制之外，其作用和外层的.gitignore 文件类似。

9. app.iml

app.iml 指 IntelliJ IDEA 项目自动生成的文件，可以不需要关心或修改其内容。

10. build.gradle

build.gradle 是 app 模块的 Gradle 构建脚本，在这个文件中会指定很多与项目构建相关的配置。

11. proguard-rules.pro

proguard-rules.pro 用于指定项目代码的混淆规则，当代码开发完成后会打成安装包文件。如果不希望代码被别人破解，通常会将代码混淆，让破解者难以阅读。

12. R.java

app 构建后，在 app/build 目录下会出现 R.java 文件，具体路径是 app/build/generated/not_namespaced_r_class_sources/debug/processDebugResources/r/com/mialab/myhello/R.java，如图 1-33 所示。R.java 是定义该模块（这里是 app）所有资源的索引文件。Android 会根据 app/src/main/res 目录的资源，生成相应的 R.java 文件。R.java 文件是只读模式，不可编辑。选中 R.java 选项，单击分别加 View→Tool Windows →Structure，打开 Structure 工具窗口，可以看到 R.java 内部类的列表，如图 1-34 所示，这些内部类对应 app 模块中不同的类型资源。

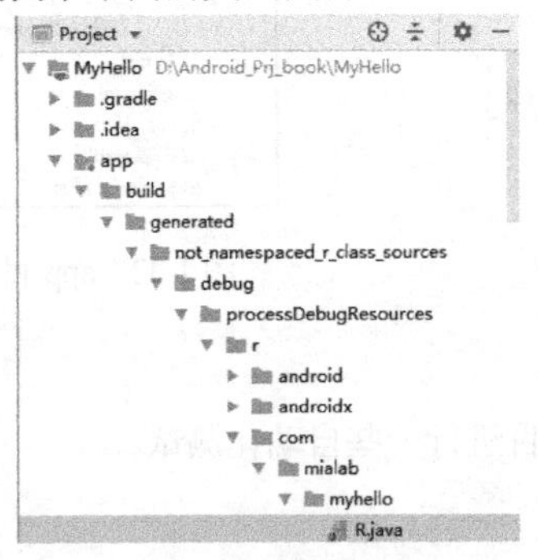

图 1-33　R.java 文件

图 1-34　R.java 内部类

1.4.4 项目全局 build.gradle 和模块 build.gradle

MyHello 工程中有两个 build.gradle 文件：一个是项目全局 build.gradle，在 MyHello 根目录下；另一个是模块级别 build.gradle，在 app 目录下。

1. 项目全局 build.gradle

项目全局 build.gradle 文件（最外层目录）的代码如下：

```
buildscript {
    //构建过程依赖的仓库
    repositories {    //repositories 闭包
        google()         //用于引用托管在 Google 公司自有仓库下的开源项目
        jcenter()        //代码托管仓库，可以引用 jcenter 上任何开源项目
    }
    dependencies {    //dependencies 闭包
     //声明 Gradle 插件 (Android Plugin for Gradle)，插件版本号为 3.4.2
        classpath 'com.android.tools.build:gradle:3.4.2'
     //NOTE: Do not place your application dependencies here; they belong
        //in the individual module build.gradle files
    }
}

//这里可配置整个项目依赖的仓库，这样每个 module 就不用配置仓库了
allprojects {
    repositories {
        google()
        jcenter()        //代码托管仓库，可以引用 jcenter 上任何开源项目
    }
}

//运行 gradle clean 时，执行此处定义的 task
//该任务继承自 Delete，删除根目录中的 build 目录
//相当于执行 Delete.delete(rootProject.buildDir)
task clean(type: Delete) {
    delete rootProject.buildDir
}
```

2. 模块 build.gradle

app 目录下 build.gradle 文件的代码如下：

```
apply plugin: 'com.android.application'        //表示一个应用程序的模块，可独立运行
//apply plugin: 'com.android.library'          //表示一个依赖库，不能独立运行
android {
    compileSdkVersion 28                        //指定项目的编译版本
    buildToolsVersion "29.0.0"                  //指定项目构建工具的版本
    defaultConfig {
        applicationId "com.mialab.myhello"      //指定该模块的应用编号，即 App 的包名
        minSdkVersion 15                        //指定最低兼容 Android 系统的版本
        //指定目标版本，表示在该 Android 系统版本已经做过充分的测试
        targetSdkVersion 28
        versionCode 1                           //版本号
        versionName "1.0"                       //版本名称
        testInstrumentationRunner "android.support.test.runner.AndroidJUnitRunner"
    }
    buildTypes { //指定生成安装文件的配置，常有两个子包：release 和 debug
```

```
                //注：直接运行的都是 debug 安装文件
        release { //用于指定生成正式版安装文件的配置
            minifyEnabled false  //指定是否对代码进行混淆，true 表示开启混淆
            //指定混淆时使用的规则文件，proguard-android.txt 指所有项目通用的混淆规则
            //proguard-rules.pro 指当前项目特有的混淆规则
            proguardFiles getDefaultProguardFile('proguard-android-optimize.txt'),
'proguard-rules.pro'
        }
    }
}

//指定 app 编译的依赖信息
dependencies { //指定当前 app 的所有依赖关系：本地依赖、库依赖、远程依赖
    implementation fileTree(dir: 'libs', include: ['*.jar'])      //本地依赖
    implementation 'com.android.support:appcompat-v7:28.0.0'      //远程依赖，
    //com.android.support 是域名部分，appcompat-v7 是组名称，28.0.0 是版本号
    //指定编译 Android 的高版本支持库，如 AppCompatActivity 必须指定 appcompat-v7 库
    implementation 'com.android.support.constraint:constraint-layout:1.1.3'
    testImplementation 'junit:junit:4.12'     //指定单元测试编译用的 junit 版本号
    androidTestImplementation 'com.android.support.test:runner:1.0.2'
    androidTestImplementation 'com.android.support.test.espresso:espresso-core:
3.0.2'
}
```

1.5 Android 应用的基本组件

1．Activity：应用表示层（基类 Activity）

一个活动表示一个可视化的用户界面。虽然多个活动一起工作可形成一个整体的用户界面，但同一应用中的每个 Activity 都是相互独立的。每一个活动都可作为 Activity 的一个子类。应用程序中每个屏幕都是通过继承和扩展 Activity 基类来实现的。

Activity 利用 View 来实现应用中的 GUI（用户直接通过 GUI 和应用程序交互）。 Activity 窗口内的可见内容通过 View 提供，使用 Activity.setContentView()方法可设置当前 Activity 中的 View 对象。

2．Service：虽没有可见的用户界面，但可长时间运行于后台（基类 Service）

Service（服务）是一个（没有用户界面的）在后台运行执行耗时操作的应用组件。其他应用组件能够启动 Service，并且当用户切换到另外的应用场景时，Service 能持续在后台运行。另外，一个组件还能够绑定到一个 Service 与之交互（IPC 机制），如一个 Service 可能会进行网络操作、播放音乐、操作文件 I/O 或与内容提供者（content provider）交互处理，所有这些活动都是在后台进行的。Service 与 Activity 都存在于当前进程的主线程中，所以一些阻塞 UI 的操作，如耗时操作就不能放在 Service 里进行。

3．Broadcast Receiver：用户接收广播通知的组件（基类 BroadcastReceiver）

广播接收者仅是接受广播公告并做出相应的反应。许多广播源自系统代码，如公告时区的改变、电池电量低、已采取图片、用户改变了语言偏好。应用程序也可以发起广播，如为了让其他程序知道某些数据已经下载到设备且可以使用时。一个应用程序可以有任意数量的广播接收者去反应它认为重要的公告。所有的接收者都继承自 BroadcastReceiver。

BroadcastReceiver 自身并不实现图形用户界面，但是当它收到某个通知后，就可以启动 Activity

作为响应，或者通过 NotificationMananger 提醒用户。BroadcastReceiver 是对发送出来的 Broadcast 进行过滤接收并响应的一类组件。

4. Content Provider：应用程序间数据的通信、共享（基类 ContentProvider）

内容提供者能使一个应用程序的指定数据集提供给其他应用程序。这些数据可以存储在文件系统、一个 SQLite 数据库中，或以任何其他合理的方式存储。内容提供者继承自 ContentProvider 并实现了一个标准的方法集，使得其他应用程序可以检索和存储数据。然而，应用程序并不直接调用这些方法，而是使用一个 ContentResolver 对象替代并调用其方法。ContentResolver 能与任何内容提供者通信，它与提供者合作来管理参与进来的进程间通信。

5. Intent：连接组件的纽带通信

Intent 能在不同的组件之间传递消息，将一个组件的请求意图传给另一个组件。因此，Intent 是包含具体请求信息的对象。针对不同的组件，Intent 所包含的消息内容有所不同，且不同组件的激活方式也不同。

Intent 是一种运行时绑定（runtime binding）机制，它能够在程序运行的过程中连接两个不同的组件。通过 Intent，程序可以向 Android 表达某种请求或意愿，Android 会根据意愿的内容选择适当的组件来处理请求。

1.6 Activity 的生命周期

Activity 主要用于和用户进行交互，一个 App 允许有多个 Activity。开发 Activity 的时候，一般需要重写 onCreate()方法。onCreate(Bundle)是初始化 Activity 的地方，可以看作是这个页面的入口函数，在这儿通常可以调用 setContentView(int)设置在资源文件中定义的 UI。在系统中的 Activity 被一个 Activity 栈（任务栈）管理。当一个新的 Activity 启动时，将被放置到栈顶，成为运行中的 Activity，前一个 Activity 保留在栈中，不再放到前台。当按下 Back 键或调用 finish()方法去销毁一个 Activity 时，处于栈顶的活动会出栈，这时前一个入栈的 Activity 就会重新处于栈顶的位置。系统总是会显示处于栈顶的 Activity 给用户。

1. Activity 的 4 种状态：

（1）运行状态或活动状态（Running or Active）：在屏幕的前台（Activity 栈顶）用户可见，并可获得焦点。

（2）暂停状态（Paused）：如果一个 Activity 失去焦点，但是依然可见（一个新的非全屏的 Activity 或一个透明的 Activity 被放置在栈顶），这时该 Activity 就处于暂停状态。一个暂停状态的 Activity 依然保持活力，但是在系统内存极低的时候将被销毁。

（3）停止状态（Stopped）：如果一个 Activity 被另外的 Activity 完全覆盖，这时该 Activity 就处于停止状态。它依然保持所有状态和成员信息，但是不再可见，当系统内存需要被用在其他地方的时候，停止状态的 Activity 将被销毁。

（4）销毁状态（Destroyed）：如果一个 Activity 是暂停或停止状态，系统可以将该 Activity 从内存中销毁，Android 系统采用两种方式进行销毁：要求该 Activity 结束或直接销毁其进程。当该 Activity 再次显示给用户时，必须重新开始和重置前面的状态。

2. Activity 被系统回调的方法

（1）onCreate()：这个方法会在 Activity 第一次被创建时调用。在该方法中完成 Activity 的初始化操作，如加载布局、初始化布局控件、绑定按钮事件等。

（2）onStart()：这个方法在 Activity 中由不可见变为可见时，进行调用。

（3）onResume()：这个方法在 Activity 准备好与用户交互时调用。此时的 Activity 一定位于返回栈的栈顶，并且处于运行状态。

（4） onPause()：这个方法在系统准备去启动或恢复另一个 Activity 时调用，或者说在当前 Activity 画面无法完全显示时（如半遮状态）调用。

（5）onStop()：这个方法在 Activity 完全不可见时调用。它和 onPause()方法的主要区别在于，如果启动的新 Activity 是对话框，则 onPause()方法会得到执行，但 onStop()方法并不会被执行。

（6）onDestroy()：这个方法在销毁 Activity 时被回调，且只会被调用一次。

（7）onRestart()：这个方法在 Activity 由停止状态变为运行状态之前调用，也就是 Activity 被重新启动了。

Activity 的生命周期如图 1-35 所示。

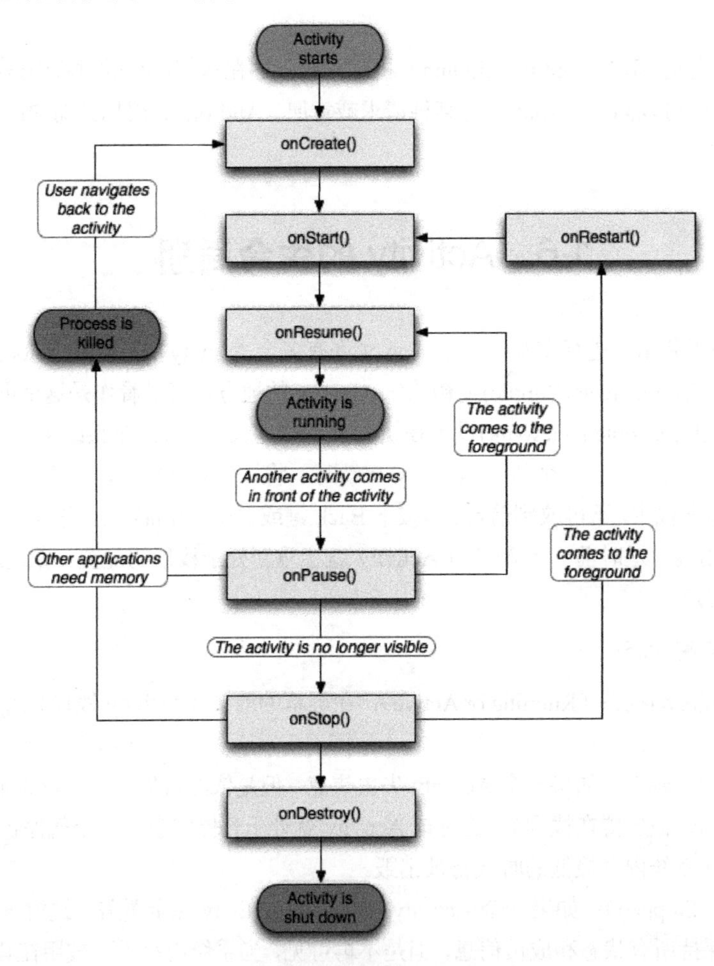

图 1-35　Activity 的生命周期

3．Activity 的 3 个关键循环

（1）全生命周期：从 onCreate()开始到 onDestroy()结束。Activity 在 onCreate()中设置所有的“全局”状态，在 onDestroy()中释放所有的资源。如果某个 Activity 有一个在后台运行的线程，用于从网络下载数据，则该 Activity 就可以在 onCreate()中创建线程，在 onDestroy()中停止线程。

（2）可见生命周期：从 onStart()开始到 onStop()结束。在这段时间，可以看到 Activity 在屏幕上，但有可能不在前台，也不能和用户交互。在这两个接口之间，需要保持显示给用户的 UI 数据和资源等，如可以在 onStart()中注册一个 BroadcastReceiver 组件来监听数据变化导致 UI 的变动，当不再需要显示时，可以在 onStop()中注销。onStart()、onStop()都可以被多次调用，因为 Activity 随时可以在可见和隐藏之间转换。

（3）前台生命周期：从 onResume()开始到 onPause()结束。在这段时间里，该 Activity 处于所有 Activity 的前面，能和用户进行交互。Activity 可以经常在前台和非前台状态之间切换，因此 onResume() 和 onPause()也会被反复调用。

除了上述的 Activity 生命周期事件回调函数，还有 onRestoreInstanceState()和 onSaveInstanceState()两个函数经常会被使用，用于保存和恢复 Activity 的临时数据。

小贴士：Android 系统的回收机制会在未经用户主动操作的情况下销毁 Activity，为了避免系统回收 Activity 时导致数据丢失，Android 提供了 onSaveInstanceState(Bundle outState)和 onRestoreInstance-State(Bundle savedInstanceState)用于保存和恢复数据。

① onSaveInstanceState(Bundle outState)在什么时机会被调用？

在 Activity 有可能被系统回收的情况下，而且是在 onStop()之前。注意是有可能，如果是已经确定会被销毁的，如用户已点击“返回”按钮，或者调用了 finish()方法销毁 Activity，则 onSaveInstanceState()不会被调用。

也可以说，此方法只有在 Activity 被异常终止的情况才会被调用。

onSaveInstanceState(Bundle outState)会在以下情况被调用：

① 当用户按 Home 键时；

② 从最近应用中选择运行其他程序时；

③ 按下电源键（关闭屏幕显示）时；

④ 从当前 Activity 启动一个新的 Activity 时；

⑤ 屏幕方向切换时（无论竖屏切横屏，还是横屏切竖屏都会调用）。

在前 4 种情况下，当 Activity 的生命周期为 onPause→onSaveInstanceState→onStop 时，同学们可自行验证。

② onRestoreInstanceState(Bundle savedInstanceState)在什么时机会被调用？

onRestoreInstanceState(Bundle savedInstanceState)只有在 Activity 确实是被系统回收，重新创建 Activity 的情况下才会被调用。

如在上述第 5 种情况下，Activity 的生命周期为 onPause→onSaveInstanceState→onStop→onDestroy→onCreate→onStart→onRestoreInstanceState→onResume 时。

在这里 onRestoreInstanceState()被调用，因为屏幕切换时原来的 Activity 确实被系统回收了，又重新创建了一个新的 Activity。

按 Home 键返回桌面，再点击“应用”图标回到原来页面时，Activity 的生命周期为 onPause→onSaveInstanceState→onStop→onRestart→onStart→onResume，因为 Activity 没有被系统回收，因此 onRestoreInstanceState()没有被调用。如果 onRestoreInstanceState()被调用，则页面就会被回收，而 onSaveInstanceState()也必然会被调用。

③ onCreate()也有 Bundle 参数可以用来恢复数据，它和 onRestoreInstanceState()有什么区别呢？

因为 onSaveInstanceState() 不一定会被调用，所以 onCreate()中的 Bundle 参数可能为空，如果使用 onCreate()来恢复数据，一定要做非空判断。

onRestoreInstanceState 的 Bundle 参数一定不会是空值，因为它只有在上次 Activity 被回收了才会被调用，且 onRestoreInstanceState 是在 onStart()之后被调用的。有时也需要在 onCreate()中进行一些初始化后再恢复数据，这时用 onRestoreInstanceState 就会比较方便。

1.7 Android 日志工具 Log

Android 的日志工具类是 Log，Log 类在 android.util 包中，可以使用它将运行过程的信息输出到 IDE，直接查看程序运行的过程。Log 类提供了若干静态方法来打印日志。

```
Log.v(String tag, String msg);
Log.d(String tag, String msg);
Log.i(String tag, String msg);
Log.w(String tag, String msg);
Log.e(String tag, String msg);
```

其中，v、d、i、w、e 分别对应 Verbose、Debug、Info、Warn 和 Error。参数 tag 是一个标识，可以是任意字符串，主要用来在查看日志时提供一个筛选条件，对打印信息进行过滤，可将其视作过滤器。参数 msg 表示具体想打印的内容。

Log 类 5 种静态方法的功能说明如下：

（1）Log.v()：用于打印最烦琐的、意义最小的日志信息，对应级别为 Verbose，是 Android 日志中级别最低的一种。

（2）Log.d()：用于打印一些调试信息，对应级别为 Debug，比 Verbose 高一级。

（3）Log.i()：用于打印一些比较重要的数据，这些数据可以帮助分析用户的行为。对应级别为 Info，比 Debug 高一级。

（4）Log.w()：用于打印一些警告信息，提示程序在这个地方可能会有潜在风险，应尽快修复出现警告的地方。对应级别为 Warn，比 Info 高一级。

（5）Log.e()：用于打印程序中的错误信息，如程序进入到 catch 语句中。当有错误信息打印出来时，表示程序中有严重错误信息，必须尽快修复。对应级别为 Error，比 Warn 高一级。

每种方法都有不同的重载。它们的重载形式如下：

```
public static int v(String tag, String msg) {…}
public static int d(String tag, String msg) {…}
public static int i(String tag, String msg) {…}
public static int w(String tag, String msg) {…}
public static int e(String tag, String msg) {…}
```

【示例】 在 Ch1 工程 LifeCycle 模块的 MainActivity.java 中，部分示例代码如下：

```
private static final String TAG ="LogTest";
@Override
protected void onCreate(Bundle savedInstanceState) {
   super.onCreate(savedInstanceState);
   setContentView(R.layout.activity_main);
   Log.d(TAG,"----- onCreate() method called -----");
}
```

在 Android Studio 中打开 Ch1 工程，通过选择“View”→“Tool Windows”→“Logcat”，弹出 Logcat 调试窗口，如图 1-36 所示。点击 Logcat 调试窗口右侧的倒黑三角▾，弹出过滤器选择清单，如图 1-37 所示。选择“Edit Filter Configuration”选项，弹出“Create New Logcat Filter”对话框，如图 1-38 所示。

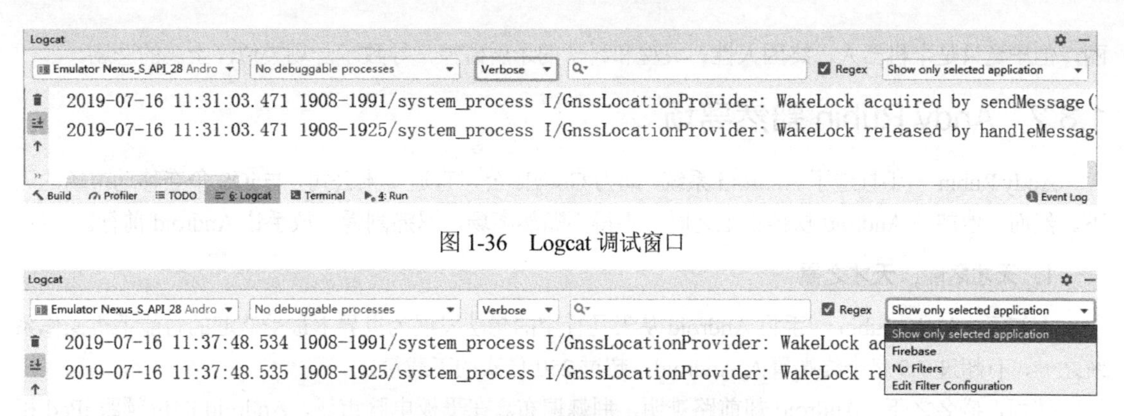

图 1-36　Logcat 调试窗口

图 1-37　Logcat 调试窗口的过滤器选择清单

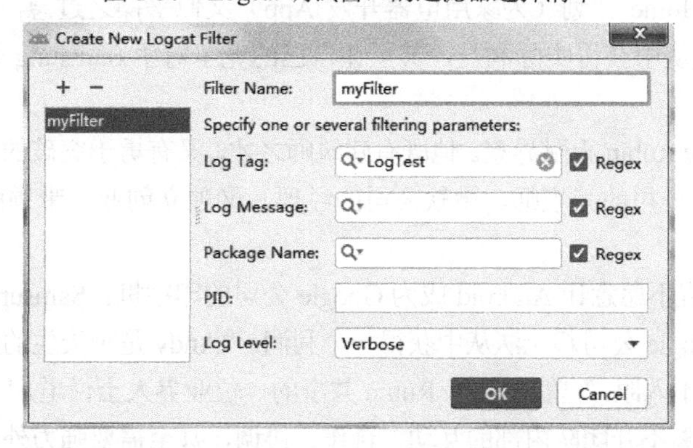

图 1-38　创建 Logcat Filter

在图 1-38 中，创建一个新的 Logcat Filter。在 Filter Name 中输入“myFilter”，并在 LogTag 中输入“LogTest”，再点击“OK”按钮。运行 LifeCycle 应用，在 Logcat 调试窗口右侧选择新创建的“myFilter”过滤器，对输出信息进行过滤，如图 1-39 所示。

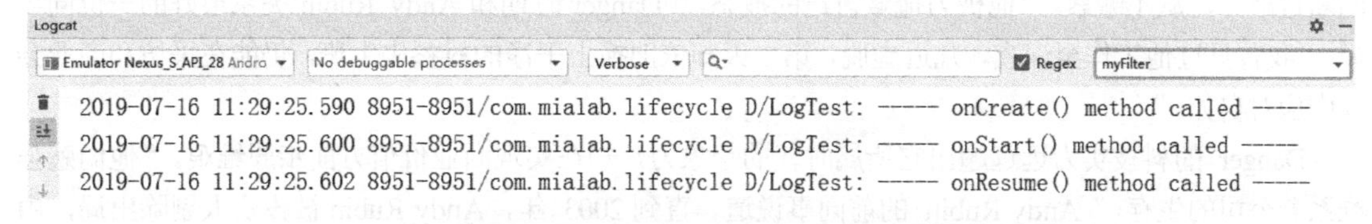

图 1-39　选择新创建的 myFilter 对 LifeCycle 输出信息进行过滤

小贴士：有一种方法可以把程序运行过程信息的输出当作程序运行的一部分，如使用 Toast Notification 将输出信息显示在界面中。当然这些只是调试代码，在发布程序时需要将其去掉。还有一种方法很有效，即直接将运行过程的信息以文件的方式存储，并在程序运行后打开文件，查看输出的信息。在一些复杂的工具中，采用这种日志文件的方法来记录文件运行的过程。

1.8　相关阅读：Andy Rubin 与 Android

1.8.1　未来的夏娃

“Android”这一词最先出现在法国作家利尔・亚当在 1886 年发表的科幻小说《未来的夏娃》中，作者将外表像人类的机器起名为 Android。

这部书描写一位和爱迪生同名的大发明家，利用电学原理制作了一个完美的女人，虽然她聪明美丽，但毕竟只是个机器人，终因人性、灵魂和科学的矛盾碰撞，导致了一场类似浮士德的悲剧。

1.8.2 Andy Rubin 黯然离场

Andy Rubin 一手打造了 Android 系统，助力 Google 公司直追苹果公司，与 iOS 争夺移动市场的天下。然而，他却在 Android 扶摇直上之时，不得不黯然离场，忍痛割爱，放手让 Android 前行。

1. 天才之能，天才之累

十年间，Andy Rubin 一手将 Android 从天马行空的想法发展为世界上最受欢迎的智能手机操作系统之一，有超过 7.5 亿个终端置入 Android，超过 250 亿次的下载量。

然而，盛名之下，Android 却前路难明，荆棘遍布。在平板电脑市场，Android 的份额跟 iPad 相去甚远，而 Android@Home 计划（为家用电器开发 App）又似乎操之过急。尽管 Google 公司坚持开放系统理念，却无法从中获得实际利益，甚至让自己的竞争对手 Samsung 公司抢占了移动设备市场，赚得盆满钵满。

没有人会否认 Andy Rubin 天赋异禀。他既有建筑师之才，又有勇于突破的黑客精神。进入 Google 公司之前，他做过苹果公司的工程师、微软公司的经理，又独立创业，带领两支团队走向成功，即 Danger 和 Android。

但 Andy Rubin 一直不愿意让 Android 成为 Google 公司的印钞机。Samsung 公司制造出远销海外的 Android 智能机，Google 公司却无法从中获得一分利润。"Andy 是个天生的独舞者。他不断奔向心中所想，无谓的人事皆不入眼。"曾与 Andy Rubin 共事的一位业界人士说道，"但是 Android 必定会成长到某种规模，那时它离不开团队内部的互动、合作、协调，甚至需要强力外部资源的支撑。事实证明，Android 成长到这个阶段时，Andy Rubin 发现自己失去了掌控力。他开始举棋不定，看不清航向。"

2. Danger 往事

Andy Rubin 和他创立的 Danger 被看成是首台智能机之父。在 Danger 的 4 年中，Andy Rubin 从未间断过战斗，从不喊累。"他极力掩藏自己的疲态，" Danger 时期和 Andy Rubin 关系很好的一位同事说，"我曾见过他工作至深夜，几近虚脱，第二天却依旧换上干净的衬衫去上班，仍能在采访和商业会议中应对自如。"

Danger 的科技实力远远超出它所属时代的想象力，却在实现商业价值方面步履维艰。"他们勉强维系着公司的生存，" Andy Rubin 的前同事说道。直到 2003 年，Andy Rubin 被投资人剔除出局，由通信业老手、精明的生意人 Hank Nothhaft 接过权杖。

"当时我们有一个超棒的产品即将推出市场，因此我们找来 Hank Nothhaft。只有他知道如何处理这么大的生意，如何同合伙人打好交道。" Danger 的前投资人 Rees 说，"只有 Hank Nothhaft 才是带领这家企业上一个台阶的 CEO"。

离开 Danger 不久，Andy Rubin 就开始着手 Android 的研发。

3. 在 Google 公司的岁月

2005 年，Larry Page 买下仍属雏形的 Android，Andy Rubin 就此加盟 Google 公司，但 Sergey Brin 和 Eric Schmidt 一直同 Android 保持距离，他们不懂移动领域。Page 却对 Andy Rubin 开放移动操作系统的远见赞赏不已，他甚至觉得 Google 公司可以推动全球范围的手机革新浪潮，哪怕困难重重。

Larry Page 坚信 Android 对 Google 公司意义非凡，Andy Rubin 却半信半疑，他感受到 Google 公司内部结构的松散，企业文化的疯狂，以及最初的不适。

在外界看来，Google 公司快速收购 Android 的举动似乎预示着他们进军无线领域的决心，但 Google 公司自身并不愿坦露过多细节。Andy Rubin 进入 Google 公司的领导层，担任移动数字部门高级副总裁。他带领产品研发，寻找合作伙伴，同乔布斯成为盟友。

4．交出权杖

2013年3月14日，Andy Rubin辞去Android业务主管一职，Chrome及应用高级副总裁Sundar Pichai将接管Android业务。经过这次人事变动，Google公司的移动软件、应用程序和Chrome浏览器业务都统一于Sundar Pichai的领导之下。

Google公司对Andy Rubin的去处缄默不言，但外界的揣测纷纷指向Google X研发实验室。这位乐于发明的Android之父，已拥有11项专利权，以及一沓专利申请书。他对硬件有天生的热爱，尤其是可以拍照的硬件。在微软公司的时候，他总是偷偷走开，回到自己位于洛杉矶的实验室，在氖灯管的照耀下敲敲打打。

Andy Rubin离开后，Android被交到高级副总裁Sundar Pichai手中，他同时负责Chrome和Apps两个项目。融合Chrome OS和Android的猜想由来已久，如果真能如此，Google公司就能将PC端和移动端的软件统一起来，创造出类似Pixel这种可以触摸的笔记本。Google公司也表示这两个操作系统未来可能融合，但目前却不是最佳时机。

Larry Page希望Google公司的产品线能连贯统一，这样不仅有利于研发，也有利于市场。但对Chrome OS和Android这两支团队而言，融合之路困难重重。它们的风格不一，目标不同。Android团队习惯飞快的工作节奏，他们必须不间断地升级软件、修复bug，才能赶上移动互联网迅猛发展的步调。

而Pichai则兼具冷静和战斗力。作为产品经理，Larry Page把Chrome浏览器一路做到行业顶尖，同时能兼顾Google公司的专业产品Apps。若Android和Chrome OS团队合并，必要经历阵痛才能达到稳定之态。对此时的Google公司而言，融合绝非明智之选。Android换将之后，Sundar Pichai令Android走向盈利的任务呼之欲出，他需要先完成Larry Rubin未竟的任务。

1.8.3 Andy Rubin早年岁月

Andy Rubin 1963年生于纽约州Chappaqua镇，父亲是学心理学的，经营一家电子玩具直销公司，那些销售样品拍照后放进销售目录，就属于Andy Rubin的了，他的房间里满是各种最新的电子玩具。反复拆装这些玩具是他最爱做的事情之一，他的Geek（极客）基因由此种下。

大学毕业后，Andy Rubin加入以光学仪器出名的卡尔·蔡司公司担任机器人工程师，主要从事数字通信网络的研发，后来他还在瑞士一家机器人公司工作过，工作得很开心。然而，一个偶然事件改变了他的一生。

1989年，Andy Rubin到开曼群岛旅游，清晨独自在沙滩漫步时遇到一个人可怜地睡在躺椅上——由于和女朋友吵架被赶出了海边别墅。Andy Rubin给他找了住处，作为回报，这位老兄答应引荐Andy Rubin到自己所在的公司工作。原来，此人是正处在第一个全盛时期的苹果公司的一名工程师。

不平凡的硅谷经历让Andy Rubin在苹果公司如鱼得水，桌面系统Quadra和历史上第一个软件Modem都是他的作品。他也不忘展示一下Geek本色：对公司的内部电话系统进行了重新编程，伪装CEO打电话给人事部，指示要给自己组里的工程师同事进行股票奖励。当然，为此信息部门免不了来找他的麻烦。

1990年，苹果公司的手持设备部门独立出来，成立了General Magic公司。两年后，Andy Rubin认定这个领域会大有作为，便选择加入。在这里，他完全投入到公司的工程师文化中，和同事们在自己的小隔间上方搭起了床，几乎24小时吃住在办公室。他们开发的产品是具有突破性意义的基于互联网的手机操作系统和界面Magic Cap，在市场上也曾经取得短暂的成功，1995年公司甚至因此上市，而且第一天股票就实现了翻番。但是好景不长，这款产品太超前了，运营商的支持完全跟不上，很快被市场判了死刑。

此后，Andy Rubin 又加入苹果公司员工创办的 Artemis Research 公司，继续吃住在办公室，追逐互联网设备的梦想。这次他参与开发的产品是交互式互联网电视 WebTV，创造了多项通信专利。产品获得了几十万用户认可，成功实现盈利，年收入超过一亿美元。1997 年，公司被微软公司收购。Andy Rubin 也随之加入，雄心勃勃地开始了他的超级机器人项目。他开发的互联网机器人在微软公司四处游荡，随时记录所看所闻。不料，有一天控制机器人的计算机被黑客入侵，激怒了微软公司的安全官员。不久，Andy Rubin 就离开微软公司在 Palo Alto 租了一个商店，与他的工程师朋友们继续研究各种机器人和新设备，构思各种新产品的奇思妙想。这就是 Danger 的前身。

创办 Danger 并担任 CEO 的过程中，使 Andy Rubin 完成了从工程师到管理者的转变。更为重要的是，他和同事们一起找到了将移动运营商和手机制造商结合起来的利益模式，这与 iPhone 非常类似。但是，公司的运营并不理想，Andy Rubin 接受了董事会的决定，辞职并有些失望地离开了公司。Danger 后来被微软公司收购，2010 年这个部门发布了很酷但很快又失败了的产品 Kin 系列手机。

1.9 本章小结

本章主要介绍了 Android 的系统架构、Android 开发环境的搭建（包括 Android Studio 安装）、Android 项目结构分析（包括 Project 工具窗口、工程目录说明、app 目录结构、项目全局 build.gradle 和模块 build.gradle 等）、Android 应用基本组件介绍、Activity 的生命周期，以及 Android 日志工具 Log 的使用。学会使用 SDK Manager 和 AVD Manager 下载 Android SDK 和创建 Android 虚拟设备（如手机模拟器等），学会如何编写一个简单的 Android 应用 MyHello，运行 Android 应用并在模拟器中查看运行结果。学会使用 LogCat 工具查看 Activity 生命周期中一些方法的回调。

习 题 1

1. Android 的体系结构是怎样的？请简要加以说明。
2. 如何搭建 Android 开发环境？有哪些注意事项？
3. Android 应用的程序结构是怎样的？请简要加以分析。
4. Android 应用有哪些基本组件？请简要加以说明。
5. 如何使用 Android 的日志工具类 Log 对打印信息进行过滤？请编程并加以说明。
6. Activity 的生命周期是怎样的？请编程并加以说明。

第2章 用户界面设计

本章导读

如何评价一款App的好坏，用户体验是非常重要的。所谓用户体验，最直接的感受就是界面的美观与否，其次才是功能。用户界面相当于App的门面，越来越受到人们的重视。如何设计能够兼容不同规格屏幕的用户界面是本章的重点。

本章对Android前台手机界面布局方法及常用组件进行阐述，主要知识点有：(1)视图组件和容器组件；(2)控制UI界面的3种方式；(3)基本UI组件(包括TextView、EditText、Button、ImageView、ProgressBar、ScrollView、Toast、布局管理器等)；(4)高级UI组件(包括列表类组件、对话框、菜单、标签栏等)；(5)Handler消息传递机制。

2.1 视图组件与容器组件

Android的UI界面都是由View和ViewGroup及其派生类组合而成的。其中，View是所有UI组件的基类，而ViewGroup是容纳View及其派生类的容器，它也是从View派生出来的。一般来说，开发UI界面不会直接使用View和ViewGroup，而是使用其派生类。

关于View类：

```
public class View extends Object implements Drawable.Callback,
KeyEvent.Callback, AccessibilityEventSource
```

关于ViewGroup类：

```
public abstract class ViewGroup extends View implements ViewParent, ViewManager
```

Android中的视图可分为3种：布局类(Layout)、视图容器类(View Container)和视图类(如TextView就是一个直接继承View类的视图类)。

android.view.ViewGroup是一个容器类，该类也是View的子类，所有的布局类和视图容器类都是ViewGroup的子类，而视图类直接继承自View类，如图2-1所示。

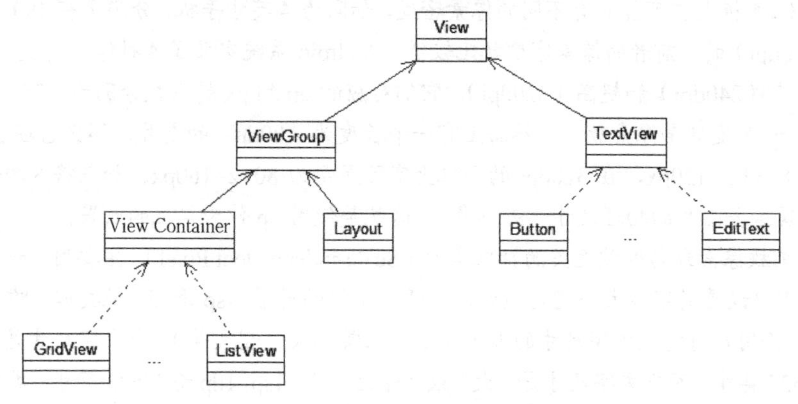

图2-1 Android中的视图组件与容器组件

在Android App中，所有的用户界面元素都是由View和ViewGroup的对象构成的。View是绘制在屏幕上的用户能与之交互的一个对象，ViewGroup 则是一个用于存放其他 View 和 ViewGroup 对象的布局容器。由于 View 和 ViewGroup 之间采用组合设计模式，可以使得“部分-整体”同等对待。ViewGroup 作为布局容器类的顶层，布局容器里面又可以有 View 和 ViewGroup。

UI 布局的层次结构如图 2-2 所示。

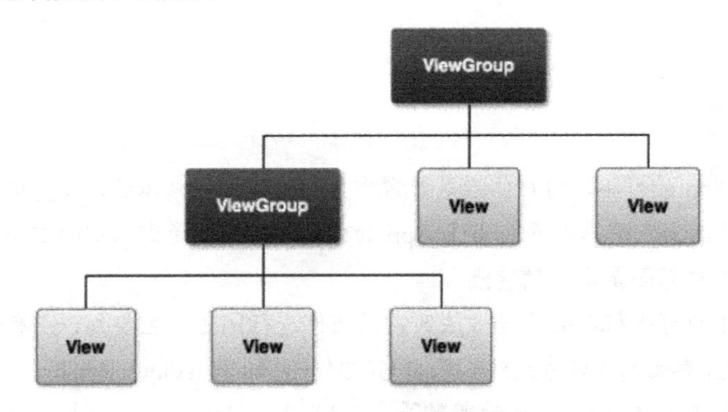

图 2-2 UI 布局的层次结构

小贴士：对于软件开发程序员来说，用户界面单位最熟悉的就是像素（px）和磅（pt）。像素是图像的基本取样单位，是最小的元素单元，每一张图片都是由若干像素组合而成的，像素的大小随着屏幕分辨率而变化，分辨率越高，像素单位越小。磅通常是印刷业常用单位，1 磅等于 1/72 英寸，通常用来定义字体的大小。

当布局用户界面的时候，UI 控件需要固定大小，如果此时选用像素为单位，就会出现不同分辨率的屏幕控件大小并不一致的情况。不同分辨率的设备对像素的显示比例不一致，低分辨率的设备，宽度为 150px 的按钮几乎占据屏幕的 1/2，而高分辨率的设备只占大概 1/3。如此看来，为了兼容不同的移动设备，像素和磅并不适合作为单位来描述视图的大小。因此，为了解决不同分辨率单位适配的问题，Android 引入了 dp 跟 sp 两种单位制度。

dp 意为与密度无关的像素，是一种基于屏幕密度的抽象单位，在不同密度的屏幕中显示比例保持一致。dp 与 px 之间的换算关系：px=dp*density/160，其中，density 表示屏幕密度。

什么是像素密度？假设有一部手机，屏幕的物理尺寸为 1.5 英寸×2 英寸，屏幕分辨率为 240×320（dpi），则可以计算出在这部手机的屏幕上，每英寸包含的像素点的数量为 240/1.5=160dpi（横向）或 320/2=160dpi（纵向），160dpi 就是这部手机的像素密度，像素密度的单位 dpi 是 Dots Per Inch 的缩写，即每英寸像素数量。横向和纵向的这个值都是相同的，原因是大部分手机屏幕使用正方形的像素点。

不同的手机/平板电脑可能具有不同的像素密度，如同为 4 英寸手机，分辨率有 480×320（dpi）的，也有 800×480（dpi）的，前者的像素密度就比较低。Android 系统定义了 4 种像素密度：低（120dpi）、中（160dpi）、高（240dpi）和超高（320dpi），它们对应的 dp 到 px 的系数分别为 0.75、1、1.5 和 2，这个系数乘以 dp 长度就是像素数。如界面上有一个长度为“80dp”的图片，那么它在 240dpi 的手机上实际显示为 80×1.5=120px，在 320dpi 的手机上实际显示为 80×2=160px。如果将这两部手机放在一起对比，会发现这个图片的物理尺寸“差不多”，这就是使用 dp 作为单位的效果。

sp 意为可缩放像素或与缩放无关的抽象像素（Scale-independent Pixel），主要用于设置字体，可根据屏幕密度和用户设置的字体大小进行调整，以适配不同的屏幕。sp 和 dp 很类似，唯一的区别是，Android 系统允许用户自定义文字尺寸的大小（小、正常、大、超大等），当文字尺寸是“正常”时，1sp=1dp=0.00625 英寸，而当文字尺寸是“大”或“超大”时，1sp>1dp=0.00625 英寸。类似在 Windows 中调整字体尺寸后的效果，即窗口大小不变，只有文字大小改变。

dip（Device Independent pixels，设备独立像素）和 dp 是一样的。在早期的 Android 版本中多使用 dip，后来为了与 sp 统一就建议使用 dp 这个名字了。

最佳实践：文字的尺寸一律用 sp，非文字的尺寸一律使用 dp，如 textSize="16sp"、layout_width="60dp"。如需要在屏幕上画一条细的分隔线时，偶尔需要使用 px，其代码如下：

```
<View layout_width="match_parent" layout_height="1px"/>
```

2.2 控制 UI 界面

Android 用户界面的搭建分静态和动态两种方式。所谓静态方式即以 XML 布局文件来定义用户界面，通过 XML 布局文件中的相关属性进行控制，这是较推荐的一种方式。动态方式是指通过 Java（或 Kotlin）代码来开发用户界面，动态地控制界面中的组件。

2.2.1 使用 XML 布局文件控制 UI 界面

利用 XML 布局文件来控制用户界面是开发人员常使用的方法。用 XML 文件来描述用户界面，并将其保存在资源文件夹/res/layout 下。这种方法可极大地简化界面设计的过程，将界面视图从 Java 代码中分离出来，将用户界面中静态部分定义在 XML 中，代替了写代码，使得程序结构更加清晰、明了。

使用 XML 布局文件控制 UI 界面的基本步骤如下：

（1）打开/res/layout 目录下的布局文件，遵照 XML 规范编写用户界面代码。保存编写后的布局文件，R.java 将自动注册该布局资源。

（2）在 Activity 中设置界面为上述布局文件描述的用户界面，Java 代码描述如下：

```
setContentView(R.layout.<布局文件名>);
```

XML 布局文件是 Android 系统中定义视图的常用方法。所有 XML 布局文件必须保存在 res/layout 目录中。

XML 布局文件的命名及定义需要注意如下 6 点：

（1）XML 布局文件的扩展名必须是 XML。

（2）Android 会根据每一个 XML 布局文件名在 R.java 中生成一个变量，这个变量名就是 XML 布局文件名，因此 XML 布局文件名（不包含扩展名）必须符合 Java 变量名的命名规则。

（3）每一个 XML 布局文件的根节点可以是任意的组件。

（4）XML 布局文件的根节点必须包含 Android 命名空间，而且命名空间的值必须是 http://schemas.android.com/apk/res/android。

（5）为 XML 布局文件中的标签指定 ID 时，需要使用格式：@+id/somestringvalue，其中@+语法标识：如果 ID 值在 R.id 类中不存在，则会新产生一个与 ID 同名的变量；如果在 R.id 类中存在该变量，则直接使用这个变量。

（6）由于每一个视图 ID 都会在 R.id 类中生成与之对应的变量，因此视图 ID 的值也要符合 Java 变量的命名规则，这一点与 XML 布局文件名的命名规则相同。

【示例】 启动 Android Studio，在 Ch2 工程中创建 FindButton 模块。

在 Ch2\FindButton\src\main\res\layout\activity_main.xml 文件中输入以下代码（将 activity_main.xml 文件的原有代码清空）：

```
<?xml version="1.0" encoding="utf-8"?>
<LinearLayout xmlns:android="http://schemas.android.com/apk/res/android"
    android:layout_width="match_parent"
```

```
        android:layout_height="match_parent"
        android:orientation="vertical">
        <TextView
            android:id="@+id/textview"
            android:layout_width="match_parent"
            android:layout_height="wrap_content"
            android:text="还有诗和远方" />
        <Button
            android:id="@+id/button"
            android:layout_width="wrap_content"
            android:layout_height="wrap_content"
            android:text="第一个按钮" />
    </LinearLayout>
```

这是一个标准XML布局文件的内容。

在MainActivity.java的onCreate()方法中，使用setContentView()方法指定XML布局文件的资源ID，其代码如下：

```
    protected void onCreate(Bundle savedInstanceState) {
        super.onCreate(savedInstanceState);
        setContentView(R.layout.activity_main);
    }
```

运行FindButton模块，其界面如图2-3所示。

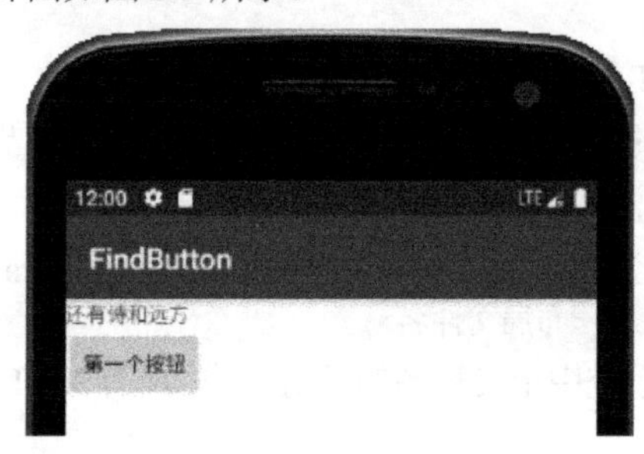

图2-3　使用XML布局文件控制UI界面

如果想要获得在activity_main.xml文件中定义的某个View，可以使用如下代码：

```
    TextView textView = (TextView)findViewById(R.id.textview);
    Button button = (Button)findViewById(R.id.button);
```

在获得XML布局文件中的视图对象时需要注意如下3点：

（1）在使用findViewById方法之前，必须先使用SetContentView方法装载XML布局文件，否则系统会抛出异常。

（2）虽然所有XML布局文件中的视图ID都在R.id类中生成了相应变量，但使用findViewById方法只能获得已经装载XML布局文件中的视图对象。

（3）在不同XML布局文件中可以有相同ID值的视图，但在同一个XML布局文件中，虽然也可以有相同ID值的视图，但通过ID值获得视图对象时，只能按定义顺序得到第一个视图对象，其他相同ID值的视图对象将无法获得。因此，在同一个XML布局文件中应尽量使视图的ID值唯一。

2.2.2　在Java代码中控制UI界面

Android允许开发者完全在Java（或Kotlin）代码中控制UI界面，其基本步骤如下：

（1）创建布局管理器，如线性布局管理器、帧布局管理器、相对布局管理器等，设置布局管理器的属性（如为布局管理器设置背景图片等），并将其设置为显示界面。

（2）创建UI组件，并调用相应的方法设置UI组件属性。

（3）调用 addView()方法，将 UI 组件添加到布局管理器中。

【示例】 启动 Android Studio，在 Ch2 工程中创建 FindButton2 模块。

Ch2\FindButton2\src\main\java\com\mialab\findbutton2\MainActivity.java 的 onCreate()方法代码如下：

```
protected void onCreate(Bundle savedInstanceState) {
        super.onCreate(savedInstanceState);
        //setContentView(R.layout.activity_main);
        LinearLayout layout = new LinearLayout(this); //创建线性布局容器对象
        setContentView(layout);
        layout.setOrientation(LinearLayout.VERTICAL);
        //设置线性布局容器中组件的排列方式为垂直方向
        TextView txt = new TextView(this);              //创建 TextView 控件
        txt.setText("还有诗和远方");                      //设置 TextView 控件显示文本
        Button btn = new Button(this);                  //创建按钮控件
        btn.setText("第一个按钮");                        //设置按钮控件显示文本
        btn.setLayoutParams(new ViewGroup.LayoutParams
                (ViewGroup.LayoutParams.WRAP_CONTENT,
                        ViewGroup.LayoutParams.WRAP_CONTENT));
        //设置按钮大小
        layout.addView(txt);                            //向 layout 容器添加文本控件
        layout.addView(btn);                            //向 layout 容器添加按钮控件
    }
```

由上述代码可以看出，完全在代码中控制用户界面，一方面不利于高层次解耦，另一方面界面中的控件需要通过 new 来创建，控件属性的设置还需要调用相应的方法，因此代码显得较臃肿，对程序开发人员来说，无论是设计还是维护都较烦琐。因此，除非必须，一般不会完全使用 Java 代码来控制 UI 界面。

2.2.3 使用 XML 布局文件和 Java 代码混合控制 UI 界面

使用 XML 布局文件和 Java 代码混合控制 UI 界面时，习惯上把变化小、行为控制比较固定的组件放在 XML 布局文件中，把变化较大、行为控制比较复杂的组件交给 Java 代码来管理。

【示例】 启动 Android Studio，在 Ch2 工程中创建 ImageDemo 模块。

Ch2\ImageDemo\src\main\res\layout\activity_main.xml 文件内容如下：

```
<?xml version="1.0" encoding="utf-8"?>
<LinearLayout xmlns:android="http://schemas.android.com/apk/res/android"
    android:id="@+id/root"
    android:layout_width="match_parent"
    android:layout_height="match_parent"
    android:orientation="vertical">
    <TextView
        android:id="@+id/textview"
        android:layout_width="match_parent"
        android:layout_height="wrap_content"
        android:text="还有诗和远方" />
    <Button
        android:id="@+id/button"
        android:layout_width="match_parent"
        android:layout_height="wrap_content"
        android:text="切换图片并用消息提示编辑框内容" />
    <EditText
        android:id="@+id/editText"
        android:layout_width="match_parent"
        android:layout_height="wrap_content"
```

```
        android:text="" />
</LinearLayout>
```

利用 Java 代码向线性布局 root 容器添加一个图像控件 ImageView，并初始化显示第一张图片；为视图中按钮控件添加点击监听事件，用于控制图片的切换显示，并显示消息提示。MainActivity.java 的 onCreate()方法主要实现代码如下：

```
protected void onCreate(Bundle savedInstanceState) {
    super.onCreate(savedInstanceState);
    setContentView(R.layout.activity_main);
    //获取 LinearLayout 布局容器
    LinearLayout main = (LinearLayout) findViewById(R.id.root);
    //创建 ImageView 控件
    final ImageView image = new ImageView(this);
    //将 ImageView 控件添加到 LinearLayout 布局容器中
    main.addView(image);
    image.setImageResource(R.drawable.pic1);
    ViewGroup.LayoutParams params = image.getLayoutParams();
    params.height = 360;
    params.width = 420;
    image.setLayoutParams(params);
    image.setScaleType(FIT_XY);
    //以 findViewById()方法取得 EditText 对象
    final EditText myEditText = (EditText)findViewById(R.id.editText);
    Button btn = findViewById(R.id.button);
    btn.setOnClickListener(new View.OnClickListener() {
        @Override
        public void onClick(View v) {
            image.setImageResource(R.drawable.pic2);
            //得到由用户输入 EditText 的文字内容
            String str = myEditText.getText().toString();
            /* 通过 Toast 的静态方法 makeText()创建一个 Toast 对象，该方法的参
            数分别为上下文、显示的文本、显示的时间长短，显示的时间还可以设置
            为 Toast.LENGTH_SHORT，这样显示的时间会相对短一些，
            然后调用 show()方法显示该 Toast  */
            Toast.makeText(MainActivity.this, "编辑框内容：" + str,
                    Toast.LENGTH_LONG).show();
        }
    });
}
```

运行 ImageDemo 模块，如图 2-4 所示。

图 2-4　切换图片并用消息提示编辑框内容

2.3 基本 UI 组件

View 是 Android 的基本视图，所有控件和布局都是由 View 类直接或间接派生而来的，故 View 类的基本属性和方法都是各控件和布局通用的。

Android 系统 UI 界面通常采用 XML 布局文件进行描述，每个控件均用特有的标签来表示，通过 Android 提供的属性来描述控件特性，其表示如下：

```
<控件标签 android属性="属性值"></控件标签>
```

其中，Android 系统为常见控件提供了一些通用属性，这些属性也是视图 Vie 在 XML 布局文件中常用的属性，如表 2-1 所示。

表 2-1 Android 控件通用属性

属性名称	属 性 值	描 述
android:id	@+id/<控件 ID>	标识控件 ID。@表示后面的字符串是 ID 资源；+表示新资源名称，如果不是新资源则可省略+
android:layout_width	wrap_content\fill_parent\match_parent	描述控件的宽（高）度。wrap_content 表示填满父控件的空白，继承父控件的大小。fill_parent\match_parent 表示大小刚好足够显示当前控件里的内容。在 Android 2.2 中推荐用 match_parent 代替 fill_parent。
android:layout_height		
android:padding	数值	描述控件内容与控件边框的距离
android:margin	数值	描述控件之间的距离
android:text	字符串	描述控件文本值

2.3.1 TextView 和 EditText

1. 文本框控件 TextView

TextView 是一种显示文本信息的控件，该控件可直接继承 View，从功能上看，类似于网页中的标签 Label 控件。TextView 类声明如下：

```
public class TextView extends View implements ViewTreeObserver.OnPreDrawListener
```

已知的子类包括：Button、CheckedTextView、Chronometer、DigitalClock、EditText、TextView，如图 2-5 所示。

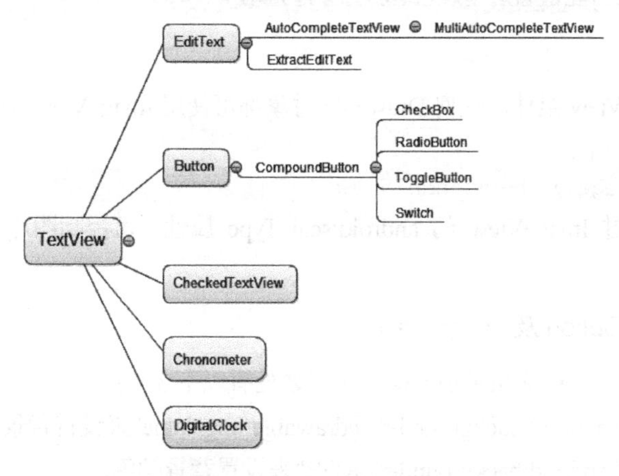

图 2-5 TextView 及其子类

TextView 控件在 XML 中的控件标签用<TextView>表示，对该控件的特性具有丰富的属性描述，可以查阅 Android 的官方 API 文档进一步学习。

2．编辑框控件 EditText

编辑框控件 EditText 是一种继承于 TextView 控件的可编辑文本框。EditText 类声明如下：

```
public class EditText extends TextView
```

EditText 控件在 XML 中的控件标签用<EditText>表示，EditText 常用属性如表 2-2 所示。

表 2-2　EditText 常用属性

属 性 名 称	描　述
android:digits	设置允许输入的字符
android:editable	设置是否可编辑
android:inputType	设置文本的类型，用于帮助输入法显示合适的键盘类型
android:hint	设置当文本框内容为空时，默认显示的提示文本
android:numeric	设置与文本框关联的数值输入法，其属性值包括以下内容。 integer：关联整数输入法； signed：允许输入符号的数值输入法； decimal：允许输入小数点的数值输入法

2.3.2　Button、ImageButton、ImageView、RadioButton 和 CheckBox

1．Button 与 ImageButton

Button 类的声明如下：

```
public class Button extends TextView
```

Button 在 XML 中的控件标签用<Button>表示，如果为按钮添加 android:background 属性，则为按钮添加了背景。

ImageButton 显示一个可以被用户点击的图片按钮，默认情况下，ImageButton 看起来像一个普通的按钮。按钮的图片可以通过<ImageButton> XML 元素的 android:src 属性或 setImageResource(int)方法指定。要删除按钮的背景，可以定义背景图片或设置背景为透明，如<ImageButton android:src= "@drawable/ok"/>。

Image Button 类的声明如下：

```
public class ImageButton extends ImageView
```

2．ImageView

ImageView 继承自 View 组件，任何 Drawable 对象都可使用 ImageView 来显示。

ImageView 类声明如下：

```
public class ImageView extends View
```

值得注意的是，通过 ImageView 的 android:scaleType 属性，可控制图片（调整大小/移动）匹配 ImageView 的 size。

3．Button、ImageButton 及 ImageView

（1）Button、ImageButton 及 ImageView 的相似之处有以下 3 点：

① 三者均可以通过 android:background="@drawable/imgName"来设置背景图片；

② 三者均可以通过 android:background="#RGB"来设置背景颜色；

③ 三者均可以定义用户触摸时的响应方法，如常用的设置点击事件监听器 setOnClickListener()对应的 onClick()方法。

（2）Button、ImageButton 及 Imageview 的不同之处有以下 3 点：

① 资源图片。Button 不支持 android:src="@drawable/imgName"属性，而 ImageButton 和 ImageView 可以通过该属性来设置资源图片（前景图较易理解）。

② 文本内容。ImageButton 和 ImageView 不支持 android:text="@string/textContent"属性，而 Button 可以通过该属性来定义按钮名称。

③ 响应方法。ImageView 能够实现的方法有限，而 Button 和 ImageButton 相对来说多一些。具体可参考教学资源包中的 Ch2 工程的 ImageButtonDemo（模块）示例（Ch2\ImageButtonDemo）。

4. RadioButton 和 CheckBox

RadioButton 和 CheckBox 分别表示可以选择单个或多个选项的控件，除了与 Button 控件外观不同，还比 Button 控件多了一个表示选中的属性 android:checked，该属性可以用于显示被选中的项。

需要注意的是，RadioButton 与 CheckBox 不同：一组单选按钮只能有一个被选中，为了表示这些单选按钮在一个"组"中，同组的单选按钮需要放置在一个单选按钮组的 RadioGroup 容器中，从而达到单选功能。

RadioButton、RadioGroup 和 CheckBox 类的声明分别如下：

```
public class RadioButton extends CompoundButton
public class RadioGroup extends LinearLayout
public class CheckBox extends CompoundButton
```

其中，RadioButton 控件在 XML 中的控件标签用<RadioButton>表示；CheckBox 控件在 XML 中的控件标签用<CheckBox>表示；RadioGroup 控件在 XML 中的控件标签用<RadioGroup>表示。相应示例参见教学资源包中的 Ch2 工程。

2.3.3 ProgressBar、ScrollView 和 Toast

1. ProgressBar

ProgressBar 是 UI 控件中用于显示某个时刻操作完成的进度控件。该控件应随时间、操作完成程度而实时变化，从而达到提醒用户任务进度的目的。

ProgressBar 类声明如下：

```
public class ProgressBar extends View
```

其中，ProgressBar 控件在 XML 中的控件标签用<ProgressBar>表示，通过 style 属性可以为 ProgressBar 指定显示风格。例如，

- @android:style/Widget. ProgressBar.Horizontal：表示水平进度条。
- @android:style/Widget. ProgressBar.Inverse：表示圆形进度条。

与此同时，为了更好地控制进度，ProgressBar 支持如表 2-3 所示的属性。

表 2-3 ProgressBar 属性

属性名称	描述
android:max	设置进度条的最大值
android:progress	设置进度条已完成的进度值
android: progressDrawable	设置进度条轨道对应的 Drawable 资源
android:indeterminate	设置是否精确显示进度
android: indeterminateDrawable	设置不显示进度的进度条 Drawable 资源
android: indeterminateDuration	设置不精确显示进度的持续时间

ProgressBar 还提供了动态控制进度条的方法如下。

- setProgress(int)：利用 Java 代码设置进度条的完成百分比。
- incrementProgressBy(int)：利用 Java 代码设置进度条的进度变化，其中，int 表示变化量，其为正数时表示增加，为负数时表示减小。

相应示例参见教学资源包中的 Ch2 工程。

2. ScrollView

ScrollView 是为普通控件添加滚动条的控件。该控件每次只能作用于一个普通控件以控制其滚动。ScrollView 类声明如下：

```
public class ScrollView extends FrameLayout
```

ScrollView 控件在 XML 中的控件标签用<ScrollView>表示。该滚动条是一个垂直滚动条，与之相反，Android 还提供水平滚动条控件 HorizontalScrollView，其使用方法与 ScrollView 相同。相应示例参见教学资源包中的 Ch2 工程。

3. Toast

在某些情况下需要向用户弹出提示消息，显示此类消息的目的仅仅是提醒用户，如显示操作完成状况、收到短消息等，Android 提供了一类消息提示框 Toast，它会随着设定的时间自动消失而不会获得焦点。Toast 类声明如下：

```
public class Toast extends Object
```

Toast 可以通过 makeText()方法创建 Toast 对象并设置相关属性，以及调用 show()方法显示提示，其示例代码如下：

```
Toast.makeText(MainActivity.this, "提示的内容", Toast.LENGTH_LONG).show();
```

其中，makeText()方法的第一个参数是上下文对象，第二个参数是显示的内容，第三个参数是显示的时间，只有 LONG 和 SHORT 这两种会生效。

2.3.4 日期时间类组件

1. AnalogClock（模拟时钟）

AnalogClock 类声明如下：

```
public class AnalogClock extends View
```

【示例】 启动 Android Studio，在 Ch2 工程中创建 AnalogClockDemo 模块。

Ch2\AnalogClockDemo\src\main\res\layout\activity_main.xml 文件内容如下：

```
<LinearLayout xmlns:android="http://schemas.android.com/apk/res/android"
   android:layout_width="match_parent"
   android:layout_height="match_parent"
   android:orientation="vertical">
   <AnalogClock
      android:id="@+id/clock"
      android:layout_width="wrap_content"
      android:layout_height="wrap_content" />
</LinearLayout>
```

运行 AnalogClockDemo 模块，其界面如图 2-6 所示。

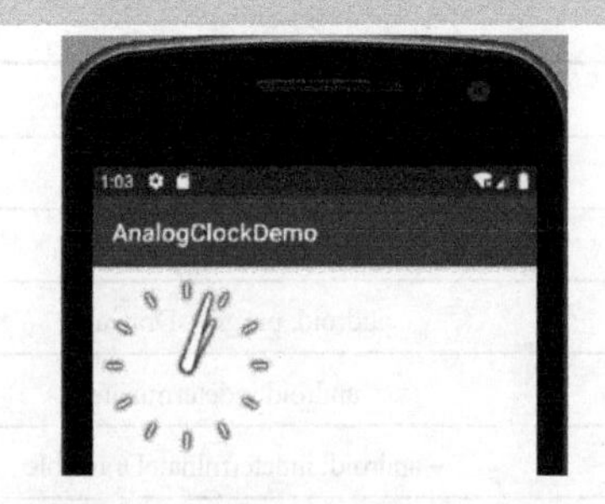

图 2-6 运行 AnalogClockDemo 模块的界面

2. DatePicker（日期选择器）

DatePicker 类声明如下：

```
public class DatePicker extends FrameLayout
```

【示例】 启动 Android Studio，在 Ch2 工程中创建

DatePickerDemo 模块。

Ch2\DatePickerDemo\src\main\res\layout\activity_main.xml 文件内容如下：

```
<?xml version="1.0" encoding="utf-8"?>
<LinearLayout xmlns:android="http://schemas.android.com/apk/res/android"
  android:layout_width="match_parent"
  android:layout_height="match_parent"
  android:orientation="vertical">
  <DatePicker
     android:id="@+id/datePicker"
     android:layout_width="wrap_content"
     android:layout_height="wrap_content"
     android:datePickerMode="spinner" />
</LinearLayout>
```

DatePickerDemo 模块的 MainActivity.java 主要代码如下：

```
public class MainActivity extends AppCompatActivity implements DatePicker.
OnDateChangedListener {
  @Override
  protected void onCreate(Bundle savedInstanceState) {
     super.onCreate(savedInstanceState);
     setContentView(R.layout.activity_main);
     DatePicker datePicker = (DatePicker) findViewById(R.id.datePicker);
     Calendar calendar = Calendar.getInstance();
     int year = calendar.get(Calendar.YEAR);
     int monthOfYear = calendar.get(Calendar.MONTH);
     int dayOfMonth = calendar.get(Calendar.DAY_OF_MONTH);
     datePicker.init(year, monthOfYear, dayOfMonth, this);
  }
  @Override
  public void onDateChanged(DatePicker view, int year, int monthOfYear, int
dayOfMonth) {
     Toast.makeText(MainActivity.this,
        "您选择的日期是: " + year + "年" + (monthOfYear + 1) + "月" +
      dayOfMonth + "日!", Toast.LENGTH_LONG).show();
  }
}
```

运行 DatePickerDemo 模块，其界面如图 2-7 所示。

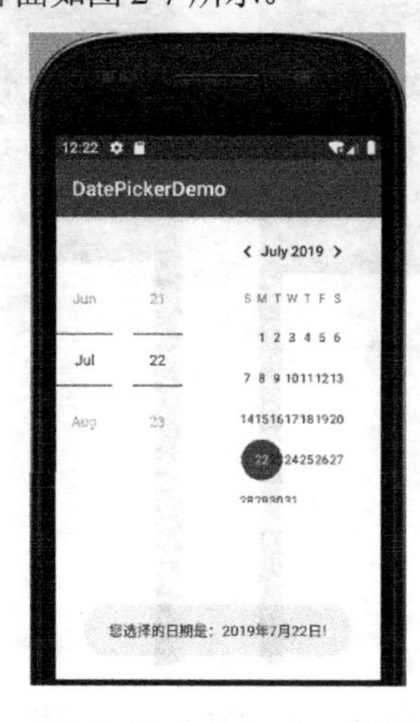

图 2-7　运行 DatePickerDemo 模块的界面

3. TimePicker（时间选择器）

TimePicker 类声明如下：

```
public class TimePicker extends FrameLayout
```

【示例】 启动 Android Studio，在 Ch2 工程中创建 TimePickerDemo 模块。

Ch2\TimePickerDemo\src\main\res\layout\activity_main.xml 文件内容如下。

```
<LinearLayout xmlns:android="http://schemas.android.com/apk/res/android"
    android:layout_width="match_parent"
    android:layout_height="match_parent"
    android:orientation="vertical">
    <TimePicker
        android:id="@+id/timePicker"
        android:layout_width="wrap_content"
        android:layout_height="wrap_content"
        android:timePickerMode="clock" />
    <!--android:timePickerMode="spinner" 或 clock, 设置显示样式-->
</LinearLayout>
```

TimePickerDemo 模块的 MainActivity.java 主要代码如下：

```
public class MainActivity extends AppCompatActivity {
    @Override
    protected void onCreate(Bundle savedInstanceState) {
        super.onCreate(savedInstanceState);
        setContentView(R.layout.activity_main);
        TimePicker timePicker = (TimePicker) findViewById(R.id.timePicker);
        timePicker.setOnTimeChangedListener(new
      TimePicker.OnTimeChangedListener() {
            @Override
            public void onTimeChanged(TimePicker view, int hourOfDay, int minute) {
                Toast.makeText(MainActivity.this,"您选择的时间是："+
              hourOfDay+"时"+minute+"分!",Toast.LENGTH_LONG).show();
            }
        });
    }
}
```

运行 TimePickerDemo 模块，其界面如图 2-8 所示。

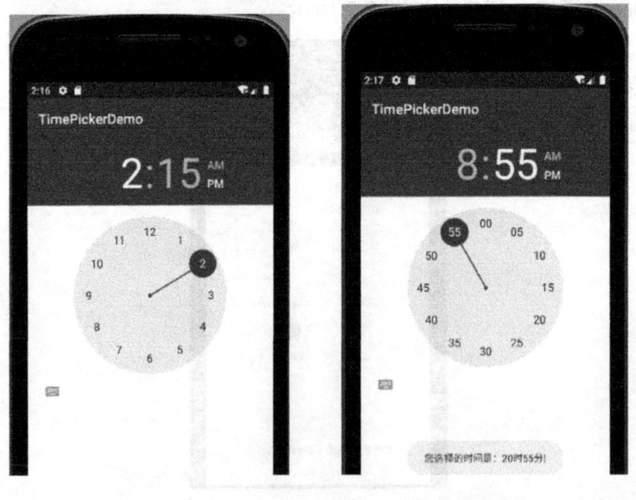

图 2-8　运行 TimePickerDemo 模块的界面

2.3.5 布局管理器

Android 布局管理器本身就是一个 UI 控件，所有的布局管理器都是 ViewGroup 的子类。

布局管理器可以包含 UI 组件，也可以包含其他布局管理器，因此，可以将其看成是一个 ViewGroup 对象。通过多层布局的嵌套能够完成一些比较复杂的界面实现。

LayoutParams 表示布局参数，子 View 通过 LayoutParams 告诉父容器（ViewGroup）应该如何放置自己。LayoutParams 与 ViewGroup 是息息相关的。

LayoutParams 携带了子控件针对父控件的信息，告诉父控件如何放置自己。

LayoutParams 类也只是简单描述了宽和高，宽和高都可以设置成以下 3 种值：

（1）一个确定的值；

（2）MATCH_PARENT，即填满（和父容器大小一样）；

（3）WRAP_CONTENT，即包裹住组件就好。

小贴士：LayoutParams 继承了 Android.View.ViewGroup.LayoutParams，相当于一个 Layout 的信息包，它封装了 Layout 的位置、高、宽等信息。假设屏幕上的一块区域是由一个 Layout 占领的，若将一个 View 添加到一个 Layout 中，应将一个认可的 LayoutParams 对象传递进去。事实上，每个 ViewGroup 的子类都有自己对应的 LayoutParams 类，典型的如 LinearLayout.LayoutParams 和 FrameLayout.LayoutParams 等，可以看出，LayoutParams 是对应 ViewGroup 子类的内部类。基础的 LayoutParams 是定义在 ViewGroup 中的静态内部类，封装着 View 的宽度和高度信息，对应着 XML 中的 layout_width 和 layout_height 属性。布局管理器一般提供了相应的 LayoutParams 内部类，该内部类用于控制其子元素支持指定 android:layout_gravity 属性，该属性设置该子元素在父容器中的对齐方式。与 android:layout_gravity 相似的属性还有 android: gravity 属性，该属性用于控制其子元素的对齐方式。

1. 线性布局

LinearLayout 类声明如下：

```
public class LinearLayout extends ViewGroup
```

线性布局 LinearLayout 是 Android 系统中最基础的一种布局。采用自上而下或从左往右的方式一个元素接着一个元素排列，当排列的元素超出屏幕范围时，超出部分将做隐藏处理，如表 2-4 所示。

表 2-4 LinearLayout 的常用 XML 属性及对应方法

属性名称	对应方法	描述
android:baselineAligned	setBaselineAligned(boolean)	设置为 false，将会阻止该布局管理器与其子元素的基线对齐
android:gravity	setGravity(int)	设置布局管理器内组件的对齐方式。该属性支持对齐的方式有 top、bottom、left、right、center_vertical、fill_vertical、center_horizontal、fill_horizontal、center、fill、clip_vertical、clip_horizontal。它可以只有一种对齐方式，也可以是多种对齐方式的组合，如 top|left 表示左上对齐
android:orientation	setOrientation(int)	设置排列方式，有 horizontal（水平排列）和 vertical（垂直排列）两种

在线性布局中有一个非常重要的属性，即 android:layout_weight。该属性的作用是分配空间，设置控件在水平（垂直）方向和宽度（高度）的比例。这个属性允许使用比例的方式来指定控件的大小，它在手机屏幕的适配性方面起到了非常重要的作用。

layout_weight 指定当前视图的宽或高占上级线性布局的权重，其并非在当前 LinearLayout 节点中设置属性，而是在下级视图的节点中设置。如果 layout_weight 指定的是当前视图在宽度上占的权重，layout_width 就要同时设置为 0dp；如果 layout_weight 指定的是当前视图在高度上占的权重，layout_height 就要同时设置为 0dp。

线性布局 LinearLayout 应掌握的主要内容如图 2-9 所示，相应示例参见教学资源包中的 Ch2 工程。

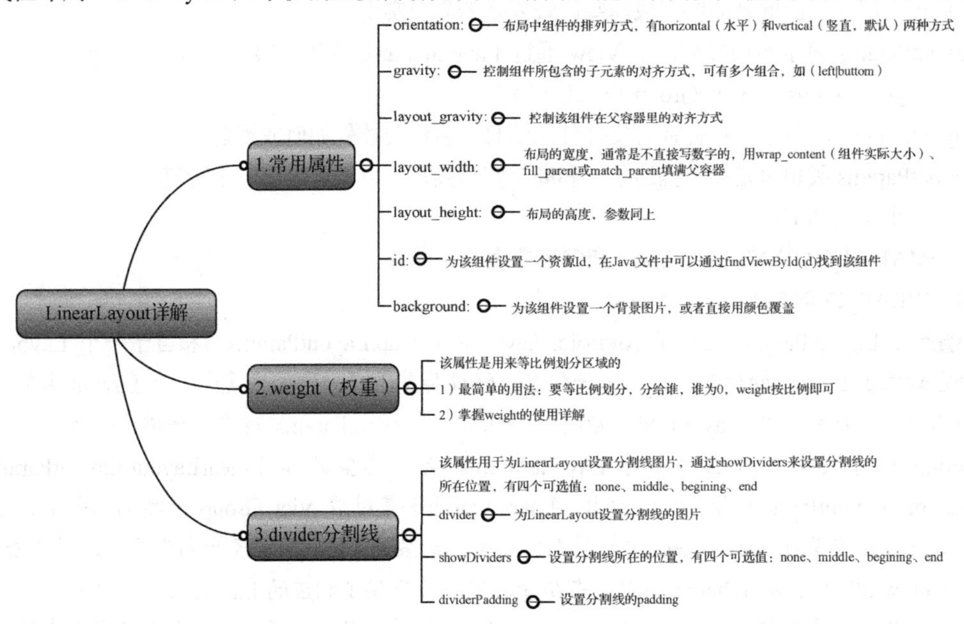

图 2-9　线性布局 LinearLayout 应掌握的主要内容

2. 相对布局

RelativeLayout 类声明如下：

```
public class RelativeLayout extends ViewGroup
```

相对布局 RelativeLayout 是位置相对于兄弟组件或布局容器来布局的一种布局管理器。Android 为相对于布局容器提供了 7 个属性，如表 2-5 所示。

表 2-5　相对于布局容器属性

属 性 名 称	描　述
android:layout_centerHorizontal	相对于布局容器水平居中
android:layout_centerVertical	相对于布局容器垂直居中
android:layout_centerInParent	相对于布局容器中心位置
android:layout_alignParentBottom	相对于布局容器底端对齐
android:layout_alignParentLeft	相对于布局容器左侧对齐
android:layout_alignParentRight	相对于布局容器右侧对齐
android:layout_alignParentTop	相对于布局容器顶部对齐

Android 为相对于兄弟组件提供了 8 个属性，如表 2-6 所示，相应示例参见教学资源包中的 Ch2 工程。

表 2-6　相对于兄弟组件属性

属 性 名 称	描　述
android:layout_toRightOf	在给出 ID 组件的右侧

续表

属 性 名 称	描　述
android:layout_toLeftOf	在给出 ID 组件的左侧
android:layout_above	在给出 ID 组件的上方
android:layout_below	在给出 ID 组件的下方
android:layout_alignLeft	与给出 ID 组件的左侧边界对齐
android:layout_alignRight	与给出 ID 组件的右侧边界对齐
android:layout_alignTop	与给出 ID 组件的上部边界对齐
android:layout_alignBottom	与给出 ID 组件的底部边界对齐

3．帧布局

FrameLayout 类声明如下：

```
public class FrameLayout extends ViewGroup
```

帧布局 FrameLayout 是最简单的布局方式。帧布局向屏幕绘制了一个空白的矩形区域，所有添加到这个布局中的视图都会以层叠的方式显示。第一个添加到框架布局中的视图显示在底层，最后一个被放在顶层，上一层的视图会覆盖下一层的视图，因此帧布局类似于堆栈布局，相应示例参见教学资源包中的 Ch3 工程。

4．表格布局

TableLayout 类声明如下：

```
public class TableLayout extends LinearLayout
```

表格布局 TableLayout 是以表格形式布局界面的一种管理器，它与传统所理解的表格略有区别，表格布局本质上是一种线性布局，它继承了 LinearLayout，通过 TableRow 来表示表格中行的概念，一个 TableRow 表示一行，每一行中有多少个组件即表示有多少列，当一个组件单独放在 TableLayout 中表示独占一整行。

在表格布局中，列的宽度由该列中最宽的那个单元格指定，而表格的宽度由父容器指定。在 TableLayout 中，可以为列设置 3 种属性。

- Shrinkable：如果一个列被标识为 Shrinkable，则表示该列的宽度可以收缩，以使表格能够适应其父容器的大小。
- Stretchable：如果一个列被标识为 Stretchable，则表示该列的宽度可以拉伸，以填满表格中的空闲空间。
- Collapsed：如果一个列被标识为 Collapsed，则表示该列会被隐藏。

注意：一个列可以同时具有 Shrinkable 属性和 Stretchable 属性，在这种情况下，该列的宽度将任意拉伸或收缩以适应父容器。

TableLayout 还支持如下属性，如表 2-7 所示，相应示例参见教学资源包中的 Ch2 工程。

表 2-7　TableLayout 支持的属性

属 性 名 称	描　述
android:collapseColumns	设置指定列号的列属性为 Collapsed
android:shrinkColumns	设置指定列号的列属性为 Shrinkable
android:stretchColumns	设置指定列号的列属性为 Stretchable

5. 网格布局

GridLayout 类声明如下：

```
public class GridLayout extends ViewGroup
```

网格布局 GridLayout 是 Android 4.0 以后新增的一种布局管理器，该布局将屏幕中某个矩形区域划分成排列整齐的若干个单元格，与表格布局相比，网格布局更接近于人们所理解的表格。对于一个网格布局来说，需要明确划分成几行几列，为此，Android 为网格布局提供了多个属性来完善网格布局，如表 2-8 所示，相应示例参见教学资源包中的 Ch2 工程。

表 2-8　GridLayout 属性

属 性 名 称	描　述
android:columnCount	设置网格布局的列数
android:rowCount	设置网格布局的行数
android:alignmentMode	设置网格布局的对齐模式
android:useDefaultMargins	是否使用默认的边距

Android 还为网格布局的单元格提供了一些属性来增加网格布局的功能，常用属性有 android:rowSpan 和 android:columnSpan，这两个属性类似于表格中合并行、合并列的功能。

6. 约束布局

ConstraintLayout 类声明如下：

```
public class ConstraintLayout extends ViewGroup
```

约束布局 ConstraintLayout 是一个 ViewGroup，可以在 API 9 以上的 Android 系统中使用，它的出现主要是为了解决布局嵌套过多的问题，以灵活的方式定位和调整小部件。从 Android Studio 2.3 起，官方的模板默认使用 ConstraintLayout。

ConstraintLayout 可以有效地解决布局嵌套过多的问题。复杂的布局往往伴随着多层嵌套，而嵌套越多，程序性能也就越差。ConstraintLayout 使用约束的方式来指定各个控件的位置和关系，有些类似于 RelativeLayout，但远比 RelativeLayout 更强大。

传统的 Android 开发，界面基本是靠编写 XML 代码完成的，虽然 Android Studio 也支持使用可视化方式来编写界面，但是操作起来并不方便。ConstraintLayout 适合使用可视化方式编写界面，但并不太适合使用 XML 方式。约束布局的缺点是实际使用时比较烦琐。为了使用 ConstraintLayout，需要在模块 build.gradle 文件中添加 ConstraintLayout 的依赖，其代码如下：

```
implementation 'com.android.support.constraint:constraint-layout:1.1.3'
```

2.4　高级 UI 组件

2.4.1　列表类组件

1. Adapter 接口

Android 里的 Adapter 接口是 View 视图与 data 数据之间的桥梁，Adapter 接口提供对数据的访问，也负责为每一项数据产生一个对应的视图。Adapter 接口声明如下：

```
public interface Adapter
```

Adapter 接口及其实现类的继承关系如图 2-10 所示。

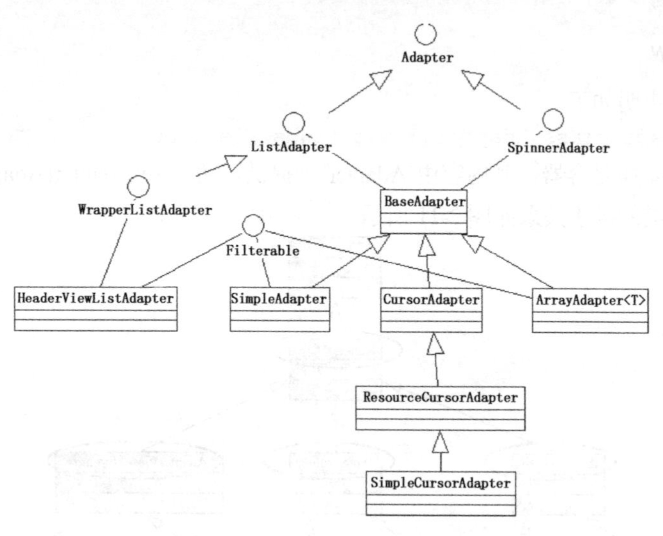

图 2-10　Adapter 接口及其实现类的继承关系

Adapter 接口派生了 ListAdapter 接口和 SpinnerAdapter 接口，ListAdapter 接口为 AbsListView 提供列表项，SpinnerAdapter 接口为 AbsSpinner 提供列表项。Adapter 接口常用的实现类如下。

（1）ArrayAdapter。

ArrayAdapter 通常由数组或 List 集合表示列表项数据源，将其绑定给 ListView，从而以列表项形式显示在界面上。ArrayAdapter 类声明如下：

```
public class ArrayAdapter extends BaseAdapter implements Filterable
```

（2）SimpleAdapter。

SimpleAdapter 是在 ArrayAdapter 基础上，能够更丰富地显示列表项信息的一种适配器，列表的每一项可以包含更多的内容，通过配置文件布局，使界面样式更多样化。

SimpleAdapter 类声明如下：

```
public class SimpleAdapter extends BaseAdapter implements Filterable
```

（3）BaseAdapter

BaseAdapter 类声明如下：

```
public abstract class BaseAdapter extends Object implements ListAdapter
SpinnerAdapter
```

BaseAdapter 通常被扩展，扩展后的 BaseAdapter 可以自定义适配器，实现对各列表项进行最大限度的设定，需要重写的 4 个方法如下。

- getCount()：返回列表项的数目。
- getItem(int position)：返回值决定第 position+1 项的列表项内容。
- getItemId(int position)：获取第 position 处列表项的 ID。
- getView(int position, View convertView , ViewGroup parent)：返回值决定第 position 处的列表项组件，即第 position 处的列表项视图界面。

（4）SimpleCursorAdapter。

SimpleCursorAdapter 是一个简单的适配器，将游标的数据映射到布局文件的 TextView 控件或 ImageView 控件中。

SimpleCursorAdapter 封装了 Cursor 提供的数据。

SimpleCursorAdapter 类声明如下：

```
public class SimpleCursorAdapter extends ResourceCursorAdapter
```

相应示例参见教学资源包中的 Ch3 工程。

2. AdapterView

AdapterView 类声明如下：

```
public abstract class AdapterView extends ViewGroup
```

AdapterView 的本质是容器，其内容由 Adapter 来提供，通过 setAdapter(Adapter)的方法来设置。AdapterView 及其子类的继承关系如图 2-11 所示。

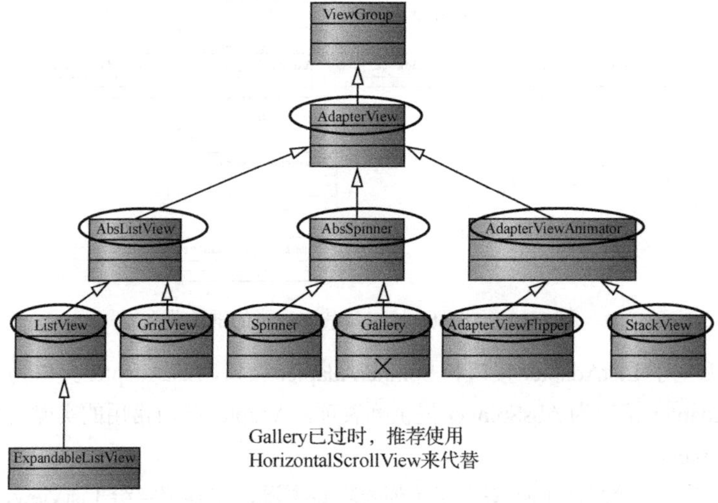

图 2-11　AdapterView 及其子类的继承关系

3. Spinner（下拉框）

Spinner 类声明如下：

```
public class Spinner extends AbsSpinner implements DialogInterface.OnClickListener
```

Spinner 是一个下拉列表框，需要给它指定对应的数据源。Spinner 数据源通过指定 android:entries 属性或 Adapter 方式设定数据源。

【示例】 启动 Android Studio，在 Ch2 工程中创建 SpinnerDemo 模块。

Ch2\SpinnerDemo\src\main\res\layout\activity_main.xml 文件内容如下。

```
<LinearLayout xmlns:android="http://schemas.android.com/apk/res/android"
    android:orientation="vertical"
    android:padding="10dip"
    android:layout_width="fill_parent"
    android:layout_height="wrap_content">
    <TextView
        android:layout_width="fill_parent"
        android:layout_height="wrap_content"
        android:layout_marginTop="10dip"
        android:text="@string/planet_prompt"
        android:textColor="@android:color/holo_red_dark"
        android:textSize="18sp"
        android:textStyle="bold" />
    <Spinner
        android:id="@+id/spinner"
        android:layout_width="fill_parent"
        android:layout_height="wrap_content"
        android:prompt="@string/planet_prompt"
        android:spinnerMode="dropdown" />
</LinearLayout>
```

在 Ch2\SpinnerDemo\src\main\res\values\strings.xml 文件中添加字符串和字符串数组。

```
<resources>
   <string name="planet_prompt">Choose a planet</string>
   <string-array name="planets_array">
      <item>Mercury</item>
      <item>Venus</item>
      <item>Earth</item>
      <item>Mars</item>
      <item>Jupiter</item>
      <item>Saturn</item>
      <item>Uranus</item>
      <item>Neptune</item>
   </string-array>
</resources>
```

SpinnerDemo 模块的 MainActivity.java 主要代码如下：

```
public class MainActivity extends AppCompatActivity {
   @Override
   protected void onCreate(Bundle savedInstanceState) {
      super.onCreate(savedInstanceState);
      setContentView(R.layout.activity_main);
      Spinner spinner = (Spinner) findViewById(R.id.spinner);
      ArrayAdapter<CharSequence> adapter = ArrayAdapter.createFromResource(
            this,                                   R.array.planets_array,
android.R.layout.simple_spinner_item);

adapter.setDropDownViewResource(android.R.layout.simple_spinner_dropdown_item);
      spinner.setAdapter(adapter);
      spinner.setOnItemSelectedListener(new MyOnItemSelectedListener());
   }
   public class MyOnItemSelectedListener implements AdapterView.OnItemSelected
Listener {
      public void onItemSelected(AdapterView<?> parent, View view, int pos, long
id) {
         Toast.makeText(parent.getContext(), "The planet is " +
               parent.getItemAtPosition(pos).toString(),
Toast.LENGTH_LONG).show();
      }
      public void onNothingSelected(AdapterView parent) {
         //Do nothing.
      }
   }
}
```

运行 SpinnerDemo 模块，其界面如图 2-12 所示。

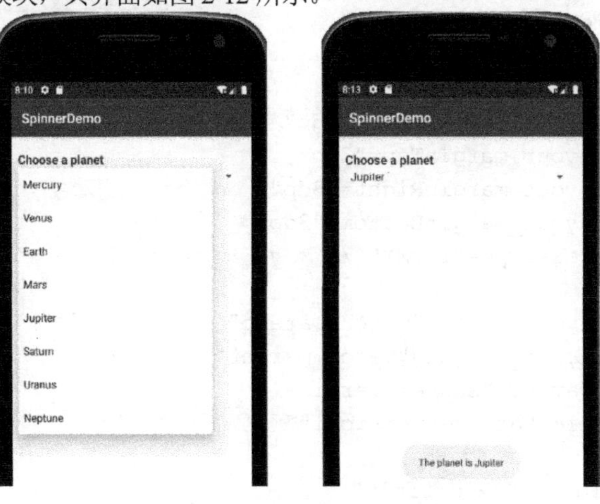

图 2-12　运行 SpinnerDemo 模块的界面

4. ListView

ListView 类声明如下：

```
public class ListView extends AbsListView
```

ListView 是手机中使用非常频繁的一类控件，以垂直的方式显示列表项，使显示信息更加清晰明了。Android 为 ListView 提供了多种属性，如表 2-9 所示。

表 2-9 ListView 的常用属性

属性名称	描述
android:divider	设置列表项之间的分隔条
android:dividerHeight	设置分隔条的高度
android:entries	设置列表项的信息（数组资源）
android:footerDividersEnabled	设置是否在 footer 项前绘制分隔条
android:headerDividersEnabled	设置是否在 header 项后绘制分隔条

【示例】 启动 Android Studio，在 Ch2 工程中创建 ListViewDemo 模块。

Ch2\ListViewDemo\src\main\res\layout\activity_main.xml 文件内容如下：

```
<?xml version="1.0" encoding="utf-8"?>
<LinearLayout xmlns:android="http://schemas.android.com/apk/res/android"
   android:layout_width="match_parent"
   android:layout_height="match_parent"
   android:orientation="vertical">
   <ListView
      android:id="@+id/listview"
      android:layout_width="match_parent"
      android:layout_height="match_parent"
      android:layout_marginTop="5sp"
      android:divider="#FFCC00"/>
</LinearLayout>
```

Ch2\ListViewDemo\src\main\res\layout\goods_item.xml 文件内容如下：

```
<?xml version="1.0" encoding="utf-8"?>
<LinearLayout xmlns:android="http://schemas.android.com/apk/res/android"
   android:layout_width="match_parent"
   android:layout_height="match_parent">
   <ImageView
      android:id="@+id/imageView"
      android:layout_width="80dp"
      android:layout_height="60dp"
      android:layout_marginLeft="6dp"
      android:layout_marginTop="3dp"
      android:layout_marginRight="8dp"
      android:layout_marginBottom="3dp"
      android:scaleType="fitXY" />
   <LinearLayout
      android:layout_width="match_parent"
      android:layout_height="match_parent"
      android:gravity="center_vertical"
      android:orientation="vertical">
      <TextView
         android:id="@+id/title"
         android:layout_width="match_parent"
```

```
            android:layout_height="wrap_content"
            android:text="Title"
            android:textSize="18sp"
            android:textStyle="bold" />
        <TextView
            android:id="@+id/desc"
            android:layout_width="match_parent"
            android:layout_height="wrap_content"
            android:text="Description"
            android:textColor="@android:color/holo_red_dark"
            android:textSize="24sp"
            android:textStyle="bold" />
    </LinearLayout>
</LinearLayout>
```

Ch2\ListViewDemo\src\main\java\com\mialab\listviewdemo\GoodsAdapter.java 的主要代码如下：

```
public class GoodsAdapter extends BaseAdapter {
    private Context context;
    private ListView listView;
    private List<GoodsInfo> goodsList;
    public GoodsAdapter(Context context, ListView listView) {
        this.context = context;
        this.listView = listView;
    }
    public void setData(List<GoodsInfo> data){
        goodsList = data;
    }
    @Override
    public int getCount() {
        return goodsList.size();
    }
    @Override
    public Object getItem(int position) {
        return null;
    }
    @Override
    public long getItemId(int position) {
        return 0;
    }
    @Override
    public View getView(int position, View convertView, ViewGroup parent) {
        convertView = View.inflate(context,R.layout.goods_item,null);
        ViewHolder viewHolder = new ViewHolder(convertView);
        GoodsInfo goodsInfo = goodsList.get(position);
        viewHolder.img.setImageResource(goodsInfo.getImgId());
        viewHolder.title.setText(goodsInfo.getTitle());
        viewHolder.desc.setText(goodsInfo.getDesc());
        return convertView;
    }
    static class ViewHolder {
        View baseView;
        TextView title;
        TextView desc;
        ImageView img;
        public ViewHolder(View itemView) {
```

```
            this.baseView = itemView;
            this.title = (TextView) baseView.findViewById(R.id.title);
            this.desc = (TextView) baseView.findViewById(R.id.desc);
            this.img = (ImageView)baseView.findViewById(R.id.imageView);
        }
    }
}
```

Ch2\ListViewDemo\src\main\java\com\mialab\listviewdemo\bean\GoodsInfo.java 的主要代码如下：

```
public class GoodsInfo {
    private String title;
    private String desc;
    private int imgId;
    public String getTitle() {
        return title;
    }
    public void setTitle(String title) {
        this.title = title;
    }
    public String getDesc() {
        return desc;
    }
    public void setDesc(String desc) {
        this.desc = desc;
    }
    public int getImgId() {
        return imgId;
    }
    public void setImgId(int imgId) {
        this.imgId = imgId;
    }
}
```

ListViewDemo 模块的 MainActivity.java 主要代码如下：

```
public class MainActivity extends AppCompatActivity {
    private List<GoodsInfo> goodsList;
    private ListView listView;
    private GoodsAdapter goodsAdapter;
    @Override
    protected void onCreate(Bundle savedInstanceState) {
        super.onCreate(savedInstanceState);
        setContentView(R.layout.activity_main);
        initData();
        listView = (ListView) findViewById(R.id.listview);
        goodsAdapter = new GoodsAdapter(MainActivity.this,listView);
        goodsAdapter.setData(goodsList);
        listView.setAdapter(goodsAdapter);
    }
    private void initData() {
        goodsList = new ArrayList<GoodsInfo>();
        for (int i = 0; i < 12; i++) {
            GoodsInfo goodsInfo = new GoodsInfo();
            goodsInfo.setTitle("宽带(100Mbps 带宽)");
            goodsInfo.setDesc("50 元/月");
            goodsInfo.setImgId(R.drawable.phone);
```

```
            goodsList.add(goodsInfo);
        }
    }
}
```

运行 ListViewDemo 模块，其界面如图 2-13 所示。

图 2-13 运行 ListViewDemo 模块的界面

5. RecyclerView

RecylerView 类声明如下：

```
public class RecyclerView  extends ViewGroup
implements ScrollingView, NestedScrollingChild
```

从 Android 5.0 开始，Google 公司推出了一个用于大量数据展示的、更加强大和灵活的新控件 RecyclerView，可以用来代替传统的 ListView。RecyclerView 能起到增强型列表的作用。

为了让 RecyclerView 在所有 Android 版本上能使用，Android 团队将 RecyclerView 定义在 support 库中。若使用 RecyclerView 控件，需要在 build.gradle 文件的 dependencies 闭包（节点）中加入以下代码，以导入 RecyclerView 库。

```
dependencies {
    implementation 'com.android.support:recyclerview-v7:28.0.0'
}
```

在使用 RecyclerView 时，必须指定一个适配器 Adapter 和一个布局管理器 LayoutManager。适配器继承 RecyclerView.Adapter 类，具体实现类似于 ListView 的适配器，取决于数据信息及展示的 UI。布局管理器用于确定 RecyclerView 中 Item 的展示方式，以及决定何时复用已经不可见的 Item，避免重复创建及执行高成本的 findViewById()方法。

（1）为 RecyclerView 准备一个适配器。

RecyclerView 有专门的适配器类，即 RecyclerView.Adapter。在调用 RecyclerView 的 setAdapter 方法前，需先实现一个从 RecyclerView.Adapter 派生而来的数据适配器，用来定义列表项的布局与具体操作。

创建适配器标准实现的步骤如下。

① 创建 Adapter：创建一个继承 RecyclerView.Adapter<ViewHolder>的 Adapter 类（或<ViewHolder>用<自定义 Adapter 类名.VH>代替）。

② 创建 ViewHolder：在自定义的 Adapter 中创建一个继承 RecyclerView.ViewHolder 的静态内部类，记为 VH。ViewHolder 的实现和 ListView 的 ViewHolder 实现几乎一样。

③ 在自定义 Adapter 中实现有 3 个方法：

- onCreateViewHolder()：先得到对应于列表 Item 的 View，再创建 ViewHolder 实例（实际上就是列表项组件），把 View 封装在 ViewHolder 中，然后返回 ViewHolder 实例，ViewHolder 类（实际是扩展 RecyclerView.ViewHolder 的子类）需要自己编写。使用该方法所创建的组件会被自动缓存。
- onBindViewHolder()：用于适配渲染数据到 View 中，为列表项组件绑定数据。
- getItemCount()：类似于 BaseAdapter 的 getCount()方法，即获得列表项的数目。

可以看出，RecyclerView 将 ListView 中 getView()的功能拆分成了 onCreateViewHolder()和 onBindViewHolder()。

（2）使用 LayoutManager 布局管理器。

LayoutManager 负责 RecyclerView 的布局，RecyclerView 在使用过程中要比 ListView 多一个

setLayoutManager 步骤，LayoutManager 用于控制 RecyclerView 的最终展示效果。

LayoutManager 是 RecyclerView 的内部类，其声明如下：

```
public static abstract class RecyclerView.LayoutManager extends Object
```

LayoutManager 只是抽象类，其实现类有 LinearLayoutManager、GridLayoutManager、StaggeredGridLayoutManager，这也是 RecyclerView 提供的 3 种具体布局管理器。

```
public class LinearLayoutManager extends RecyclerView.LayoutManager implements
ItemTouchHelper.ViewDropHandler {}
public class GridLayoutManager extends LinearLayoutManager {}
public class StaggeredGridLayoutManager  extends RecyclerView.LayoutManager {}
```

- LinearLayoutManager：以垂直或水平列表方式展示 Item。
- GridLayoutManager：以网格方式展示 Item。
- StaggeredGridLayoutManager：以瀑布流方式展示 Item。

可以看出，RecyclerView 与 ListView 相比，在使用上的主要区别如下：

- ViewHolder 的编写规范化了，不再由手工编写。
- RecyclerView 复用 Item 的工作已由 Google 公司实现，不再需要 ListView 自己调用 setTag。
- RecyclerView 使用多出的 LayoutManager 设置工作。

【示例】 启动 Android Studio，在 Ch2 工程中创建 RecyclerViewDemo 模块。

Ch2\RecyclerViewDemo\src\main\res\layout\activity_main.xml 文件内容如下：

```
<?xml version="1.0" encoding="utf-8"?>
<LinearLayout xmlns:android="http://schemas.android.com/apk/res/android"
   android:layout_width="match_parent"
   android:layout_height="match_parent"
   android:orientation="vertical">
   <android.support.v7.widget.RecyclerView
       android:id="@+id/recyclerView"
       android:layout_width="match_parent"
       android:layout_height="wrap_content" />
</LinearLayout>
```

com.mialab.recyclerviewdemo 包中的 GoodsAdapter.java 文件内容如下：

```
public class GoodsAdapter extends RecyclerView.Adapter<GoodsAdapter. ItemHolder>
{
   private List<GoodsInfo> mGoodsList;
   private Context mContext;
   public GoodsAdapter(Context context, List<GoodsInfo> goodsList) {
       this.mContext = context;
       this.mGoodsList = goodsList;
   }
   @Override
   public ItemHolder onCreateViewHolder(ViewGroup parent, int viewType) {
       View view = LayoutInflater.from(mContext).inflate(R.layout.goods_item,
parent, false);
       ItemHolder itemHolder = new ItemHolder(view);
       return itemHolder;
   }
   @Override
   public void onBindViewHolder(@NonNull ItemHolder itemHolder, int position)
{
       GoodsInfo goodsInfo = mGoodsList.get(position);
       //ItemHolder itemHolder = (ItemHolder) viewHolder;
       itemHolder.img.setImageResource(goodsInfo.getImgId());
```

```
        itemHolder.title.setText(goodsInfo.getTitle());
        itemHolder.desc.setText(goodsInfo.getDesc());
    }
    @Override
    public int getItemCount() {
        return mGoodsList.size();
    }
    //定义列表项的视图持有者
    static class ItemHolder extends RecyclerView.ViewHolder {
        TextView title;
        TextView desc;
        ImageView img;
        public ItemHolder(View view) {
            super(view);
            this.title = (TextView) view.findViewById(R.id.title);
            this.desc = (TextView) view.findViewById(R.id.desc);
            this.img = (ImageView) view.findViewById(R.id.imageView);
        }
    }
}
```

com.mialab.recyclerviewdemo 包中的 MainActivity.java 文件内容如下：

```
public class MainActivity extends AppCompatActivity {
    private List<GoodsInfo> goodsList;
    private GoodsAdapter goodsAdapter;
    @Override
    protected void onCreate(Bundle savedInstanceState) {
        super.onCreate(savedInstanceState);
        setContentView(R.layout.activity_main);
        initData();
        RecyclerView recyclerView = (RecyclerView)findViewById(R.id.recycler
View);
        //创建一个垂直方向的线性布局管理器
        LinearLayoutManager layoutManager = new
            LinearLayoutManager(this,LinearLayout.VERTICAL,false);
        //设置 recyclerView 的布局管理器
        recyclerView.setLayoutManager(layoutManager);
        goodsAdapter = new GoodsAdapter(MainActivity.this, goodsList);
        recyclerView.setAdapter(goodsAdapter);
    }

    private void initData() {
        goodsList = new ArrayList<GoodsInfo>();
        for (int i = 0; i < 12; i++) {
            GoodsInfo goodsInfo = new GoodsInfo();
            goodsInfo.setTitle("宽带(200Mbps 带宽)");
            goodsInfo.setDesc("60 元/月");
            goodsInfo.setImgId(R.drawable.phone);
            goodsList.add(goodsInfo);
        }
    }
}
```

运行 RecyclerViewDemo 模块，其界面类似于图 2-13。

小贴士：RecyclerView 封装了 ViewHolder 的回收复用，也就是说，RecyclerView 标准化了 ViewHolder，编写 Adapter 面向的是 ViewHolder 而不再是 View 了，复用的逻辑被封装了，写起来更加简单。直接省去了 ListView 中 convertView.setTag(holder)和 convertView.getTag()这些烦琐的步骤。

RecyclerView 提供了一种插拔式的体验，即高度解耦、异常灵活，针对一个 Item 的显示，RecyclerView 专门抽取出了相应的类，来控制 Item 的显示，使其扩展性更强。

RecyclerView 设置 LinearLayoutManager、GridLayoutManager、StaggeredGridLayoutManager 布局管理器以控制 Item 的布局方式，包括线性布局（纵向或横向）、网格布局、瀑布流布局等方式。

如果想实现控制横向或者纵向滑动列表的效果，可以通过 LinearLayoutManager 这个类来进行控制。与 GridView 效果对应的是控制类 GridLayoutManager，与瀑布流效果对应的是控制类 StaggeredGridLayoutManager。RecyclerView 不再拘泥于 ListView 的线性展示方式，它可以实现线性布局（纵向或横向）、网格布局、瀑布流布局等多种效果。

RecyclerView 可设置 Item 的间隔样式（可绘制）。通过继承 RecyclerView 的 ItemDecoration 这个类，就可针对自己的业务需求去编写代码了。通过 ItemAnimator 这个类，也可以控制 Item 实现增、删的动画效果。

但是关于 Item 的点击和长按事件，需要用户自己去实现。相应示例参见教学资源包中的 Ch2 工程。

2.4.2 对话框

1. Dialog

Dialog 就是在屏幕上弹出一个让用户做出选择或者输入额外信息的对话框，一个对话框并不会占满整个屏幕，通常用于模态事件（modal events），需要用户做出一个决定后才会继续执行。

Dialog 类声明如下：

```
public class Dialog  extends Object implements DialogInterface, Window.Callback,
KeyEvent.Callback, View.OnCreateContextMenuListener
```

Dialog 有很多的子类实现，所以要定义一个对话框，使用其子类来实例化一个即可，而不要直接使用 Dialog 这个父类来构造。Dialog 的常用子类（直接子类 Direct Subclasses 和间接子类 Indirect Subclasses）如下：

```
public class AlertDialog extends Dialog implements DialogInterface {}
public class ProgressDialog extends AlertDialog {}
public class DatePickerDialog extends AlertDialog implements
DialogInterface.OnClickListener, DatePicker.OnDateChangedListener {}
public class TimePickerDialog  extends AlertDialog implements
DialogInterface.OnClickListener, TimePicker.OnTimeChangedListener {}
```

2. AlertDialog

AlertDialog 是一款功能强大的对话框控件，可以显示标题、图标、按钮（最多 3 个）、可选择的列表项或者一个普通的布局视图。Android 将该控件划分的区域有图标、标题、内容和按钮，通过分域区控制对话框的内容，可使界面变得灵活多变。

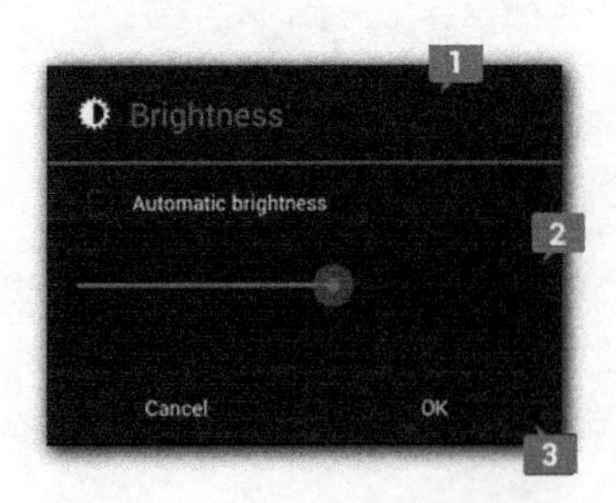

图 2-14　AlertDialog 对话框界面布局

AlertDialog 对话框界面布局如图 2-14 所示。

① 区域 1（标题区+图标区）：定义弹出框的头部信息，包括标题和一个图标。

② 区域 2（内容区）：AlertDialog 对话框的 Content（内容）部分，在这里可以设置一些提示信息、定义一组选择框、自己设定的布局等。

③ 区域 3（按钮区）：定义 Action Buttons（操作按钮）。

使用 AlertDialog 创建对话框的主要步骤如下：

● 创建 AlertDialog.Builder 对象，该对象是 AlertDialog 的创建器。
● 调用 AlertDialog.Builder 的 setIcon()、setTitle()或 setCustomTitle()、setMessage()等方法为对话框设置图标、标题、内容等。
● 调用 AlertDialog.Builder 的 create()方法创建 AlertDialog 对话框。
● 调用 AlertDialog 的 show()方法显示对话框。

AlertDialog.Builder 提供了 APIs，使得创建 AlertDialog 对话框更为容易，其示例代码如下：

```
//1. Instantiate an AlertDialog.Builder with its constructor
AlertDialog.Builder builder = new AlertDialog.Builder(getActivity());
//2. Chain together various setter methods to set the dialog characteristics
builder.setMessage(R.string.dialog_message).setTitle(R.string.dialog_title);
//3. Get the AlertDialog from create()
AlertDialog dialog = builder.create();
```

在 AlertDialog.Builder 中，设置按钮都是通过 setXXXButton 方法来完成的，一共有 3 种不同的 Action Buttons 可选择，但弹出对话框时每种 Button 最多只能出现一个。

● setPositiveButton(CharSequence text, DialogInterface.OnClickListener listener)：这是一个相当于确定操作（OK）的按钮，
● setNegativeButton (CharSequence text, DialogInterface.OnClickListener listener)：这是一个相当于取消操作的按钮。
● setNeutralButton (CharSequence text, DialogInterface.OnClickListener listener)：这是一个相当于忽略操作的按钮。

AlertDialog.Builder 添加 Button 的示例代码如下：

```
AlertDialog.Builder builder = new AlertDialog.Builder(getActivity());
//Add the buttons
builder.setPositiveButton(R.string.ok, new DialogInterface.OnClickListener() {
        public void onClick(DialogInterface dialog, int id) {
            //User clicked OK button
        }
    });
builder.setNegativeButton(R.string.cancel, new DialogInterface. OnClickListener() {
        public void onClick(DialogInterface dialog, int id) {
            //User cancelled the dialog
        }
    });
//Set other dialog properties
…
//Create the AlertDialog
AlertDialog dialog = builder.create();
```

为了能够灵活设定对话框的内容，AlertDialog.Builder 提供了以下 6 种方法。

● setMessage：设置对话框内容为普通文本。
● setItems，设置对话框内容为简单列表项。
● setSingleChoiceItems：设置对话框内容为单选列表项。
● setMultiChoiceItems：设置对话框内容为多选列表项。
● setAdapter：设置对话框内容为自定义列表项。
● setView：设置对话框内容为自定义 View。

使用 AlertDialog 创建对话框的示例可参见教学资源包中 Ch2 工程的 DialogDemo 模块。

2.4.3 菜单

Android 中的菜单（Menu）主要有 OptionMenu（选项菜单）、ContextMenu（上下文菜单）、SubMenu（子菜单）和 PopupMenu（弹出菜单）。

1. OptionMenu

选项菜单是 Android 中最常见的菜单。在选项菜单中，应当包括与当前 Activity 上下文相关的操作和其他选项，如搜索、撰写电子邮件和设置等。选项菜单在屏幕上的显示位置取决于开发应用所适用的 Android 版本。

选项菜单经过了 3 个阶段：①在 Android 2.3.x 或者更低版本，因为该阶段大部分机型都是带有 Menu 键的，此阶段可通过点击 Menu 键弹出菜单。②在 Android 3.0 或者更高的版本，则是引入 ActionBar 中的 Setting 菜单。③在 Android 5.0 以上的版本，则是在 ToolBar 中，通过点击手机屏幕右上角⋮（垂直方向 3 个点）弹出一个溢出式的菜单样式。

使用 OptionMenu 需要重写（Override）以下两个方法。

- public boolean onCreateOptionsMenu(Menu menu)：调用 OptionMenu，在这里完成菜单初始化。
- public boolean onOptionsItemSelected(MenuItem item)：菜单项被选中时触发，在这里完成事件处理。

加载菜单的方式有两种：①直接通过编写菜单 XML 文件，然后调用 getMenuInflater().inflate(R.menu.menu_main, menu)加载菜单。②通过代码动态添加，通过 onCreate OptionsMenu(Menu menu)中的参数 menu，调用其 add (int groupId, int itemId, int order, CharSequence title)方法添加菜单项。

2. ContextMenu

上下文菜单是通过长按某个视图（View）组件后出现的菜单，该组件需注册上下文菜单。在 Windows 中经常用鼠标右键点击弹出的菜单就是上下文菜单。

使用 ContextMenu 的步骤如下：

- 重写 onCreateContextMenu()方法；
- 为 View 组件注册上下文菜单，使用 registerForContextMenu()方法，其参数是 View；
- 重写 onContextItemSelected()方法为菜单项指定事件监听器。

ContextMenu 与 OptionMenu 的主要区别在于：

- OptionMenu 对应的是 Activity，一个 Activity 只能拥有一个选项菜单；
- ContextMenu 对应的是 View，每个 View 都可以设置上下文菜单；

一般情况下，ContextMenu 常用于 ListView 或者 GridView。

3. SubMenu

SubMenu 代表子菜单，可包含 1～*N* 个 MenuItem。子菜单不支持嵌套，所谓子菜单只是在<item>标签中又嵌套了一层<menu>而已，其示例代码如下：

```
<item android:id="@+id/submenu" android:title="一级菜单">
      <menu>
          <group android:checkableBehavior = "none">
              <item android:id="@+id/one" android:title = "子菜单一"/>
              <item android:id="@+id/two" android:title = "子菜单二"/>
          </group>
      </menu>
</item>
```

4. PopupMenu

弹出菜单是锚定到 View 的模态菜单。如果空间足够，它将显示在定位视图下方，否则显示在其上方。如果使用 XML 定义菜单，则创建弹出菜单的步骤如下：

- 实例化 PopupMenu 及其构造函数，该函数将提取当前应用的 Context 及菜单应锚定到的 View。如调用 PopupMenu(Context context, View anchor)构造器创建下拉菜单，其中 anchor 代表激发该弹出菜单的组件。
- 使用 MenuInflater 的 inflate()方法将菜单资源填充到 PopupMenu 实例调用 getMenu()方法返回的 Menu 对象中。
- 调用 PopupMenu 实例的 show()方法。

例如，使用 android:onClick 属性显示弹出菜单的按钮，其示例代码如下：

```
<ImageButton android:layout_width="wrap_content"
         android:layout_height="wrap_content"
         android:src="@drawable/ic_overflow_holo_dark"
         android:contentDescription="@string/descr_overflow_button"
         android:onClick="showPopup" />
```

稍后，Activity 可按照如下方式显示弹出菜单：

```
public void showPopup(View v) {
   PopupMenu popup = new PopupMenu(this, v);
   MenuInflater inflater = popup.getMenuInflater();
   inflater.inflate(R.menu.actions, popup.getMenu());
   popup.show();
}
```

在 API 14 及更高版本中，可以将两行合并在一起，使用 PopupMenu.inflate() 扩充菜单。当用户选择菜单项或触摸菜单以外的区域时，系统即会清除此菜单。

相关示例参见教学资源包中 Ch2 工程的 MenuDemo 模块和 ContextMenuDemo 模块。

2.4.4 标签栏

在 Android 开发中，经常使用 Tab 进行主界面的布局。由于手机屏幕尺寸的限制，合理使用 Tab 可以极大地利用屏幕资源，给用户带来良好的体验。学会 Tab 的使用方法已经成为 Android 开发必不可少的技能了，如微信、QQ 就是使用 Tab 的方式进行主界面布局的。

Tab 选项卡（标签栏）几乎成为了 App 必备的一个功能，它的实现方法多种多样，常见标签栏的实现方式有以下几种类型。

1. 传统的 ViewPager 实现

通过 ViewPager+ViewAdapter 实现，是比较常见的方式。

单纯使用 ViewPager 方式可以实现左右滑动切换页面和点击 Tab 切换页面的效果。但是这种方式需要 Activity 完成所有的代码实现，包括初始化 Tab 及其对应页面的初始化控件、数据、事件及业务逻辑的处理。这样就会使 Activity 看起来非常臃肿，进而造成代码的可读性和可维护性变得极差。

Google 公司在 Android 3.0 中推出了 Fragment，可以分别使用 Fragment 来管理每个 Tab 对应的页面的布局及功能的实现。然后将 Fragment 与 Android 关联，这样 Android 只需要管理 Fragment 就行了，起到调度器的作用，不用再关心每个 Fragment 里的内容及功能实现是什么了。这样就极大地解放了 Activity，使代码变得简单、易读。

单纯使用 ViewPager 实现 Tab 标签栏的示例可参见教学资源包中的 Tab 工程。

2. FragmentManager+Fragment 实现

使用 Fragment 实现了 Activity 与 Tab 对应的页面分离，特别是当 Tab 对应的页面布局和逻辑比较复杂时，更能体会到使用 Fragment 的好处。但是单纯使用 Fragment 只能通过点击 Tab 来切换页面，并不能实现左右滑动进行切换。

使用 Fragment 实现 Tab 标签栏的示例可参见教学资源包中的 Tab 工程。

Tab 工程中示例模块（TabFragmentDemo）的运行界面如图 2-15（a）所示。

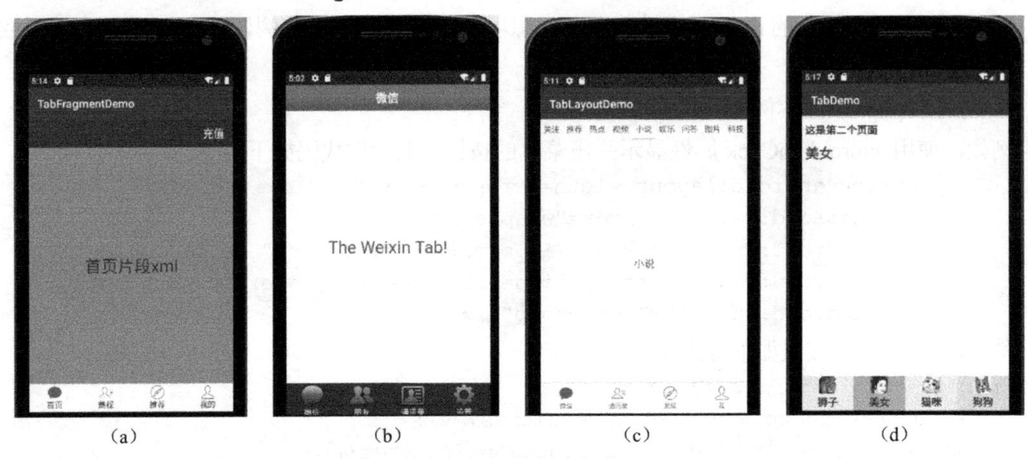

（a）（b）（c）（d）

图 2-15 Tab 工程各模块的运行界面

3. ViewPager +Fragment 实现

使用 ViewPager + Fragment 的方式，综合了各自的优势，既能使用 Fragment 管理 Tab 对应页面的布局及业务逻辑的实现，使得 Activity 与 Tab 对应的页面分离，又能使用 ViewPager 实现左右滑动切换页面的效果。

这种方式需要为 ViewPager 设置 FragmentPagerAdapter 适配器。

使用 ViewPager + Fragment 的方式实现 Tab，也是开发中使用比较广泛的，该示例可参见教学资源中的 Tab 工程。

Tab 工程中示例模块（TabFragmentDemo2）的运行界面如图 2-15（b）所示。

4. TabLayout + ViewPager 实现

Google 公司在 2015 的 IO 大会上，给开发者们带来了全新的 Android Design Support Library，里面包含了许多新控件，这些新控件有许多是把以前的一些第三方开源库官方化，实现起来更为简便，其中的 TabLayout 控件可用于设计 App 的 Tab 标签栏，而且 TabLayout 用法更加简单。使用 TabLayout 与 ViewPager 结合起来实现 Tab 标签栏成为了一个固定的套路。

由于 TabLayout 是 design 库中的控件，在使用前还需要在 build.gradle 的 dependencies 节点添加导入 design 库的代码，其代码如下：

```
dependencies {
   implementation 'com.android.support:design:28.0.0'
}
```

使用 TabLayout 实现 Tab 标签栏的示例可参见教学资源包中的 Tab 工程。

Tab 工程中示例模块（TabLayoutDemo）的运行界面，如图 2-15（c）所示。

5. TabActivity+TabHost+Activity 实现

TabActivity 类声明如下：

```
public class TabActivity extends ActivityGroup
```

TabHost 类声明如下：

```
public class TabHost extends FrameLayout implements ViewTreeObserver.OnTouchModeChangeListener
```

TabHost 是一个标签窗口的容器。一个 TabHost 对象包含两个子元素对象：一个对象是 Tab 标签集合（TabWidget），用户通过点击它们来选择一个特定的标签；另一个是 FrameLayout 对象，展示当前页的内容。

XML 布局文件中的 TabHost 的写法是固定的。TabHost 及其包含的 TabWidget、FrameLayout，它们的 ID 必须是固定的，其示例如下：

```
<Tabhost android:id="@android:id/tabhost" …/>
<TabWidget android:id="@android:id/tabs" …/>
<FrameLayout android:id="@android:id/tabcontent" …/>
```

注意：TabActivity 这个类已经被标注为 This class was deprecated in API level 13。

使用 TabActivity 实现 Tab 标签栏的示例可参见教学资源包中的 Tab 工程。

Tab 工程中示例模块（TabDemo）的运行界面，如图 2-15（d）所示。

6．ActivityGroup+Activity 实现

ActivityGroup 类声明如下：

```
public class ActivityGroup extends Activity
```

它是指在一个屏幕里包含及运行多个嵌入的 Activity。从类的继承结构可以看出，ActivityGroup 也是一个 Activity，其使用方法同 Activity 类似。

使用 ActivityGroup 实现 Tab 标签栏的示例可参见教学资源包中的 Tab 工程。

2.5 Handler 消息传递机制

在 Android 的设计机制里，为了避免多个线程同时操作 UI 造成线程安全问题，只允许主线程（程序第一次启动时所启动的线程，也称为 UI 线程）对 UI 进行修改等操作，这时问题就出现了，如果只允许主线程修改 UI，那么如果新线程的操作需要修改原来的 UI 时该如何进行呢？这时候就引入了 Handler，借助 Handler 消息传递机制在新线程和主线程（UI 线程）之间传递消息来实现。

1．Handler

Handler 是一个消息处理类，主要用于异步消息的处理：当发出一条消息之后，首先进入一个消息队列，发送消息的函数即刻返回，而另外部分从消息队列中将消息逐个取出，然后再进行处理，也就是说，发送消息和接收消息不是同步处理的。

因此，Handler 在整个消息传递机制中主要扮演两个角色：

- 在新启动的线程中发送消息；
- 在主线程中获取和处理消息。

Handler 通过如下方法来发送、处理消息：

- void handleMessage(Message msg)：处理消息的方法。线程通过重写这个方法来处理消息。
- final boolean hasMessage(int what)：检查消息队列中是否包含 what 属性为指定值的消息。
- final boolean hasMessage(int what，Object object)：检查消息队列中是否包含 what 属性为指定值和指定对象的消息。
- Message obtainMessage()：获取消息，可被多种方式重载。
- sendEmptyMessage(int what)：发送空消息。

- final boolean sendEmptyMessageDelayed(int what, long delayMillis)：指定多少毫秒之后发送空消息。
- final boolean sendMessage(Message msg)：携带消息立即发送消息。
- final boolean sendMessageDelayed(Message msg, long delayMillis)：指定多少毫秒之后携带消息发送消息。

2．Looper

循环对象 Looper 用于为一个线程开启一个消息队列，循环等待其他线程发送消息，当有消息时会唤起线程来处理消息，直到线程结束为止。通常情况下 Android 中并不会为新线程开启消息循环，不会用到 Looper。但主线程除外，系统会自动为主线程创建一个 Looper 对象及消息队列，所以主线程会一直运行，处理用户事件直至退出。

当需要一个线程时，这个线程要能够循环处理其他线程发来的消息事件，或者需要长期与其他线程进行复杂的交互，这时就需要用到 Looper 来给线程建立消息队列。

Looper 对象提供了以下 3 个方法：

- prepare()：用于初始化 Looper，prepare()方法保证每个线程最多只有一个 Looper 对象。
- loop()：用于开启消息循环，当调用了 loop()方法后，Looper 线程就真正开始工作了，它会从消息队列中获取消息并交给对应的 Handle 对象处理消息。
- quit()：用于结束 Looper 线程的消息循环。

3．Message

消息类 Message 是能够发送给 Handler 并由 Handle 处理的消息对象，它包含了消息的描述和任意数据对象。一个 Message 对象拥有 5 个属性，如表 2-10 所示。

表 2-10　Message 对象属性

属 性 名 称	类　型	描　述
arg1	int	用于存放整型数据
arg2	int	用于存放整型数据
obj	Object	用于存放发送给接收器的 Object 对象
replyTo	Message	用来指定此 Message 发送到何处的可选 Message 对象
what	int	用于指定用户自定义的消息代码

4．消息传递机制

利用 Handler 消息传递的过程：使用 Handler 发送消息，该消息被传送到指定的 MessageQueue。为了保证正常工作，当前线程必须有 MessageQueue，而 MessageQueue 是由 Looper 对象来管理的。因此要求当前线程必须有一个 Looper 对象，根据不同类型的线程，处理情况不同，主要分为两类：

- 主线程：系统已经为其初始化了 Looper 对象，因此可以直接创建 Handler 对象，并由该 Handler 对象发送、处理消息。
- 非主线程：需要先调用 Looper 的 prepare()方法创建一个 Looper 对象，然后再调用 Looper 的 loop()方法启动它。

当线程有了 Looper 对象之后，先要创建 Handler 子类的实例，再重写 handleMessage()方法处理消息，最后由 loop()启动 Looper 对象，如图 2-16 所示。

基于 Handler、Looper 机制传递消息主要包括以下 7 个步骤：

（1）目标线程调用 Looper.prepare()创建 Looper 对象和消息队列；

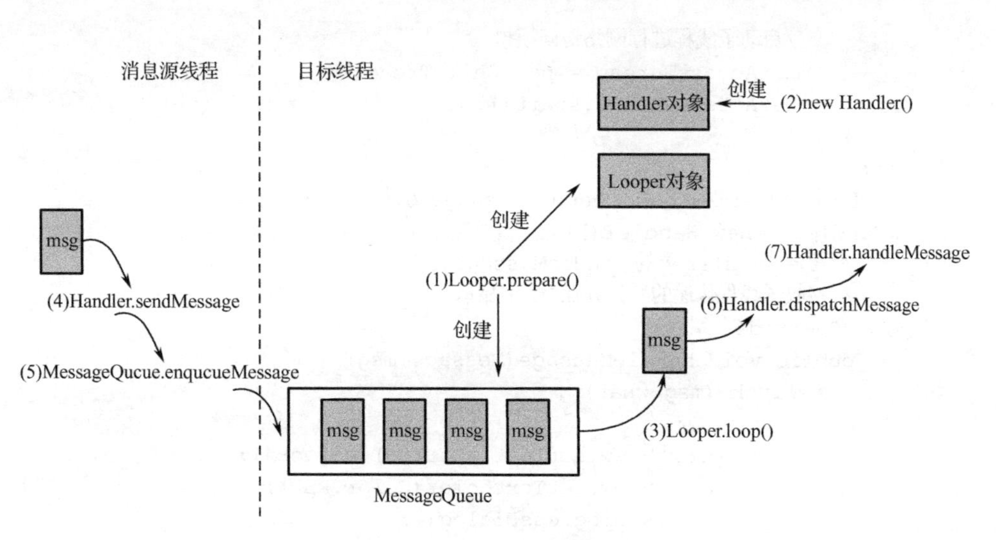

图 2-16　Handler 消息传递机制分析

（2）目标线程通过 new Handler()创建 Handler 对象，将 Handler、Looper、消息队列三者关联起来，并覆盖其 HandleMessage 函数；

（3）目标线程调用 Looper.loop() 监听消息队列；

（4）消息源线程调用 Handler.sendMessage 发送消息；

（5）消息源线程调用 MessageQueue.enqueueMessage 将待发消息插入消息队列；

（6）目标线程的 loop() 检测到消息队列有消息插入，将其取出；

（7）目标线程将取出消息通过 Handler.dispatchMessage 派发给 Handler.handleMessage 进行消息处理；

到这里整个 Android 的 Handler 、Looper 机制传递消息原理就分析完毕了。

【示例】 模拟手机客户端下载网络数据，并在手机界面显示新数据。

分析业务逻辑：①UI 线程获得用户请求；②启动子线程完成网络数据下载（网络下载过程通过强制子线程休眠若干秒来模拟）；③子线程将下载的数据返回 UI 线程并显示。

具体操作过程：启动 Android Studio，在 Ch2 工程中创建 HandlerDemo 模块。

HandlerDemo 模块的 MainActivity.java 主要代码如下：

```
public class MainActivity extends AppCompatActivity {
  private Button mButton;
  private TextView mTextView;
  private Handler mHandler;
  private Thread mNetAccessThread;
  private ProgressDialog mProgressDialog;
  private int mDownloadCount = 0;
  @Override
  protected void onCreate(Bundle savedInstanceState) {
     super.onCreate(savedInstanceState);
     setContentView(R.layout.activity_main);
     mButton = (Button) findViewById(R.id.begin);
     mTextView = (TextView) findViewById(R.id.dataText);
     //设置按钮的点击事件监听器
     mButton.setOnClickListener(new View.OnClickListener() {
        @Override
        public void onClick(View v) {
           showProgressDialog("","正在下载…");
```

```
                //启动子线程进行网络访问模拟
                mNetAccessThread = new ChildTread();
                mNetAccessThread.start();
            }
        });
        //继承 Handler 类并覆盖其 handleMessage 方法
        mHandler = new Handler(){
            //覆盖 Handler 类的 handleMessage 方法
            //接收子线程传递的数据并在 UI 中显示
            @Override
            public void handleMessage(Message msg) {
                switch (msg.what) {
                    case 1:
                        mTextView.setText((String) msg.obj);
                        mTextView.setTextColor(Color.RED);
                        dismissProgressDialog();
                        break;
                    //可以添加其他情况，如网络传输错误
                    //case…
                    default:
                        break;
                }
            }
        };
    }
    class ChildTread extends Thread {
        @Override
        public void run() {
            try {
                Thread.sleep(6000);  //休眠 6s，模拟网络访问延迟
            } catch (InterruptedException e) {
                e.printStackTrace();
            }
            //将结果通过消息返回主线程
            Message msg = new Message();
            msg.what = 1;
            mDownloadCount ++;
            msg.obj = new String("第"+mDownloadCount+"次从网上下载的数据");
            mHandler.sendMessage(msg);
        }
    };
    protected void showProgressDialog(String title,String content) { //开启
progressDialog
        mProgressDialog = new ProgressDialog(this);
        if(title != null)
            mProgressDialog.setTitle(title);
        if(content != null)
            mProgressDialog.setMessage(content);
        mProgressDialog.show();
    }
    protected void dismissProgressDialog() { //关闭 progressDialog
        if(mProgressDialog != null)
        {
            mProgressDialog.dismiss();
```

```
            }
        }
    }
```

运行 HandlerDemo 模块，其界面如图 2-17 所示。

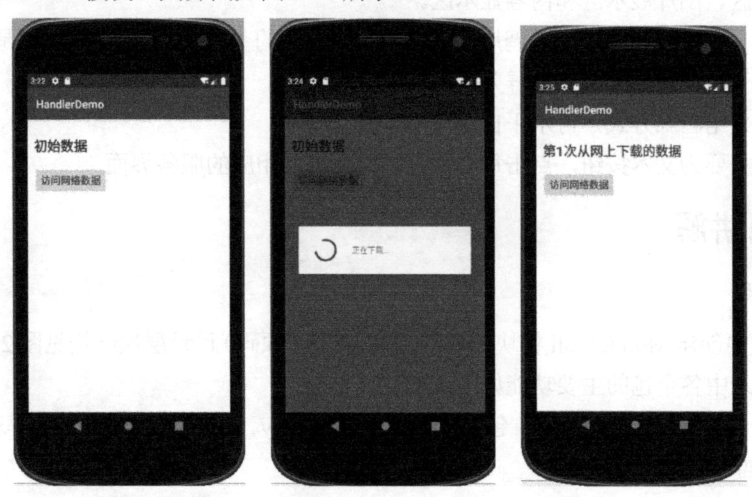

图 2-17　运行 HandlerDemo 模块的界面

小贴士：Thread 和 Looper 是一一对应关系，Thread 和 MessageQueue 也是一一对应关系。因此，一个 Thread 只有一个 Looper 对象和一个 MessageQueue 对象。一个 Looper 却可以与多个 Handler "绑定"，但是一个 Handler 只能与一个 Looper "绑定"。通过阅读 Handler 的构造函数的源码可以找到原因，由于 Looper 分发消息是阻塞式的，即一个消息分发完成之后才会处理下一个消息。Looper 就是一个 While 循环，不停地从 MessageQueue 读取 Message 并分发给 Handler 处理，再调用 Looper.loop() 来启动循环。

2.6　项目实战："移动商城"（一）

"移动商城"是一款针对 XX 网络运营商开发的电子商务平台手机应用程序，通过该 App，用户可以购买手机、流量及订制宽带。"移动商城"（一）运行主界面如图 2-18 所示，宽带列表界面如图 2-19 所示，MobileMall 目录结构如图 2-20 所示。

图 2-18　运行主界面

图 2-19　宽带列表界面

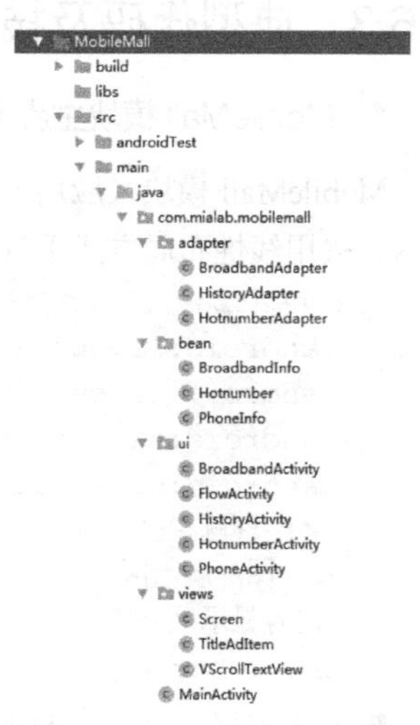

图 2-20　MobileMall 目录结构

2.6.1 任务说明

本例主要完成“移动商城”（一）用户界面的搭建，包括主界面和宽带列表界面等。主界面包含3个部分：标题栏区、图片展示区和内容显示区。

标题栏区有3个部分：返回原先调用（上层模块）界面的“Home”按钮（这里是退出）、查看历史信息的“历史”按钮和具有向上翻屏效果的文字信息栏。

图片展示区采用画廊方式，可水平自动切换多张广告。

内容显示区主要为文本按钮，点击每个按钮，即可进入相应的服务界面。

2.6.2 项目讲解

1. 总体框架

在Ch2工程中创建MobileMall模块，MobileMall模块（源码）目录层次结构见图2-20。MobileMall模块（源码）目录中各个包的主要功能如下：

（1）com.mialab.mobilemall.ui包：包含BroadbandActivity、HotnumberActivity等表示具体商品服务的Activity界面。

（2）com.mialab.mobilemall.adapter包：包含BroadbandAdapter等多个对应于RecyclerView控件的适配器。

（3）com.mialab.mobilemall.bean包：包含程序中用到的BroadbandInfo.java等多个实体类。

（4）com.mialab.mobilemall.views包：此包为视图包，主要包含获取屏幕信息的Screen.java类和自定义翻屏滚动文本的视图类VscrollTextView.java。

（5）com.mialab.mobilemall包：此包名是该App的标识，被称为applicationId，在build.gradle文件里可以看到applicationId的声明。这个包只包含MainActivity.java这一个类。

2. 程序主要流程

程序的入口是MainActivity的onCreate方法。进入该方法后，依次初始化生成翻屏滚动的文本和水平切换的图片。调用setTitle()方法来设置待滚动的文本信息，并将其传给由VscrollTextView设置的滚动文本信息视图，从而实现翻屏滚动的文本。调用initAdvert()方法设置初次显示的图片及画廊，从而实现水平依次切换图片。当点击用线性布局制作的宽带文本区域时，便可打开不同宽带产品的列表界面。宽带列表界面是使用RecyclerView控件制作的，与相应的适配器BroadbandAdapter进行绑定。

2.6.3 典型代码及技术实现

1. MobileMall模块主界面MainActivity.java

MobileMall模块（应用）的主界面MainActivity分为3个部分：标题栏区、图片展示区和内容显示区，采用线性布局方式垂直排列，其主界面布局文件代码如下：

```
<LinearLayout xmlns:android="http://schemas.android.com/apk/res/android"
    android:layout_width="fill_parent"
    android:layout_height="fill_parent"
    android:background="#EBEBEB"
    android:orientation="vertical" >
    <!—标题栏区-->
   <!—图片展示区-->
 <!—内容显示区-->
 </LinearLayout>
```

● 标题栏区：它包含两个按钮和一行具备翻屏功能的文本。两个按钮分列屏幕左右两端，文本居中显示，因此，标题栏区采用相对布局方式，其中编写 Java 代码生成具备翻屏功能的文本界面，将描述界面的 Java 代码注入标题栏区域的布局代码中。

标题栏左右两侧的按钮通过设置 android:background 属性实现拥有点击和未点击两种状态，按钮的样式分别定义在资源文件夹 drawable 的 headbtn_style.xml 文件中。

屏幕切换指的是在同一个 Activity 内进行屏幕间的切换，最常见的情况就是在一个 FrameLayout 内有多个页面，如系统设置页面、个性化设置页面。

android.widget.ViewAnimator 类继承了 FrameLayout 类，ViewAnimator 类的作用是为 FrameLayout 中的 View 切换提供动画效果。实际使用时，一般不会直接使用 ViewAnimator，而是使用它的子类 ViewFlipper。ViewFlipper 可以用来指定 FrameLayout 内多个 View 之间的切换效果，并且可以一次指定也可以每次切换时都指定单独的效果。

ViewFlipper 类声明如下：public class ViewFlipper extends ViewAnimator。

ViewFlipper 类继承结构如图 2-21 所示。

```
java.lang.Object
 ↳android.view.View
   ↳android.view.ViewGroup
     ↳android.widget.FrameLayout
       ↳android.widget.ViewAnimator
         ↳android.widget.ViewFlipper
```

图 2-21　ViewFlipper 类继承结构

在 com.mialab.mobilemall.views 包中，创建 Java 类文件 VScrollTextView.java，用于实现带翻屏的文字界面。该类继承了类 ViewFlipper，其主要代码如下：

```
public class VScrollTextView extends ViewFlipper {
   private TextView tView;
   private List<TitleAdItem> vInfos;
   private int currentIndex = 0;
   int scrollTime = 9000;
   private int type = 0; //0 横向滚动，1 竖直滚动
   private TextClickListener tcListener;
    …
   private void init() {
       setFlipInterval(scrollTime);
       setInAnimation(getContext(), R.anim.push_up_in);
       setOutAnimation(getContext(), R.anim.push_up_out);
       tView = new TextView(getContext());
       …
       this.addView(tView);
   }
   public void setTxtInfos(List<TitleAdItem> infos) {
      if (infos != null && infos.size() > 0) {
        vInfos = infos;
        if(vInfos.size()==1){
            type = 0;
        }else {
            type = 1;
        }
        notifyDataSetChange();
      }
   }
   public void setTxtInfo(String msg) {
      if (msg != null) {
        TitleAdItem titleAdItem=new TitleAdItem();
        titleAdItem.setMsg(""+msg);
        titleAdItem.setParam("");
        titleAdItem.setUrl("");
```

```
            vInfos =new ArrayList<TitleAdItem>();
            vInfos.add(titleAdItem);
            type = 0;
            notifyDataSetChange();
        }
    }
    public void notifyDataSetChange() {
        currentIndex = 0;
        if (type == 0) {
            setAutoStart(false);
            stopFlipping();
            clearAnimation();
            TitleAdItem item=vInfos.get(currentIndex);
            tView.setText(item.getMsg());
        } else if (type == 1) {
            tView.setText(vInfos.get(currentIndex).getMsg());
            Animation animation = getInAnimation();
            …
            this.startFlipping();
            setAutoStart(true);
        }
        …
    }
    …
}
```

上述代码中，VScrollTextView 封装了一个带滚动翻屏效果的文本控件，实现该类的关键是以下 4 个方法。

（1）init()方法。

构造函数调用 init()方法初始化添加待滚动的文本控件，并定义滚动的动画设置。动画的定义主要调用了以下 3 个方法。

setFlipInterval(scrollTime)：设置一次滚动的时间间隔 scrollTime，以 ms 为单位。

setInAnimation(getContext(), R.anim.push_up_in)：设置文本进入屏幕时的动画。

setOutAnimation(getContext(), R.anim.push_up_out)：设置文本退出屏幕时的动画。

资源文件夹 anim 中的 push_up_in.xml 和 push_up_out.xml 分别定义了进入和退出的动画效果，代码参见教学资源包中 Ch2 工程的 MobileMall 模块。

（2）setTxtInfo()方法、setTxtInfos()方法。

这两个方法主要用于设置文本控件的值，滚动的文本都封装在一个 List<TitleAdItem>集合中，其中 TitleAdItem 为滚动文本的 JavaBean 对象。

这两个方法的差别在于：如果只有一条文本信息，则调用 setTxtInfo()；如果有多条信息，则将这些信息封装在 List 集合中，并调用 setTxtInfos ()实现。

（3）notifyDataSetChange()方法。

该方法用以实现滚动效果。此处需要两个变量 type 和 currentIndex。type 变量用来判断是否滚动，若值为 0 则表示不滚动，若值为 1 则表示滚动。currentIndex 表示屏幕中显示的文本在 List 集合中的索引号。当 type 为 0 时，表示滚动的文本只有一条，无须翻屏，因此调用 ViewFlipper 类中的相关方法来停止动画。当 type 为 1 时，则表示有多条文本信息，可根据 currentIndex 的值确定当前显示的文本信息，然后修改 currentIndex 的值，表示翻屏显示下一条信息的索引号。

● 图片展示区：它主要通过水平切换图片展示产品或者广告，此处使用画廊来实现图片切换的功能。图片展示区的用户界面定义在布局文件 mobilemall_advert.xml 中，通过<include>导入到主界面。

● 内容显示区：它垂直线性地排列了 4 个内容，分别是靓号套餐、手机、宽度、流量。这 4 个内容界面布局类似，布局文件代码参见教学资源包中 Ch3 工程的 MobileMall 模块。

主界面 MainActivity.java 启动应用程序。除了需要设置显示的界面布局文件，还需要设置翻屏滚动的文本信息及图片切换，其主要代码如下：

```
int[] imageAdv;
public Handler handler;
private int hasad;
protected void onCreate(Bundle savedInstanceState) {
    super.onCreate(savedInstanceState);
    setContentView(R.layout. activity_main);
    imageAdv = new int[] { R.drawable.suda_lake, R.drawable.suda_snow,
            R.drawable.suda_park2, R.drawable.suda_bridge1,
            R.drawable.suda_bridge2, R.drawable.suda_park1 };
    handler = new Handler() {
        @Override
        public void handleMessage(Message msg) {
            super.handleMessage(msg);
            if (msg.what == 3) {
                if (tempGallery != null) {
                    if (msg.getData() != null) {
                        int selectedIndex = msg.getData().getInt("pos");
                        if (selectedIndex > -1 && tempGallery != null) {
                            tempGallery.setSelection(selectedIndex);
                        }
                    }
                }
            }
        }
    };
    setTitle();
        initAdvert();
}
```

上述代码中，setTitle()方法封装了设置翻屏滚动的文本信息，initAdvert()方法实现了载入图片并能水平切换的功能。

setTitle()方法的主要功能是设置用于翻屏的文本信息列表，通过调用 setTxtInfo(s)将其传入翻屏文本控件 tvName，其主要代码如下：

```
private VScrollTextView tvName;
private void setTitle() {
tvName = (VScrollTextView) findViewById(R.id.grid_title_name);
tvName.setTextSize(this, 20);
tvName.setTextColor(Color.WHITE);
tvName.setTxtInfo("移动商城");
hasad=1;
if(hasad==1){
        new Thread(){
            public void run(){
                final List<TitleAdItem> tItems = new ArrayList<TitleAdItem>();
                TitleAdItem adItem1 = new TitleAdItem();
                adItem1.setMsg("移动商城");
```

```
            TitleAdItem adItem2 = new TitleAdItem();
            adItem2.setMsg("这是一个测试");
            TitleAdItem adItem3 = new TitleAdItem();
            adItem3.setMsg("手机靓号不想要吗");
            tItems.add(adItem1);
            tItems.add(adItem2);
            tItems.add(adItem3);
            if (tItems != null && tItems.size() >0) {
                handler.post(new Runnable() {
                    @Override
                    public void run() {
                        tvName.setTxtInfos(tItems);
                    }
                });
            }else{
                handler.post(new Runnable() {
                @Override
                public void run() {
                        tvName.setTxtInfo("移动商城");
                }});
            }
        }
      }.start();
    }else{
        tvName.setTxtInfo("移动商城");
    }
}
```

initAdvert()方法主要由 initDefaultAdvert()和 initGallerPanel()这两个方法组成，分别用于初始化图片及画廊，其代码如下：

```
private void initAdvert() {
    View advert_panel = this.findViewById(R.id. advert);
    ImageView advert_view = (ImageView)
            advert_panel.findViewById(R.id.advert_deafult_image);
    initDefaultAdvert(advert_view);
    initGallerPanel(advert_panel);
}
```

其中 initDefaultAdvert()方法需要调用获取屏幕信息，并设置图片展示区的大小，因此，首先创建 Java 文件 Screen.java，用于获取移动设备的屏幕信息，其代码参见教学资源包中 Ch2 工程 MobileMall 模块的 Screen.java。

initDefaultAdvert()方法根据屏幕宽度设置图片大小，以及第一次打开应用程序显示的图片，其代码如下：

```
private void initDefaultAdvert(ImageView advert_view) {
    int screent_width = (int) Screen.getInstance(this).getScreenWidth();
    android.widget.LinearLayout.LayoutParams tempLayoutParams =
        (android.widget.LinearLayout.LayoutParams)
advert_view.getLayoutParams();
    tempLayoutParams.height = screent_width / 2;
    advert_view.setLayoutParams(tempLayoutParams);
    advert_view.setImageResource(R.drawable.suda_lake);
}
```

initGallerPanel()方法主要完成以下工作：

- 设置展示区的圆点图片，将其放置在 images[]数组中；
- 设置画廊大小，以及将显示图片 imgadv[]作为数据源利用 BasedAdapter 绑定给画廊；
- 设置画廊列表项，选择事件监听，从而更改圆点图片；
- 设置画廊 Touch 事件监听，当 Touch 画廊区域时，则停止切换动画。

initGallerPanel()方法代码如下：

```
private Gallery tempGallery = null;
private void marketcity_initGallerPanel(final View advert_panel) {
   View galler_panel = advert_panel.findViewById(R.id.gallery_panel);
   tempGallery = (Gallery) galler_panel.findViewById(R.id.gallery);
   tempGallery.setOnTouchListener(new OnTouchListener() {
      @Override
      public boolean onTouch(View v, MotionEvent event) {
          stopPlay();
          return false;
      }
   });
   …
   ImageAdapter  imageAdapter = new ImageAdapter (MainActivity.this);
   tempGallery.setAdapter(imageAdapter);
   imageAdapter.setData(imageAdv);
   tempGallery.setOnItemSelectedListener(new OnItemSelectedListener() {
      @Override
      public void onItemSelected(AdapterView<?> arg0, View arg1, int arg2, long
arg3) {
          int size_color = imageAdv.length;
          for (int i = 0; i < size_color; i++) {
             if (i == arg2) {
               images[arg2].setImageResource(R.drawable. advert_selected);
             } else {
               images[i].setImageResource(R.drawable. advert _default);
             }
          }
      }
      startPlay();
   }
```

通过调用 startPlay()和 stopPlay()这两个方法，来实现动画的开始和停止，其代码如下：

```
private Timer autoGallery;
private TimerTask timerTask;
private void startPlay() {
   if (timerTask == null) {
        timerTask = new TimerTask() {
            @Override
            public void run() {
                if (tempGallery != null && tempGallery.getAdapter() != null) {
                    int count = tempGallery.getAdapter().getCount();
                    int    galleryposition    =    tempGallery.galleryposition
Position();
                    if (galleryposition == count - 1) {
                        galleryposition = 0;
                    } else {
                        galleryposition++;
                    }
```

```
                        Message msg = new Message();
                        Bundle date = new Bundle();
                        date.putInt("pos", galleryposition);
                        msg.setData(date);
                        msg.what = 3;
                        handler.sendMessage(msg);
                    }
                }
            };
        }
        if (autoGallery == null) {
            autoGallery = new Timer();
            autoGallery.schedule(timerTask, 3000, 3000);
        }
    }
    private void stopPlay() {
        if (timerTask != null) {
            timerTask.cancel();
            timerTask = null;
        }
        if (autoGallery != null) {
            autoGallery.cancel();
            autoGallery = null;
        }
    }
```

Timer 是 jdk 中提供的一个定时器工具，使用时会在主线程之外创建一个单独的线程执行指定的计划任务，可以指定执行一次或者反复执行多次。TimerTask 是一个实现了 Runnable 接口的抽象类，代表一个可以被 Timer 执行的任务。Timer 和 TimerTask 使用起来都很简单，可先定义 Timer 和 TimerTask，再调用 Timer 的 schedule 方法，将 TimerTask 传进去即可。

2. 宽带列表界面 BroadbandActivity.java 和适配器 BroadbandAdapter.java

BroadbandAdapter.java 主要代码如下：

```
    public class BroadbandAdapter extends RecyclerView.Adapter<BroadbandAdapter.
ItemHolder> {
        private List<BroadbandInfo> mBroadbandList;
        private Context mContext;
        public BroadbandAdapter(Context context, List<BroadbandInfo> broadbandList)
{
            this.mContext = context;
            this.mBroadbandList = broadbandList;
        }
        @Override
        public ItemHolder onCreateViewHolder(ViewGroup parent, int viewType) {
            View view = LayoutInflater.from(mContext).inflate(R.layout.broadband_
item, parent, false);
            ItemHolder itemHolder = new ItemHolder(view);
            return itemHolder;
        }
        @Override
        public void onBindViewHolder(ItemHolder itemHolder, int position) {
            BroadbandInfo goodsInfo = mBroadbandList.get(position);
            itemHolder.img.setImageResource(goodsInfo.getImgId());
            itemHolder.title.setText(goodsInfo.getTitle());
```

```
            itemHolder.desc.setText(goodsInfo.getDesc());
            if (position % 2 == 0) {

itemHolder.itemView.setBackgroundColor(Color.parseColor("#ebebeb"));
            } else {

itemHolder.itemView.setBackgroundColor(Color.parseColor("#FFFFFF"));
            }
        }
        @Override
        public int getItemCount() {
            return mBroadbandList.size();
        }

        //定义列表项的视图持有者
        static class ItemHolder extends RecyclerView.ViewHolder {
            TextView title;
            TextView desc;
            ImageView img;
            ImageView img_2;
            View itemView;
            public ItemHolder(View view) {
                super(view);
                this.title = (TextView) view.findViewById(R.id.title);
                this.desc = (TextView) view.findViewById(R.id.desc);
                this.img = (ImageView) view.findViewById(R.id.imageView);
                this.img_2 = (ImageView) view.findViewById(R.id.more);
                this.itemView = view;
            }
        }
    }
```

BroadbandActivity.java 主要代码如下：

```
    public class BroadbandActivity extends Activity {
        private List<BroadbandInfo> broadbandList;
        private BroadbandAdapter broadbandAdapter;
        @Override
        protected void onCreate(Bundle savedInstanceState) {
            super.onCreate(savedInstanceState);
            setContentView(R.layout.activity_broadband);
            initHead();
            initData();
            RecyclerView recyclerView = (RecyclerView)findViewById(R.id.recyclerView);
            //创建一个垂直方向的线性布局管理器
            LinearLayoutManager layoutManager = new LinearLayoutManager(this,
            LinearLayout.VERTICAL,false);
            //设置 recyclerView 的布局管理器
            recyclerView.setLayoutManager(layoutManager);
            broadbandAdapter = new BroadbandAdapter(BroadbandActivity.this, broad
bandList);
            recyclerView.setAdapter(broadbandAdapter);
        }
        private void initData() {
            broadbandList = new ArrayList<BroadbandInfo>();
            for (int i = 0; i < 12; i++) {
```

```
            BroadbandInfo broadbandInfo = new BroadbandInfo();
            broadbandInfo.setTitle("宽带(80Mbps 带宽)");
            broadbandInfo.setDesc("60 元/月");
            broadbandInfo.setImgId(R.drawable.phone2_default);
            broadbandInfo.setImgId_2(R.drawable.arrow_more);
            broadbandList.add(broadbandInfo);
        }
    }
    private void initHead() {…}
}
```

MobileMall 模块（应用）的宽带列表运行界面见图 2-19。表示宽带列表 Item 的 Bean（BroadbandInfo.java）文件，以及宽带列表 Item 的布局文件（broadband_item.xml），参见教学资源包中 Ch2 工程的 MobileMall 模块，这里限于篇幅，不再赘述。

2.7 相关阅读：ButterKnife 的使用

在 Android 编程过程中，通常会写大量的 findViewById 和点击事件，如初始 View、设置 View 监听，这样简单而重复的操作让人觉得特别麻烦，所以可以采用注解的方式去实现，这也可以视为类似依赖注入的解决方案，而 ButterKnife 则是依赖注入（注解）中相对简单易懂的开源框架。

ButterKnife 是出自 JakeWharton 的一个开源库，它通过注解的方式来替代 Android 中 View 的相关操作，可减少大量的 findViewById 及 setOnClickListener 代码，且对性能影响较小。ButterKnife 可以说是一个非常强大的视图绑定库，它可以大大简化代码，并且不会因为反射而影响 App 的性能。应用时可以在 Activity 中绑定 ButterKnife，也可以在 Fragment、Adapter 中绑定 ButterKnife。ButterKnife 的基本使用包括绑定 View、绑定资源、事件绑定、绑定监听等。

1. 基本使用介绍

使用@BindView 和 ID 注解相应的变量时，ButterKnife 就会在 layout 文件中找到对应的 View 并进行赋值。

```
class ExampleActivity extends Activity {
    @BindView(R.id.title) TextView title;
    @BindView(R.id.subtitle) TextView subtitle;
    @BindView(R.id.footer) TextView footer;
    @Override
    public void onCreate(Bundle savedInstanceState) {
        super.onCreate(savedInstanceState);
        setContentView(R.layout.simple_activity);
        ButterKnife.bind(this);
        //TODO Use fields…
    }
}
```

代替耗时的反射，通过产生代码来实现 View 的查找，可以称为委托绑定并产生可见且可调试的代码。上面的例子所产生的代码如下：

```
public void bind(ExampleActivity activity) {
    activity.subtitle = (android.widget.TextView) activity.findViewById(2130968578);
    activity.footer = (android.widget.TextView) activity.findViewById(2130968579);
    activity.title = (android.widget.TextView) activity.findViewById(2130968577);
}
```

2. 资源绑定

使用@BindBool、@BindColor、@BindDimen、@BindDrawable、@BindInt、@BindString 与一个对应的 ID 来绑定定义好的资源。

```
class ExampleActivity extends Activity {
    @BindString(R.string.title) String title;
    @BindDrawable(R.drawable.graphic) Drawable graphic;
    @BindColor(R.color.red) int red; //int or ColorStateList field
    @BindDimen(R.dimen.spacer) Float spacer; //int (for pixel size) or float (for exact value) field
    //...
}
```

3. 非 Activity 绑定

在已知 View 情况下的任意对象中，绑定该 View 中所含有的控件。

如在 Fragment 中：

```
public class FancyFragment extends Fragment {
    @BindView(R.id.button1) Button button1;
    @BindView(R.id.button2) Button button2;
    @Override
    public View onCreateView(LayoutInflater inflater, ViewGroup container, Bundle savedInstanceState) {
        View view = inflater.inflate(R.layout.fancy_fragment, container, false);
        ButterKnife.bind(this, view);
        //TODO Use fields...
        return view;
    }
}
```

在 ViewHolder 中：

```
public class MyAdapter extends BaseAdapter {
    @Override
    public View getView(int position, View view, ViewGroup parent) {
        ViewHolder holder;
        if (view != null) {
            holder = (ViewHolder) view.getTag();
        } else {
            view = inflater.inflate(R.layout.whatever, parent, false);
            holder = new ViewHolder(view);
            view.setTag(holder);
        }
        holder.name.setText("John Doe");
        //etc...
        return view;
    }
    static class ViewHolder {
        @BindView(R.id.title) TextView name;
        @BindView(R.id.job_title) TextView jobTitle;
        public ViewHolder(View view) {
            ButterKnife.bind(this, view);
        }
    }
}
```

4. View Lists

可以将所需要的控件，全部填充到一个 List 中。

```
@BindViews({ R.id.first_name, R.id.middle_name, R.id.last_name })
List<EditText> nameViews;
```

apply()方法可以对 List 中所有的 View 执行某个操作。

```
ButterKnife.apply(nameViews, DISABLE);
ButterKnife.apply(nameViews, ENABLED,  false);
```

动作和 setter 接口允许设置的简单行为:

```
static final ButterKnife.Action<View> DISABLE = new ButterKnife.Action<View>()
{
    @Override public void apply(View view, int index) {
        view.setEnabled(false);
    }
};
static final ButterKnife.Setter<View, Boolean> ENABLED = new ButterKnife.Setter
<View, Boolean>() {
    @Override public void set(View view, Boolean value, int index) {
        view.setEnabled(value);
    }
};
```

当然也可以在 apply()方法中指定一个 Android 控件的属性名。

```
ButterKnife.apply(nameViews, View.ALPHA, 0.0f);
```

5. 绑定 Listener

监听器也可以自动配置到相应的 View 上。

```
@OnClick(R.id.submit)
public void submit(View view) {
   //TODO submit data to server…
}
```

监听器函数的参数都是可选的。

```
@OnClick(R.id.submit)
public void submit() {
   //TODO submit data to server…
}
```

指定一个确定的类型，它将会被自动转换。

```
@OnClick(R.id.submit)
public void sayHi(Button button) {
   button.setText("Hello!");
}
```

还可以将一个监听器函数绑定到多个控件上。

```
@OnClick({ R.id.door1, R.id.door2, R.id.door3 })
public void pickDoor(DoorView door) {
   if (door.hasPrizeBehind()) {
      Toast.makeText(this, "You win!", LENGTH_SHORT).show();
   } else {
      Toast.makeText(this, "Try again", LENGTH_SHORT).show();
   }
}
```

6. 下载

在模块 build.gradle 文件的 dependencies 节点中添加（可以添加 latest version）:

```
implementation 'com.jakewharton:butterknife:8.8.1'
annotationProcessor 'com.jakewharton:butterknife-compiler:8.8.1'
```

这种使用 ButterKnife 的方式，是针对 App 单组件开发的模式。所谓 App 单组件开发，是指只有一个主 Moduel，所有代码不管是 Java、XML、其他资源、依赖库等都在主 Moduel 中，这个时候使用 ButterKnife 是比较简单的。

7．注意事项

（1）在 Activity 类中绑定 ButterKnife.bind(this);，必须在 setContentView();之后绑定，且父类 bind 绑定后，子类不需要再 bind。

（2）在非 Activity（如 Fragment、ViewHolder）中绑定 ButterKnife.bind(this, view);，这里的 this 不能替换成 getActivity()。

（3）在 Activity 中不需要做解绑操作，但 Fragment 中则必须做解除绑定的操作（在 onDestroyView()中）。

（4）使用 ButterKnife 修饰的方法和字段（控件），不能用 private or static 修饰，否则会报错。如错误信息：@BindView fields must not be private or static。

（5）setContentView()不能通过注解实现。

2.8 本章小结

本章首先简要地概述了视图组件与容器组件、Android 如何控制 UI 界面，并通过示例简单介绍了 UI 界面的三种搭建方式。详细介绍了 Android 常用的基本 UI 组件（包括 TextView、EditText、Button、ImageView、ProgressBar、ScrollView、Toast 以及实现控件合理排放的布局管理器等）和高级 UI 组件（包括列表类组件、对话框、菜单、标签栏等）。针对 Android 只允许主线程操作用户界面的特点，阐述了 Handler 消息传递机制。

习 题 2

1．简述 Android 平台用户界面搭建有哪几种方式，并对这几种方式进行比较。
2．Android 界面布局主要有哪些？请简要加以说明。
3．为什么要在线性布局中使用 android:layout_weight 属性？试举例说明。
4．Handler 消息传递机制是怎样的？试举例说明。
5．观察者模式是怎样的？试举例说明。
6．为什么说在 Handler 消息传递机制中使用了观察者模式？
7．循环对象 Looper 的作用是什么？请加以说明。
8．如何使用 RecylerView？试编程加以说明。
9．RecyclerView 和 ListView 的区别是什么？
10．在子线程中创建 Handler 为何会抛出异常？
11．选择合理的布局管理器，搭建一个计算器界面，效果如图 2-21 所示。

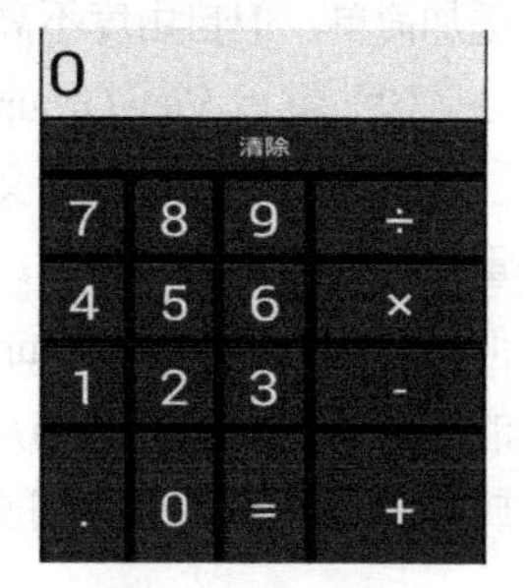

图 2-21　计算器

第3章　自定义View

View 表现为显示在屏幕上的各种视图。View 类是 Android 中各种组件的基类，如 View 就是 ViewGroup 基类。Android 中的 UI 组件都是由 View 和 ViewGroup 组成的。自定义 View 对于一个 Android 开发者来说是必须掌握的知识点，也是 Android 开发进阶的必经之路。为什么要自定义 View 呢？主要是 Android 系统内置的 View 无法完全满足业务需求，因而需要针对业务需求编制 View。本章主要内容：（1）自定义 View 的分类；（2）自定义 View 的构造函数；（3）View 的绘制流程；（4）自定义 View 示例。

3.1　自定义 View 的分类

Android 自定义 View 主要分为两大类：自定义 View 和自定义 ViewGroup。

1. 自定义 View

（1）继承特定 View。

一般用于扩展已有（特定）View 的功能，如 TextView、Button、EditText、ImageView 等。这种方法比较常见和容易实现，不需要自己支持 wrap_content 和 padding 等属性。

（2）继承 View。

主要用于实现一些不规则的效果，这种效果不方便通过布局的组合方式达到，往往需要静态或动态地显示一些不规则的图形，它需要通过绘制的方式来实现，即重写 onDraw()方法。采用这种方式需要自己支持 wrap_content 属性和 padding 属性，但不需要支持 margin 属性，因为 margin 属性是由父容器决定的。

2. 自定义 ViewGroup

（1）继承特定的 ViewGroup。

拓展某种布局的方式。在原有 ViewGroup 基础上组合，可较容易实现。与自定义 ViewGroup 相比更加简单，但自由度不高。

（2）继承 ViewGroup。

主要用于实现自定义的布局，即除了 LinearLayout、RelativeLayout、FrameLayout 等系统布局，再重新定义一种新的布局。当某种效果看起来很像几种 View 组合在一起时，可以采用这种方式实现，但需要在处理 ViewGroup 的测量（measure）和布局（layout）这两个过程时，同时处理子元素（子 View）的测量（measure）和布局（layout），过程较为复杂。这就需要自定义支持 wrap_content 属性、padding 属性和 margin 属性，可更加接近 View 的底层。

3.2 自定义 View 的构造函数

以自定义 RoundRectView 为例，这里 RoundRectView 类继承自 View 类。

```
public class RoundRectView extends View { }
```

View 有四个构造器，其区别在于参数的数目和类型不同，一般使用前三个构造器。

1. 只有一个 Context 参数的构造方法

```
public RoundRectView(Context context) {
    super(context);
    init();
}
```

如果 View 是在 Java 代码里面通过 new 关键字创建的，则调用第一个构造函数。通常在通过代码初始化控件时使用，即当在 Java 代码中直接通过 new 关键字创建这个控件时，就会调用这个方法。

2. 两个参数（Context 上下文和 AttributeSet 属性集）的构造方法

```
public RoundRectView(Context context, @Nullable AttributeSet attrs) {
    this(context, attrs, 0);
}
```

如果 View 是在.xml 中声明的，则会调用第二个构造函数。自定义属性是从 AttributeSet 参数传进来的。两个参数的构造方法通常对应布局文件中控件被映射成对象时调用（需要解析属性），即当需要在自定义控件中获取属性时，就默认调用这个构造方法。

AttributeSet 对象就是这个控件中定义的所有属性。可以通过 AttributeSet 对象的 getAttributeCount()方法获取属性的个数，通过 getAttributeName()方法获取到某条属性的名称，并通过 getAttributeValue()方法获取到某条属性的值。

注意，不管有没有使用自定义属性，都会默认调用这个构造方法。

3. 三个参数（Context 上下文、AttributeSet 属性集和 defStyleAttr 自定义属性）的构造方法

```
public RoundRectView(Context context, @Nullable AttributeSet attrs, int defStyle
Attr) {
    super(context, attrs, defStyleAttr);
    TypedArray a = context.obtainStyledAttributes(attrs, R.styleable. RoundRect
View);
    mColor = a.getColor(R.styleable.RoundRectView_roundrect_color, Color.RED);
    a.recycle();
    init();
}
```

这个构造方法不会默认调用，必须通过手动调用，它和两个参数的构造方法的唯一区别就是，这个构造方法传入了一个默认属性集。defStyleAttr 指向的是自定义属性的<declare-styleable>标签中定义的自定义属性集，在创建 TypedArray 对象时需要用到 defStyleAttr。

如果在 Code 中实例化一个 View 就会调用第一个构造函数，如果在 xml 中定义就会调用第二个构造函数，而第三个函数系统是不调用的，要由 View（自定义的或系统预定义的 View）显式调用，如这里在第二个构造函数中调用了第三个构造函数。

第三个参数的意义就如同它的名字，即默认的 Style。

3.3 View 的绘制流程

当一个应用启动时，会启动一个主 Activity，Android 系统会根据 Activity 的布局来对它进行绘制。绘制会从根视图 ViewRoot 的 performTraversals()方法开始，从上到下遍历整个视图树，每个 View 控件负责绘制自己，而 ViewGroup 还需要负责通知自己的子 View 进行绘制操作。视图绘制的过程可以分为三个步骤，分别是测量（Measure）、布局（Layout）和绘制（Draw）。View 的绘制基本由 measure()、layout()、draw()这个三个函数完成。

（1）measure()：用来测量 View 的宽和高，相关方法有 measure()、setMeasuredDimension()和 onMeasure()。

（2）layout()：用来确定 View 在父容器中的布局位置，相关方法有 layout()、onLayout()和 setFrame()。

（3）draw()：负责将 View 绘制在屏幕上，相关方法有 draw()和 onDraw()。

移动设备一般定义屏幕左上角为坐标原点，向右为 x 轴的增大方向，向下为 y 轴的增大方向。

3.3.1 Measure 过程

1．MeasureSpec

MeasureSpec 是 View 的内部类，它封装了一个 View 的尺寸，在 onMeasure()中会根据这个 MeasureSpec 值来确定 View 的宽和高。MeasureSpec 值保存在一个 32 位的 int 值中，前两位则表示模式 SpecMode，后 30 位则表示大小 SpecSize。

即 MeasureSpec = SpecMode + SpecSize。

在 MeasureSpec 当中一共存在三种 SpecMode：EXACTLY、AT_MOST 和 UNSPECIFIED。对于 View 来说，MeasureSpec 的 SpecMode 和 SpecSize 有如下意义。

（1）EXACTLY：精确测量模式。父容器已经检测出 View 所需要的精确大小，此时 View 的最终大小就是 SpecSize 所指定的值。它对应于 LayoutParams 中的 match_parent 和具体数值这两种模式。

（2）AT_MOST：最大值模式。View 的尺寸有一个最大值，View 不可以超过 MeasureSpec 当中的 SpecSize 值。它对应于 LayoutParams 中的 wrap_content。

（3）UNSPECIFIED：不指定测量模式。父容器不对 View 有任何限制，子视图可以是想要的任何尺寸，通常用于系统内部，但在应用开发中很少使用到。

获取测量模式（SpecMode）的示例代码如下：

```
int specMode = MeasureSpec.getMode(measureSpec);
```

获取测量大小（SpecModeSize）的示例代码如下：

```
int specSize = MeasureSpec.getSize(measureSpec);
```

通过 SpecModeMode 和 SpecModeSize 生成新的 MeasureSpec，其示例代码如下：

```
int measureSpec=MeasureSpec.makeMeasureSpec(size, mode);
```

子元素的 MeasureSpec 创建与父容器的 MeasureSpec 和子元素本身的 LayoutParams 有关。ViewGroup 的 getChildMeasureSpec()方法的主要代码如下：

```
 public static int getChildMeasureSpec(int spec, int padding, int childDimension)
{
      int specMode = MeasureSpec.getMode(spec);
      int specSize = MeasureSpec.getSize(spec);
      int size = Math.max(0, specSize - padding);
      int resultSize = 0;
      int resultMode = 0;
      switch (specMode) {
         //当父 View 要求一个精确值时，为子 View 赋值
```

```
        case MeasureSpec.EXACTLY:
            //如果子View有自己的尺寸，则使用自己的尺寸
            if (childDimension >= 0) {
                resultSize = childDimension;
                resultMode = MeasureSpec.EXACTLY;
                //当子View是match_parent，将父View的大小赋值给子View
            } else if (childDimension == LayoutParams.MATCH_PARENT) {
                resultSize = size;
                resultMode = MeasureSpec.EXACTLY;
                //如果子View是wrap_content，设置子View的最大尺寸为父View
            } else if (childDimension == LayoutParams.WRAP_CONTENT) {
                resultSize = size;
                resultMode = MeasureSpec.AT_MOST;
            }
            break;

        //父布局给子View一个最大界限
        case MeasureSpec.AT_MOST:
            if (childDimension >= 0) {
                //如果子View有自己的尺寸，则使用自己的尺寸
                resultSize = childDimension;
                resultMode = MeasureSpec.EXACTLY;
            } else if (childDimension == LayoutParams.MATCH_PARENT) {
                //父View的尺寸为子View的最大尺寸
                resultSize = size;
                resultMode = MeasureSpec.AT_MOST;
            } else if (childDimension == LayoutParams.WRAP_CONTENT) {
                //父View的尺寸为子View的最大尺寸
                resultSize = size;
                resultMode = MeasureSpec.AT_MOST;
            }
            break;

        //父布局对子View没有做任何限制
        case MeasureSpec.UNSPECIFIED:
            if (childDimension >= 0) {
                //如果子View有自己的尺寸，则使用自己的尺寸
                resultSize = childDimension;
                resultMode = MeasureSpec.EXACTLY;
            } else if (childDimension == LayoutParams.MATCH_PARENT) {
                //因父布局没有对子View做出限制，当子View为MATCH_PARENT时则大小为0
                resultSize = View.sUseZeroUnspecifiedMeasureSpec ? 0 : size;
                resultMode = MeasureSpec.UNSPECIFIED;
            } else if (childDimension == LayoutParams.WRAP_CONTENT) {
                //因父布局没有对子View做出限制，当子View为WRAP_CONTENT时则大小为0
                resultSize = View.sUseZeroUnspecifiedMeasureSpec ? 0 : size;
                resultMode = MeasureSpec.UNSPECIFIED;
            }
            break;
    }
    return MeasureSpec.makeMeasureSpec(resultSize, resultMode);
}
```

小贴士：针对不同的父容器和View本身不同的LayoutParams，View可以有多种MeasureSpec。

（1）当View采用固定宽和高时，不管父容器的MeasureSpec是什么，View的MeasureSpec都是精确模式，并且其大小遵循LayoutParams中的大小。

（2）当View的宽和高是match_parent时，如果父容器的模式是精确模式，那么View也是精确模式，并且其大小是父容器的剩余空间；如果父容器的模式是最大值模式，那么View也是最大值模式并且其大小不会超过父容器的剩余空间。

（3）当View的宽和高是wrap_content时，不管父容器的模式是精确还是最大值，View的模式总是最大值模式，并且其大小不能超过父容器的剩余空间。

2. onMeasure()

整个测量过程的入口位于View的Measure方法当中，该方法做了一些参数的初始化之后调用了onMeasure方法（Measure方法是一个final类型的方法，子类不能重写此方法）。

```
    protected void onMeasure(int widthMeasureSpec, int heightMeasureSpec) {
        setMeasuredDimension(getDefaultSize(getSuggestedMinimumWidth(),
widthMeasureSpec),
                getDefaultSize(getSuggestedMinimumHeight(), heightMeasureSpec));
    }
```

其中，

（1）setMeasuredDimension(int measuredWidth, int measuredHeight)，该方法用来设置View的宽和高，在自定义View时会经常用到。

（2）getDefaultSize(int size, int measureSpec)，该方法用来获取View默认的宽和高，可结合源码来进行分析。

直接继承View的自定义控件需要重写onMeasure方法，并设置wrap_content时的自身大小，否则在布局中使用wrap_content就相当于使用match_parent。

Measure过程会因为布局的不同或者需求的不同而呈现不同的形式，使用时还是要根据业务场景来具体分析。

小贴士：ViewGroup的测量（Measure）过程与View有些不同，其本身是继承自View，它没有对View的Measure方法及onMeasure方法进行重写。ViewGroup除了要测量自身的宽和高，还需要测量各个子View的大小，不同的布局，测量方式也都不同（可参考LinearLayout及FrameLayout），所以没有办法统一设置。因此它提供了测量子View的measureChildren()和measureChild()的方法，以帮助对子View进行测量。通过阅读measureChildren()及measureChild()的源码可以知道，其大致流程就是遍历所有的子View，然后再调用View的measure()方法，让子View测量自身大小。

3.3.2 Layout过程

布局（Layout）过程对于View来说，就是用来计算View的位置参数，而对于ViewGroup来说，除了要测量自身位置，还需要测量子View的位置。

layout()方法是整个Layout流程的入口。

```
    //这里的四个参数l、t、r、b分别代表View的左、上、右、下四个边界相对于其父View的距离。
    public void layout(int l, int t, int r, int b) {
        …
        int oldL = mLeft;
        int oldT = mTop;
        int oldB = mBottom;
        int oldR = mRight;
        //通过setFrame或setOpticalFrame方法确定View在父容器当中的位置。
```

```
        boolean changed = isLayoutModeOptical(mParent) ? setOpticalFrame(l, t, r, b) :
setFrame(l, t, r, b);
        //调用 onLayout 方法。onLayout 方法是一个空实现，不同的布局会有不同的实现。
        if (changed || (mPrivateFlags & PFLAG_LAYOUT_REQUIRED) == PFLAG_LAYOUT_
REQUIRED) {
            onLayout(changed, l, t, r, b);
        }
    }
```

由上述内容可知，在 layout()方法中已经通过 setOpticalFrame(l, t, r, b)或 setFrame(l, t, r, b)方法对 View 自身的位置进行了设置。所以，onLayout(changed, l, t, r, b)方法主要是 ViewGroup 对子 View 的位置进行计算。

3.3.3 Draw 过程

Draw 流程也就是 View 绘制到屏幕上的过程，整个流程的入口在 View 的 draw()方法之中，其过程可以分为 6 个步骤：

① 如果需要，绘制背景（drawBackground(canvas)）；

② 如果有必要，保存当前 canvas；

③ 绘制 View 的内容（onDraw(canvas)）；

④ 绘制子 View（dispatchDraw(canvas)）；

⑤ 如果有必要，绘制边缘、阴影等效果；

⑥ 绘制装饰，如滚动条等。

onDraw()方法用于绘制自己（自身 View），dispatchDraw()方法用于绘制子 View，它们的使用说明如下：

（1）protected void onDraw(Canvas canvas)。

绘制 View 的内容。该方法是一个空的实现，在各个业务中自行处理。

（2）protected void dispatchDraw(Canvas canvas)。

绘制子 View。该方法在 View 当中是一个空的实现，在各个业务中自行处理。

在 ViewGroup 中对 dispatchDraw 方法做了实现，主要是遍历子 View，并调用子类的 draw 方法，一般不需要重写该方法。

3.4 自定义 View 示例

3.4.1 实现一个基本的自定义 View

【示例】 实现一个基本的自定义 View。

启动 Android Studio，在 Ch3 工程中创建 RoundRectDemo 模块，并创建 MainActivity.java、activity_main.xml，以及自定义 View：RoundRectView.java。

RoundRectDemo 模块的 RoundRectView.java 主要代码如下：

```
public class RoundRectView extends View {
   private int mColor = Color.RED;
   private Paint mPaint = new Paint(Paint.ANTI_ALIAS_FLAG);  //设置无锯齿
   public RoundRectView(Context context) {
      super(context);
      init();
   }
```

```
    public RoundRectView(Context context, AttributeSet attrs) {
        super(context, attrs);
        init();
    }
    public RoundRectView(Context context, AttributeSet attrs, int defStyleAttr)
{
        super(context, attrs, defStyleAttr);
        init();
    }
    private void init() {
        mPaint.setColor(mColor);
    }
    @Override
    protected void onDraw(Canvas canvas) {
        super.onDraw(canvas);
        int width = getWidth();
        int height = getHeight();
        int rectWidth = 220;
        int rectHeight = 160;
        RectF rectF = new RectF((width - rectWidth) / 2, (height - rectHeight) /
2, (width + rectWidth) / 2,
                (height + rectHeight) / 2);
        canvas.drawRoundRect(rectF, 30, 30, mPaint);
    }
}
```

RoundRectDemo 模块的 activity_main.xml 内容如下。

```
<LinearLayout xmlns:android="http://schemas.android.com/apk/res/android"
    android:layout_width="match_parent"
    android:layout_height="match_parent"
    android:background="#ffffff"
    android:orientation="vertical">
    <com.mialab.roundrectdemo.ui.RoundRectView
        android:id="@+id/roundRectView1"
        android:layout_width="wrap_content"
        android:layout_height="80dp"
        android:background="#6D6767" />
</LinearLayout>
```

运行 RoundRectDemo 的界面如图 3-1（a）所示。在 ui.RoundRectView 中添加一行代码如下。

```
android:layout_margin="20dp"
```

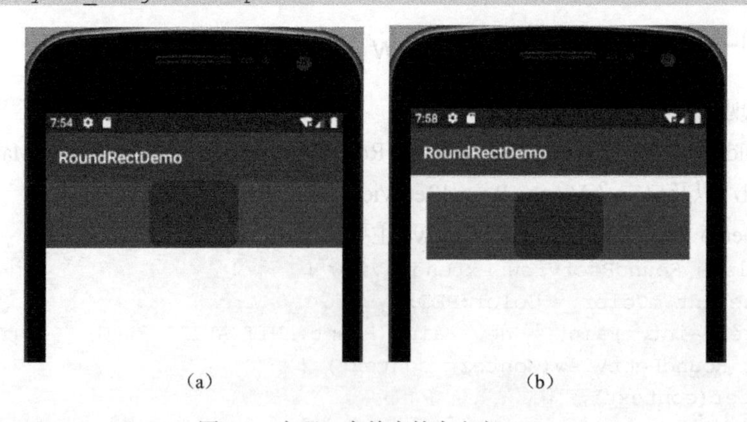

（a）　　（b）

图 3-1　实现一个基本的自定义 View

再重新运行 RoundRectDemo 模块，界面如图 3-1（b）所示。由于 margin 属性是由父容器控制的，所以 margin 属性生效了。

仍在 com.mialab.roundrectdemo.ui.RoundRectView 中添加一行 padding 属性的代码如下：

```
android:padding = "10dp"
```

重新运行 RoundRectDemo 模块，界面仍如图 3-1（a）所示，没有什么变化。这说明 padding 属性未生效。

无论如图 3-1（a）所示，还是如图 3-1（b）所示，都证明了 wrap_content 属性亦未生效。

3.4.2 支持 wrap_content 属性和 padding 属性

【示例】 支持 wrap_content 属性和 padding 属性。

启动 Android Studio，在 Ch3 工程中创建 RoundRectDemo2 模块，并创建 MainActivity.java、activity_main.xml、RoundRectView.java。

为了支持 padding 属性，RoundRectView.java 的 onDraw()方法代码改变如下。

```
protected void onDraw(Canvas canvas) {
    super.onDraw(canvas);
    final int paddingLeft = getPaddingLeft();
    final int paddingRight = getPaddingRight();
    final int paddingTop = getPaddingTop();
    final int paddingBottom = getPaddingBottom();
    int width = getWidth() - paddingLeft - paddingRight;
    int height = getHeight() - paddingTop - paddingBottom;
    int rectWidth = 220;
    int rectHeight = 160;
    RectF rectF = new RectF(paddingLeft + (getWidth() - rectWidth) / 2,
            paddingTop + (getHeight() - rectHeight) / 2,
            (getWidth() + rectWidth) / 2 - paddingRight,
            (getHeight() + rectHeight) / 2 - paddingBottom);
    canvas.drawRoundRect(rectF, 30, 30, mPaint);
}
```

运行 RoundRectDemo2 模块的界面如图 3-2（a）所示，这说明 padding 属性生效了。

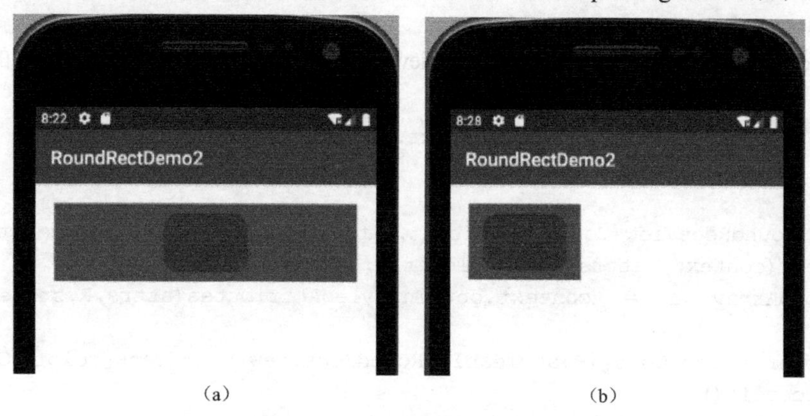

（a）　　　　（b）

图 3-2　支持 padding 和 wrap_content 属性

为了支持 wrap_content 属性，重写 RoundRectView.java 的 onMeasure()方法，其代码如下：

```
@Override
protected void onMeasure(int widthMeasureSpec, int heightMeasureSpec) {
    super.onMeasure(widthMeasureSpec, heightMeasureSpec);
    int widthSpecMode = MeasureSpec.getMode(widthMeasureSpec);
```

```
        int widthSpecSize = MeasureSpec.getSize(widthMeasureSpec);
        int heightSpecMode = MeasureSpec.getMode(heightMeasureSpec);
        int heightSpecSize = MeasureSpec.getSize(heightMeasureSpec);
        if (widthSpecMode == MeasureSpec.AT_MOST&& heightSpecMode == MeasureSpec.
AT_MOST) {
            setMeasuredDimension(240, 160);
        } else if (widthSpecMode == MeasureSpec.AT_MOST) {
            setMeasuredDimension(240, heightSpecSize);
        } else if (heightSpecMode == MeasureSpec.AT_MOST) {
            setMeasuredDimension(widthSpecSize, 160);
        }
    }
```

重新运行 RoundRectDemo2 模块的界面如图 3-2（b）所示，这说明 wrap_content 属性生效了。

3.4.3 自定义属性

系统自带的 View 可以在 XML 中配置属性，对于写好的自定义 View 同样可以在 XML 中配置属性，为了使自定义的 View 的属性可以在 XML 中配置，一般需要以下 4 个步骤：

- 通过<declare-styleable>为自定义 View 添加属性；
- 在 XML 中为相应的属性声明属性值；
- 在运行时（一般为构造函数）获取属性值；
- 将获取到的属性值应用到 View。

【示例】 为自定义 View 提供自定义属性。

启动 Android Studio，在 Ch3 工程中创建 RoundRectDemo3 模块，并创建 MainActivity.java、activity_main.xml、RoundRectView.java。

（1）在 RoundRectDemo3 模块 res\values 中创建 attrs.xml 文件，其代码如下：

```
<?xml version="1.0" encoding="utf-8"?>
<resources>
    <declare-styleable  name = "RoundRectView">
        <attr name="roundrect_color"  format="color" />
    </declare-styleable>
</resources>
```

（2）在 RoundRectDemo3 模块的 RoundRectView 类的构造方法中须解析自定义属性的值，并做相应处理。

```
    public RoundRectView(Context context, AttributeSet attrs) {
        this(context, attrs, 0);
    }
    public RoundRectView(Context context, AttributeSet attrs, int defStyleAttr) {
        super(context, attrs, defStyleAttr);
        TypedArray  a  =  context.obtainStyledAttributes(attrs,R.styleable.Round
RectView);
        mColor = a.getColor(R.styleable.RoundRectView_roundrect_color, Color.RED);
        a.recycle();
        init();
    }
```

context 通过调用 obtainStyledAttributes()方法来获取一个 TypeArray，然后由该 TypeArray 对属性进行设置。

```
    TypedArray  a  =  context.obtainStyledAttributes(attrs,  R.styleable.RoundRect
View);
```

调用结束后务必要用 recycle()方法释放资源。

（3）在 activity_main.xml 文件中使用自定义属性。

还须在布局文件 activity_main.xml 添加 schemas 声明。

```
xmlns:app = "http://schemas.android.com/apk/res-auto"
```

com.mialab.roundrectdemo3.ui.RoundRectView 中添加对于自定义属性的使用。

```
app:roundrect_color = "@color/color_blue"
```

运行 RoundRectDemo3 模块的界面如图 3-3 所示，这说明自定义属性生效了。

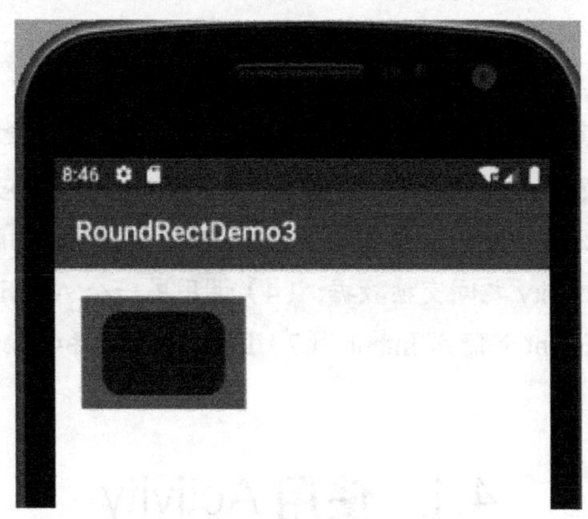

图 3-3　为 View 提供自定义属性

3.5　本章小结

本章主要介绍了自定义 View 的相关内容。首先概述了自定义 View 的分类和自定义 View 的构造函数，然后阐述了 View 的绘制流程，最后给出一个自定义 View 的示例。

习　题　3

1. 自定义 View 的分类有哪些？
2. View 的绘制流程是怎样的？
3. 如何自定义 View？试编程举例说明。
4. 如何自定义 ViewGroup？试编程举例说明。
5. View 的事件传递机制是怎样的？可上网查找相关资料并加以说明。
6. ViewGroup 的事件传递机制是怎样的？可上网查找相关资料并加以说明。

第 4 章　Activity 和 Intent

Activity 是 Android 系统中最基本的组件。每一个 Activity 都可以通过布局文件、Java 程序或者二者相结合的方式来绘制用户界面。一个应用程序由多个 Activity 组成，它们之间通过组件通信技术形成一个有机的整体，通常是松耦合关系。本章主要内容:（1）Activity 的配置;（2）Activity 的启动和关闭;（3）使用 Bundle 在 Activity 之间交换数据;（4）调用另一个 Activity 并返回结果;（5）Activity 的四种启动模式;（6）显式 Intent 和隐式 Intent;（7）Intent 过滤器和 Intent 的属性。

4.1　使用 Activity

一个应用程序通常由一系列组件组成，这些组件以进程为载体，相互协作来实现 App 功能。Activity 不仅负责 App 的视图，还负责用户的交互功能。在 Android 中每个界面都可视作一个 Activity，切换界面操作其实是多个不同 Activity 之间的实例化操作。

Activity 类声明如下：

```
    public class Activity extends ContextThemeWrapper implements LayoutInflater.
Factory2, Window.Callback,
    KeyEvent.Callback, View.OnCreateContextMenuListener, ComponentCallbacks2
```

AppCompatActivity 类声明如下：

```
    public class AppCompatActivity extends FragmentActivity implements AppCompat
Callback,
    TaskStackBuilder.SupportParentable, ActionBarDrawerToggle.DelegateProvider
```

Activity 类和 AppCompatActivity 类的继承结构如图 4-1 所示。

```
java.lang.Object
 ↳ android.content.Context
     ↳ android.content.ContextWrapper
         ↳ android.view.ContextThemeWrapper
             ↳ android.app.Activity
                 ↳ android.support.v4.app.FragmentActivity
                     ↳ android.support.v7.app.AppCompatActivity
```

图 4-1　Activity 类和 AppCompatActivity 类的继承结构

小贴士：Activity 生命周期的四个重要状态。（1）运行状态：当前的 Activity，位于 Activity 栈顶，用户可见，并且可以获得焦点。（2）暂停状态：失去焦点的 Activity，仍然可见，但是在内存低的情况下，不能被系统销毁。（3）停止状态：该 Activity 被其他 Activity 所覆盖，不可见，但是它仍然保存所有的状态和信息。在内存低的情况下，它将会被系统销毁。（4）销毁状态：该 Activity 结束，或 Activity 所在的进程结束。

4.1.1　Activity 的配置

Android 应用要求所有的组件都必须显式地在 AndroidManifest.xml 中进行配置。在配置文件<application>元素的子元素<activity>中来进行 Activity 的配置，修改该元素即可修改该 Activity 的配置。

在配置 Activity 时通常指定如下几个属性。

- name：Activity 实现类的类名。类名必须明确指明实现类所在的包和类名，如 org.suda.app.activitydemo。如果在<mainifest>中明确定义了应用程序的包，即 package= "org.suda.app"，那么<activity>的 name 可以简写为.activitydemo。
- icon：Activity 对应的图标，通常和标签一起显示在窗口上。
- label：Activity 对应的标签。
- exported：是否允许其他应用程序启动该 Activity。如果设为 true，则允许启动。
- launchMode：指定该 Activity 的加载模式，该属性支持 standard、singleTop、singleTask、singleInstance 这四种加载模式。

<activity>元素还包含有<intent-filter>子元素，该元素用于指定该 Activity 可响应的 Intent。

4.1.2 Activity 的启动和关闭

1. 启动 Activity

用 Activity 启动其他 Activity 有如下两种方法：

（1）startActivity(Intent intent)。

启动 Activity，其中参数 intent 描述了待启动 Activity 的信息。

Activity A 启动 Activity B 的代码如下：

```
Intent intent=new Intent(A.this, B.class);
startActivity(intent);
```

（2）startActivityForResult(Intent intent,int requestCode)。

以指定请求码 requestCode 启动 Activity，而且程序将会得到新启动 Activity 的结果，该结果可凭借重写 onActivityResult()方法获取。请求码的值由开发者根据业务自行设置，用于标识请求来源。

2. 关闭 Activity

当完成任务退出窗口的时候，也有两个关闭 Activity 的方法。

（1）finish()：表示结束当前 Activity。

（2）finishActivity(int requestCode)：结束以 startActivityForResult(Intent intent,int request Code)方法启动的 Activity。

4.1.3 使用 Bundle 在 Activity 之间交换数据

Activity 与 Activity 之间相互调用时，往往需要传递一些数据。Android 将数据封装成以键值对(key-value)形式表示的 Bundle 对象。如两个 Activity 之间靠 Intent 通信，只需将 Bundle 数据放入 Intent，通过调用 Intent 对象相应的方法即可实现数据传递。

Bundle 类声明如下：

```
public final class Bundle extends BaseBundle implements Cloneable, Parcelable
```

（1）Intent 提供了多个方法来设置要传递的数据，其示例如下。

public Intent putExtras (Bundle extras)：向 Intent 中放入需要传递的 Bundle 数据 extras。

public Intent putExtra (String name, Xxx value)：向 Intent 中按 key-value 对的形式存入数据（Xxx 指代各种数据类型的名称），其中 key 为 name，value 为 val。

（2）Intent 提供了相应的取出传递过来的数据方法，其示例如下。

public Bundle getExtras ()：取出 Intent 所“携带”的数据。

（3）Bundle 类提供了多个方法，把数据封装到 Bundle 对象中，其示例如下：

public void putSerializable (String key, Serializable value)：向 Bundle 中放入一个可序列化的对象。

public void putInt (String key, int value)：向 Bundle 中放入 int 类型的数据。

（4）Bundle 类提供了从 Bundle 对象中取出数据的方法。

public Serializable getSerializable (String key)：从 Bundle 中取出一个可序列化的对象。

public int getInt (String key)：从 Bundle 中取出 Int 类型的数据。

例如，ActivityA 启动 ActivityB，并携带 key 为“name”数据传递给 ActivityB。

```
Intent intent = new Intent(ActivityA.this,ActivityB.class);
Bundle bundle = new Bundle();
bundle.putString("name","Tom");
intent.putExtras(bundle);
startActivity(intent);
```

ActivityB 获取传递过来的数据，其相应代码如下：

```
Intent intent = getIntent();
Bundle bundle = intent.getExtras();
String name = bundle.getString("name");
```

【示例】 启动 Android Studio，在 Ch4 工程中创建 BundleTest 模块，并在模块中创建注册页面（MainActivity.java）和注册信息提交页面（ResultActivity.java），以及相应的 XML 布局文件（activity_main.xml 和 activity_result.xml），还有实体类（Person.java）。

BundleTest 模块的 MainActivity.java 主要代码如下：

```
public class MainActivity extends Activity {
   @Override
   protected void onCreate(Bundle savedInstanceState) {
       super.onCreate(savedInstanceState);
       setContentView(R.layout.activity_main);
       final String[] education = {"未选择"};
       Button bn = (Button) findViewById(R.id.bn);
       bn.setOnClickListener(new View.OnClickListener() {
           @Override
           public void onClick(View v) {
               EditText name = (EditText) findViewById(R.id.name);
               EditText passwd = (EditText) findViewById(R.id.passwd);
               RadioButton male = (RadioButton) findViewById(R.id.male);
               String gender = male.isChecked() ? "男" : "女";
               CheckBox checkBox1 = (CheckBox) findViewById(R.id.checkbox1);
               CheckBox checkBox2 = (CheckBox) findViewById(R.id.checkbox2);
               CheckBox checkBox3 = (CheckBox) findViewById(R.id.checkbox3);
               List<CheckBox> checkBoxList = new ArrayList<CheckBox>();
               checkBoxList.add(checkBox1);
               checkBoxList.add(checkBox2);
               checkBoxList.add(checkBox3);
               StringBuffer sb = new StringBuffer();
               //遍历集合中的checkBox,判断是否选择，获取选中的文本
               for (CheckBox checkbox : checkBoxList) {
                   if (checkbox.isChecked()) {
                       sb.append(checkbox.getText().toString() + " ");
                   }
               }
               String hobby = sb.toString();

               //把个人注册信息封装到person对象中
               Person person = new Person(name.getText().toString(), passwd.
getText().toString(),
```

```
                gender, hobby,education[0]);
            //创建一个 Bundle 对象
            Bundle data = new Bundle();
            data.putSerializable("person", person);
            Intent  intent  =  new  Intent(MainActivity.this,  ResultActivity.
class); //创建一个 Intent
            intent.putExtras(data);
            startActivity(intent); //启动 Intent 对应的 Activity
        }
    });

    //Spinner 控件与适配器绑定
    Spinner sp = ((Spinner) findViewById(R.id.spinner));
    ArrayAdapter<CharSequence> adapter = ArrayAdapter.createFromResource(
            MainActivity.this,   R.array.eduList,   android.R.layout.simple_
spinner_item);

adapter.setDropDownViewResource(android.R.layout.simple_spinner_dropdown_item);
    sp.setAdapter(adapter);
    sp.setOnItemSelectedListener(new AdapterView.OnItemSelectedListener() {
        @Override
        public void onItemSelected(AdapterView<?> parent, View view, int pos,
long id) {
            education[0] = parent.getItemAtPosition(pos).toString();
            Toast.makeText(parent.getContext(), "您的学历为: " +
                    education[0], Toast.LENGTH_LONG).show();
        }
        @Override
        public void onNothingSelected(AdapterView<?> parent) {
        }
    });
  }
}
```

BundleTest 模块的 ResultActivity.java 主要代码如下：

```
public class ResultActivity extends Activity {
   @Override
   protected void onCreate(Bundle savedInstanceState) {
       super.onCreate(savedInstanceState);
       setContentView(R.layout.activity_result);
       TextView name = (TextView)findViewById(R.id.name);
       TextView passwd = (TextView)findViewById(R.id.password);
       TextView gender = (TextView)findViewById(R.id.gender);
       TextView hobby = (TextView)findViewById(R.id.hobby);
       TextView eduText = (TextView)findViewById(R.id.edu_text);
       Intent intent = getIntent(); //获取启动该 ResultActivity 的 Intent
       //直接通过 Intent 取出它所携带的 Bundle 数据包中的数据
       Person p = (Person)intent.getSerializableExtra("person");
       name.setText("您的用户名为: " + p.getName().toString());
       passwd.setText("您的密码为: " + p.getPass().toString());
       gender.setText("您的性别为: " + p.getGender().toString());
       hobby.setText("您的兴趣爱好为: " + p.getHobby().toString());
       eduText.setText("您的学历为: " + p.getEdu().toString());
   }
}
```

运行 BundleTest 模块，注册页面如图 4-2 所示，注册信息提交页面如图 4-3 所示。其中 activity_main.xml、activity_result.xml、Person.java 代码详见教学资源包中的 Ch4\BundleTest。

图 4-2　BundleTest 的注册页面

图 4-3　BundleTest 的注册信息提交页面

4.1.4　调用另一个 Activity 并返回结果

如果启动另一个 Activity，并且希望返回结果给当前的 Activity，那么可使用 startActivityForResult()方法来启动 Activity，返回的数据也是通过 Bundle 来进行交换的。

为了获取到被启动 Activity 的返回结果，需要执行两个步骤。

（1）当前的 Activity 要重写 onActivityResult(int requestCode, int resultCode, Intent intent)方法，其中 requestCode 代表请求码，判断是哪个 Activity 返回的结果。resultCode 代表返回的结果码，由开发人员根据业务需要自定义。

（2）被启动的 Activity 需要调用 setResult(int resultCode, Intent data)方法设置返回结果。

【示例】 启动 Android Studio，在 Ch4 工程中创建 ActivityIntentDemo 模块、MainActivity.java 和 ActivityB.java，以及相应的 XML 布局文件 activity_main.xml 和 activity_b.xml。MainActivity 的用户界面有一个按钮和一个显示返回信息的文本框，点击按钮，启动 ActivityB。ActivityB 的用户界面有一个用于接收“业务相关信息”输入的编辑框，用户在编辑框中输入“业务相关信息”，点击“返回 MainActivity”按钮，则 ActivityB 关闭并返回 MainActivity，同时 MainActivity 的文本框显示 ActivityB 返回的业务相关信息。

ActivityIntentDemo 模块的 MainActivity.java 主要代码如下：

```
public class MainActivity extends Activity {
   @Override
   protected void onCreate(Bundle savedInstanceState) {
       super.onCreate(savedInstanceState);
       setContentView(R.layout.activity_main);
       Button btn=(Button)findViewById(R.id.btnA);
       btn.setOnClickListener(new View.OnClickListener() {
           @Override
           public void onClick(View arg0) {
               Intent intent = new Intent(MainActivity.this,ActivityB.class);
               startActivityForResult(intent, 0);
           }
       });
   }

   @Override
   public void onActivityResult(int requestCode, int resultCode, Intent intent)
```

```
    {
        //当 requestCode、resultCode 同时为 0，也就是处理特定的结果
        if (requestCode == 0 && resultCode == 0)
        {
            Bundle data = intent.getExtras();  //取出 Intent 里的 Extras 数据
            String name = data.getString("name");  //取出 Bundle 中的数据
            TextView txt=(TextView)findViewById(R.id.txt);
            txt.setText(name);  //设置文本框的内容
        }
    }
}
```

ActivityIntentDemo 模块的 ActivityB.java 主要代码如下：

```
public class ActivityB extends Activity {
    @Override
    protected void onCreate(Bundle savedInstanceState) {
        super.onCreate(savedInstanceState);
        setContentView(R.layout.activity_b);
        Button btn = (Button) findViewById(R.id.btnB);
        btn.setOnClickListener(new View.OnClickListener() {
            @Override
            public void onClick(View arg0) {
                EditText txt = (EditText) findViewById(R.id.txtNameB);
                Intent intent = getIntent();
                Bundle bundle = new Bundle();
                bundle.putString("name", txt.getText().toString());
                intent.putExtras(bundle);
                ActivityB.this.setResult(0, intent);
                ActivityB.this.finish();
            }
        });
    }
}
```

运行 ActivityIntentDemo 模块的结果如图 4-4 所示。其中，activity_main.xml 文件和 activity_b.xml 文件代码参见教学资源包中的 Ch4\ActivityIntentDemo 模块。

图 4-4 运行 ActivityIntentDemo 模块的结果

4.2 Activity 的四种启动模式

Activity 的启动模式决定了 Activity 的运行方式。在默认情况下，当多次启动同一个 Activity 时，系统会创建多个实例并把它们逐一放入任务栈中，当点击 back 键时，这些 Activity 会逐一回退。

任务栈是一种“后进先出”的栈结构，每按一下 back 键就会有一个 Activity 出栈，直到栈空为止。当栈中无任何 Activity 时，系统就会回收这个任务栈。

任务栈用来存放用户开启的 Activity。在应用程序创建之初，系统会默认分配给其一个任务栈（默认一个），并存储根 Activity。对于 Activity 来说，只要不在栈顶就是 onStop 状态。

Activity 的启动模式分为四种：standard、singleTop、singleTask、singleInstance。启动模式在 AndroidManifest.xml 中，通过<activity>标签的 android:launchMode 属性进行设置，其示例代码如下：

```
<activity android:name = ".MainActivity"  android:launchMode = "standard" />
```

4.2.1 standard 模式

它是标准模式，也是默认模式。每当启动一个 Activity 时，系统就会相应地创建一个实例，不管这个实例是否已经存在。standard 模式的 Activities 在栈中的管理示意如图 4-5 所示。

可以看到，使用 standard 模式的 Activity 在栈中不管有没有实例，每次新开启一个 Activity 时都会新建一个实例对象。

4.2.2 singleTop 模式

它为栈顶复用模式。如果要启动的 Activity 处于栈的顶部，那么此时系统不会创建新的实例，而是直接打开此页面，同时它的 onNewIntent()方法会被执行，可以通过 Intent 进行传值，而且它的 onCreate()方法、onStart()方法不会被调用，因为它并没有发生任何变化。

singleTop 模式的 Activity 在栈中的管理示意如图 4-6 所示。其中，将 AActivity 的启动模式设置为 android:launchMode="singleTop"，BActivity 的启动模式设置为 android:launchMode="standard"。可以看到，使用 singleTop 模式的 Activity 在栈中，若在栈顶有该 Activity 实例，则不会重新创建，保持原有对象实例。

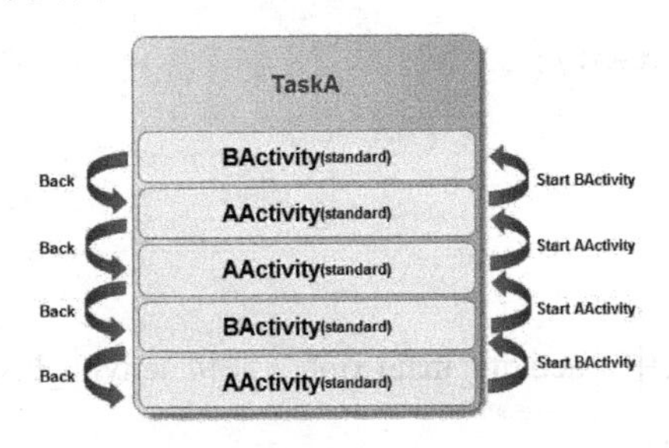

图 4-5　standard 模式

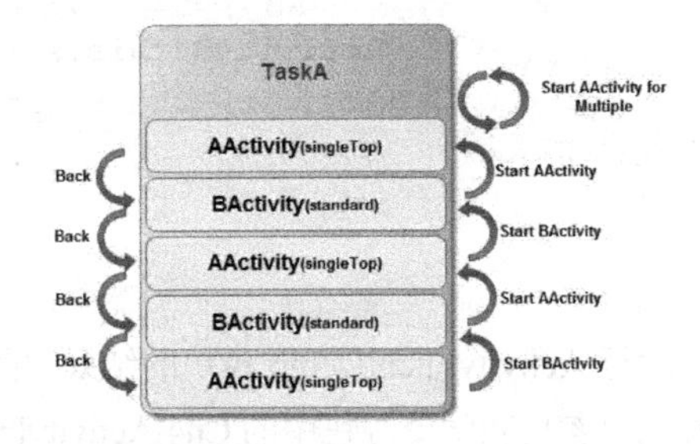

图 4-6　singleTop 模式

4.2.3 singleTask 模式

它为栈内复用模式。如果栈中存在这个 Activity 的实例就会复用这个 Activity，不管它是否位于栈顶，复用时会将它上面的 Activity 全部出栈，因为 SingleTask 本身自带 clearTop 这种功能，并且会回调该实例的 onNewIntent()方法。

singleTask 模式的 Activity 在栈中的管理示意，如图 4-7 所示。

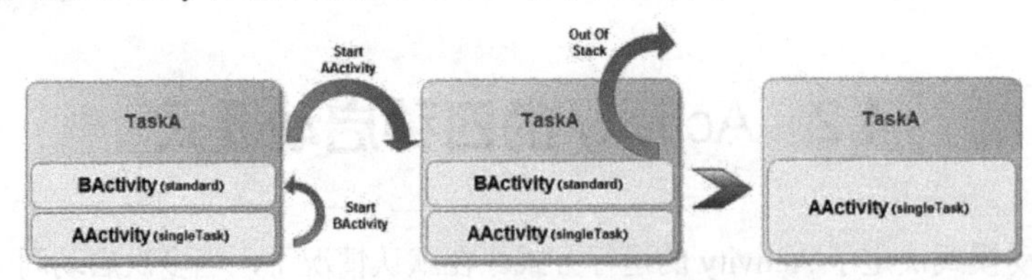

图 4-7　singleTask 模式

首先开启了 AActivity 和 BActivity 实例，然后 BActivity 再启动 AActivity，这时系统会将 BActivity 对象实例移出栈，然后恢复第一次创建 AActivity 对象实例的状态。可见，使用 singleTask 为开启模式的 Activity 在单个栈中只允许一个实例对象存在。

4.2.4 singleInstance 模式

它为单实例模式。该模式除了具备 singleTask 模式的所有特性，在这种模式下的 Activity 还会单独占用一个 Task 栈，具有全局唯一性。以 singleInstance 模式启动的 Activity 在整个系统中是单例的，如果在启动这样的 Activity 时，已经存在了一个实例，那么就会把它所在的任务调度到前台，重用这个实例。

singleInstance 模式的 Activity 在栈中的管理示意如图 4-8 所示。

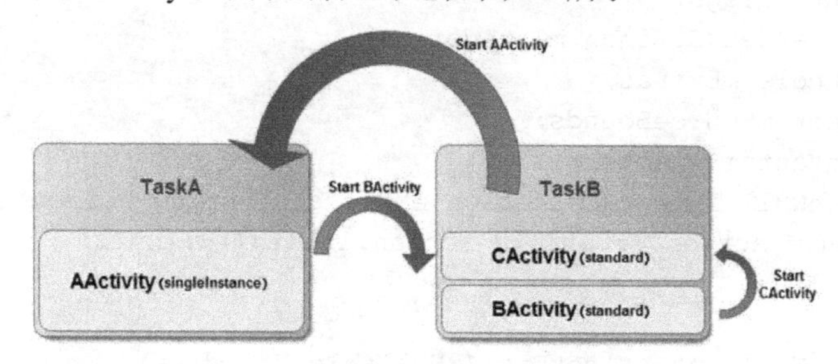

图 4-8　singleInstance 模式

使用 singleInstance 开启模式的 AActivity 在栈中只允许一个实例存在，并且该栈也只能存在一个必须是 AActivity 的实例对象。注意，使用 singleTask 和 singleInstance 为开启模式的 Activity 很相似，但它们之间是有区别的，举例来讲，当一个 Activity 使用 singleTask 模式时，装载它的 Task 中只允许一个该实例存在，不过该 Task 也可以装载其他的 Activity 实例；当一个 Activity 使用 singleInstance 模式时，装载它的 Task 中只允许一个该实例存在，并且该 Task 不允许装载其他的 Activity 实例，如要从此 Activity 跳转至其他的 Activity，系统会创建新的 Task 用于装载其他 Activity，也就是说，使用 singleInstance 模式为开启模式的 Activity（假设为 AActivity）只属于一个 Task，且该 Task 也只包含一个 Activity 实例，就是 AActivity，相关示例可参见教学资源包中的 Ch4 工程。

Activity 四种启动模式的特点如表 4-1 所示。

表 4-1　Activity 四种启动模式

启动模式名称	特　点
standard	①Activity 默认的启动模式。 ②新 Activity 在启动其 Activity 所在的栈中
singleTop	若需新建的 Activity 位于任务栈栈顶，那么此 Activity 的实例就不会重建，而是重用栈顶的实例，否则就创建该 Activity 新的实例并置于栈顶
singleTask	①查看 Activity 想要的任务栈是否存在，若不存在则重建一个任务栈，创建该 Activity 实例并置于栈中。 ②若任务栈存在，则查看 Activity 是否存在栈中，若存在栈中，则将其移出栈，使该 Activity 处于栈顶
singleInstance	直接创建一个新的任务栈，并创建该 Activity 实例放于新的任务栈中

4.3 Intent 对象

Intent 是一种轻量级消息传递机制，意在解决 Android 应用的各项组件之间的通信。它描述了应用中一次操作的动作、动作涉及数据、附加数据，Android 则根据此描述，负责找到对应的组件，将 Intent 传递给待启动的组件，并完成该组件的启动。这种消息的传递可以在同一应用程序中的不同组件之间，也可以在不同应用程序的组件之间。这类组件主要是 Activity、Broadcast、Service 等。

Intent.java 源码中有这样的代码片断如下：

```
public class Intent implements Parcelable, Cloneable {
   …
   private String mAction;
   private Uri mData;
   private String mType;
   private String mPackage;
   private ComponentName mComponent;
   private int mFlags;
   private ArraySet<String> mCategories;
   private Bundle mExtras;
   private Rect mSourceBounds;
   private Intent mSelector;
   public @NonNull Intent setAction(@Nullable String action) {
      mAction = action != null ? action.intern() : null;
      return this;
   }
   public @NonNull Intent setData(@Nullable Uri data) {
      mData = data;
      mType = null;
      return this;
   }
   public @NonNull Intent setType(@Nullable String type) {
      mData = null;
      mType = type;
      return this;
   }
   public @NonNull Intent setDataAndType(@Nullable Uri data, @Nullable String
type) {
      mData = data;
      mType = type;
      return this;
   }
   …
}
```

Intent 对象至少包括 mAction、mData、mType、mComponent、mFlags、mCategories、mExtras 等属性。

4.3.1 显式 Intent 和隐式 Intent

Intent 启动 Activity 可分为显式启动和隐式启动两种。

1. 显式启动

显式启动是指一个 Activity 通过类名明确指明要启动哪个 Activity。

显式启动 Activity 需要两个步骤：

（1）创建 Intent 对象，并初始化指明要启动的 Activity，创建方法有两种。

```
//方法一
Intent intent = new Intent(ActivityA.this, ActivityB.class);
//方法二
ComponentName componentName = new ComponentName(ActivityA.this, ActivityB.
class);
```

```
Intent intent = new Intent();
intent.setComponent(componentName);
```

ActivityA.this 表示当前上下文，ActivityB.class 表示新启动 Activity 的类名。

（2）调用启动 Activity 的方法启动新的 Activity，如 startActivity(intent)，则可完成新 Activity 的启动。

2．隐式启动

所谓隐式启动，表示不需指明要启动哪一个 Activity，只需要声明一个行为，系统会根据 Activity 配置，启动能够匹配这个行为的 Activity。隐式启动 Activity 时，系统会自动启动与 Intent 上的动作（Action）、附加数据（Category），以及数据（Data）相匹配的 Activity。

其中，Action 用于描述要完成的动作，Category 为被执行动作 Action 增加的附加信息。通常 Action 与 Category 结合使用，Data 则用于向 Action 提供操作的数据。

Intent 隐式启动分为三个步骤：

（1）创建 Intent 对象，其代码如下：

```
Intent intent=new Intent();
```

（2）设置用于匹配的 Intent 属性，Intent 属性有：动作属性、动作额外附加的类别属性及数据属性等，分别需要调用不同的方法来设置，其代码如下：

```
intent.setAction();            //设置动作属性
intent.addCategory);           //设置类别属性
intent.setData();              //设置数据属性
intent.setType();              //设置数据属性
```

（3）调用启动 Activity 的方法，启动新的 Activity，如 startActivity(intent) 。

除了上述启动应用程序中自定义的 Activity，有时还需要启动系统级的应用程序。如希望通过界面的电话号码直接点击拨打电话时，就需要启动手机中的电话程序，其代码如下：

```
Intent intent = new Intent(Intent.ACTION_DIAL); //ACTION_DIAL表示显示拨号面板
intent.setData(Uri.parse("tel:114"));
startActivity(intent);
```

4.3.2　Intent 过滤器

Android 系统的这种匹配机制是依靠过滤器 Filter 来实现的。与 Intent 属性保持一致，Intent Filter 包含动作、类别、数据等过滤内容，通过与 Intent 对象中的内容进行比较，匹配出符合 Intent 启动要求的 Activity，且必须在 AndroidManifest.xml 文件中以 Activity 的子元素<intent-filter>的方式进行声明，如表 4-2 所示。

表 4-2　<intent-filter>节点属性

节　点	属　性	描　述
<action>	android:name	指定组件所能响应的动作，以字符串形式表示，通常由 Java 类名和包的完全限定名构成
<category>	android: name	指定组件能响应的服务方式，用字符串表示
<data>	android:host	用于指定一个有效的主机名
	android:mimetype	用于指定组件能处理的数据类型
	android:path	用于指定有效的 URI 路径名
	android:scheme	用于指定所需要的特定协议

4.3.3 Intent 的属性

1. Component 属性

ComponentName 主要被用于显式 Intent 中，它明确了需要被启用的组件名称（如果是同一个应用内的其他组件，则只需具体的 ComponentName；如果在不同的应用中，则需要添加 PackageName），这样系统就会将该 Intent 发送到对应的组件中。

Intent 的 Component 属性需要接受一个 ComponentName 对象，其代码如下：

```
intent.setComponent(componentName);
```

ComponentName.java 源码中的代码片断如下：

```
public final class ComponentName implements Parcelable, Cloneable, Comparable
<ComponentName> {
    private final String mPackage;
    private final String mClass;
    public ComponentName(@NonNull Context pkg, @NonNull Class<?> cls) {
        mPackage = pkg.getPackageName();
        mClass = cls.getName();
    }
    public ComponentName(@NonNull Context pkg, @NonNull String cls) {
        if (cls == null) throw new NullPointerException("class name is null");
        mPackage = pkg.getPackageName();
        mClass = cls;
    }
    public ComponentName(@NonNull String pkg, @NonNull String cls) {
        if (pkg == null) throw new NullPointerException("package name is null");
        if (cls == null) throw new NullPointerException("class name is null");
        mPackage = pkg;
        mClass = cls;
    }
    …
}
```

ComponentName 类的构造器如表 4-3 所示。

表 4-3 ComponentName 类的构造器

构 造 器	描 述
ComponentName(String pkg, String cls)	创建 pkg 所在包下的 cls 类所对应的组件
ComponentName(Context pkg, String cls)	创建 pkg 所对应的包下的 cls 类所对应的组件
ComponentName(Context pkg, Class<?> cls)	创建 pkg 所对应的包下的 cls 类所对应的组件

上面的构造器的本质就是一个，这说明创建一个 ComponentName 需要指定包名和类名（可唯一确定的组件类），这样应用程序就可根据给定的组件类去启动特定的组件。

2. Action 属性

Action 属性是一个描述动作的普通字符串，用户可以在应用程序中通过 Intent 自定义动作字符串来隐式启动某个 Activity。

一个 Intent 对象最多只能设置一个 Action 属性。Action 要完成的动作可以自定义，也可以指定系统提供的 Action，其示例代码如下：

```
String action_str = "com.mialab.demo. SUDA_ACTION";
Intent intent = new Intent();
intent.setAction(action_str);
```

上述代码为Intent设置Action属性，但是该Intent具体启动哪个Activity，只从代码上看并无从知晓，因为它取决于AndroidManifest.xml文件中Activity的<intent-filter>元素的配置。

这条语句要启动的是包含<action>元素，并且<action>元素属性值为"com.mialab.demo.SUDA_ACTION"的Activity。

其中，Action（动作）是一个字符串，用来表示要完成的一个抽象动作。对这个动作具体由哪个组件（或许是Activity，或许是BroadcastReceiver）来完成，Action本身并不管。

Android定义了一系列的Action常量，其目标组件包括Activity和BroadcastReceiver两类。Intent类中定义的一些用于启动Activity的标准Action常量，如表4-4所示。

表4-4 Intent类中定义的一些用于启动Activity的标准Action常量

Action常量	对应字符串	描　述
ACTION_MAIN	android.intent.action.MAIN	应用程序入口
ACTION_VIEW	android.intent.action.VIEW	显示指定数据
ACTION_EDIT	android.intent.action.EDIT	编辑指定数据
ACTION_GET_CONTENT	android.intent.action. GET_CONTENT	让用户选择数据，并返回所选数据
ACTION_DIAL	android.intent.action. DIAL	显示拨号面板
ACTION_CALL	android.intent.action. CALL	直接向指定用户打电话
ACTION_SEND	android.intent.action. SEND	向其他人发送数据
ACTION_SENDTO	android.intent.action. SENDTO	向其他人发送消息
ACTION_ANSWER	android.intent.action. ANSWER	应答电话
ACTION_INSERT	android.intent.action. INSERT	插入数据
ACTION_DELETE	android.intent.action. DELETE	删除数据

3. Category属性

Category（类别）也是一个字符串，它用于为Action添加额外的附加类别信息。一个Intent对象可以设置多个Category属性，可通过调用addCategory()方法来添加Category属性。

在Intent类中也预定义了一些Category常量，常用的标准Category常量及对应的字符串如表4-5所示。

表4-5 常用的标准Category常量及对应的字符串

Category常量	对应字符串	说　明
CATEGORY_DEFAULT	android.intent.category. DEFAULT	默认的Category
CATEGORY_BROWSABLE	android.intent.category. BROWSABLE	指定该Activity能被浏览器安全调用
CATEGORY_LAUNCHER	android.intent.category. LAUNCHER	Activity显示在顶级程序列表中
CATEGORY_INFO	android.intent.category. INFO	用于提供包信息
CATEGORY_HOME	android.intent.category. HOME	设置该Activity随系统启动而运行

4. Component Data属性和Type属性

Data（数据）通常用于向Action属性提供操作数据，主要描述Intent的动作所操作数据的Uri对象及类型。setData(Uri data) 接受一个Uri对象，一个Uri对象通常通过如下形式的字符串来表示：

```
content://com.android.contacts/ contacts/1
tel:123
```

Uri对象的字符串格式如下：

```
scheme://host:port/path
```

Type（类型）用于该Data所指定Uri对象所对应的MIME类型，如setType(String type)设置Data指定Uri对象所对应的MIME类型（MIME可以为任何类型，只需满足A/B的格式即可）。

不同类型的Action会有不同的Data数据。如Action是启动拨号面板时，其封装数据则是电话号码Uri，如tel：xxxx。若启动的是输入文本内容的应用，其封装数据则是MINE类型，如text/plain。

如果为Intent先设置Data属性，后设置Type属性，则Type属性会覆盖Data属性，反之亦然。如果希望同时拥有这两个属性，可以调用Intent的setDataAndType()方法。

其示例代码如下：

```
Intent intent = new Intent();
intent.setData(Uri.parse("content://com.mialab.demo/data"));
intent.setType("abc/xyz");
```

上述代码中，Type属性后设置，因此覆盖了Data属性，最终Intent启动的Activity包含一个<data>元素，该元素的mimetype属性是指“abc/xyz”。若希望同时设置Data和Type的属性，则将代码修改如下：

```
Intent intent = new Intent();
intent. setDataAndType(Uri.parse("content://com.mialab.demo/data"), "abc/xyz");
```

5. Extra属性

Extra（额外）属性用于向Intent组件添加附加信息，通常采用键值对的形式保存附加信息。Intent对象中有一系列的putxxx()方法用于添加各种附加数据，一系列的getxxx()方法用于读取数据。Intent的Extra属性值应该是一个Bundle对象。

6. Flag属性

Intent的Flag（标记）属性用于为该Intent添加一些额外的控制标记。相关示例参见教学资源包中的Ch4工程。

4.4 本章小结

本章较为详细地介绍了Activity的基本概念及Activity的四种启动模式。由于应用程序启动一个Activity时，需要借助于Intent来实现，因此介绍了Intent对象的使用，包括Intent过滤器和Intent的属性。Intent启动Activity分隐式启动和显式启动两种，应用程序不仅可以启动自身的Activity，还可以启动系统的组件，甚至可以启动其他应用程序的Activity。

习　题　4

1．如何显式启动Activity？如何隐式启动Activity？

2．Activity的启动模式是怎样的？请编程加以说明。

3．Activity生命周期中有哪几个重要状态？并请加以解释。

4．如何使用Bundle在Activity之间交换数据？试编程加以说明。

5．创建用户拨打电话的应用程序。

6．Activity的管理机制是怎样的？可上网搜集整理相关资料并加以说明。

第5章　后台服务与广播消息

Service（后台服务）和 BroadcastReceiver（广播）均属于 Android 系统的四大组件（Activity、Service、BroadcastReceiver、ContentProvider）。Service 只能后台运行，并且可以和其他组件进行交互。它是一种广泛运用在应用程序之间传输信息的机制，是对发送出来的广播进行过滤接收并响应的一类组件。本章主要内容有：（1）Service 应用的进程内服务和进程外服务；（2）发送广播和接收广播消息；（3）有序广播。

5.1　Service 的应用

5.1.1　Service 的启动方式

Service 可以在很多应用场合中使用。如播放多媒体时用户还启动了其他的 Activity，此时程序可在后台继续播放；检测 SD 卡上文件的变化；在后台记录用户地理信息位置的改变等。总之这些服务都是藏在后台的。

Service 的启动有两种方式：context.startService() 和 context.bindService()。

1．context.startService()启动方式

```
context.startService()→onCreate()→onStartCommand()→Service running→context.stopService()→onDestroy()→Service 被关闭
```

如果 Service 还没有运行，则 Android 先调用 onCreate()方法，然后再调用 onStartCommand()方法；如果 Service 已经运行，则只调用 onStartCommand()方法，所以一个 Service 的 onStart Command ()方法可能会重复调用多次。

在调用 stopService()方法的时候会直接调用 onDestroy()方法。如果是调用者自己直接退出而没有调用 stopService()方法的话，Service 就会一直在后台运行。该 Service 的调用者在启动后可以通过 stopService()方法关闭 Service。

小贴士：如果 Service 已经启动了，当再次启动 Service 时，则不会执行 onCreate()方法，而是直接执行 onStartCommand ()方法。在 Service 每一次的开启、关闭过程中，只有 onStartCommand()方法可被多次调用（通过多次调用 startService()方法），而其他方法如 onCreate()、onBind()、onUnbind()、onDestory()，则在一个生命周期中只能被调用一次。

2．context.bindService()启动方式

```
context.bindService() → onCreate() → onBind() → Service  running → onUnbind() → onDestroy()→Service 被关闭
```

其中，onBind()将返回给客户端一个 IBind 接口实例，IBind 允许客户端回调服务的方法，如得到 Service 的实例、运行状态或其他操作。这时会把调用者 Context（如 Activity）和 Service 绑定在一起。一旦 Context 退出，Srevice 就会调用 onUnbind()→onDestroy()方法相应地退出。

5.1.2 Service 的生命周期

Service 和 Activity 一样同为 Android 的四大组件之一，并且它们都有各自的生命周期，要想掌握 Service 的用法，那就要了解 Service 生命周期的方法，以及各个方法回调的时机和作用。Service 生命周期的两种启动方式是不一样的，如图 5-1 所示。

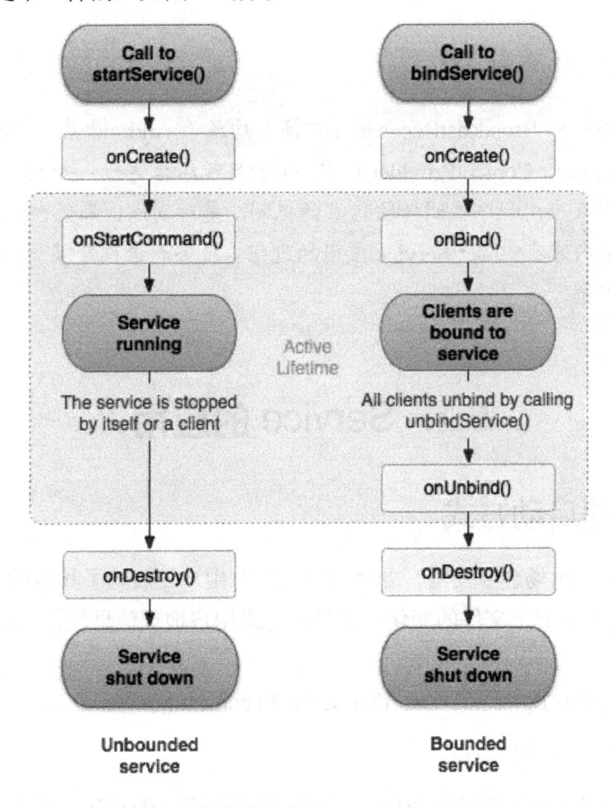

图 5-1　Service 生命周期的两种启动方法

小贴士：Service 生命周期的方法不同于 Activity，并不需要调用超类的生命周期方法，如不用调用 super.onCreate()。

1．startService 方法启动服务的生命周期

当应用组件通过 startService()方法启动 Service 时，Service 会处于启动状态。一旦服务启动，它就会在后台无限期地运行，生命周期独立于启动它的组件。即使启动它的组件已经销毁了也不受任何影响。由于启动的服务长期运行在后台，会大量消耗手机的电量。因此，在任务执行完成后应该调用 stopSelf()来停止服务，或者通过其他应用组件调用 stopService() 来停止服务。

startService()启动服务后，会执行生命周期的如下：onCreate()→onStartCommand()（可多次调用）→onStart()（现在已经废弃）→onDestroy() 。

它的生命周期方法如下。

（1）onCreate()：首次启动服务的时候，系统会调用这个方法，在 onStartCommand 方法和 onBind 方法之前，如果服务已经启动起来了，再次启动时，则不会调用此方法，因此可以在 onCreate 方法中做一些初始化的操作，如要执行耗时的操作，可以在这里创建线程；要播放音乐，也可以在这里初始化音乐播放器。

（2）onStartCommand()：当通过 startService 方法来启动服务时，在 onCreate 方法之后就会回调这个方法，服务就启动起来了。它将会在后台无限期的运行，直到通过 stopService 方法或者 stopSelf 方

法来停止服务。

（3）onDestroy()：当服务不再使用且将被销毁时，系统将调用此方法。服务应该实现此方法来清理所有资源，如线程、注册的侦听器、接收器等。这是服务接收的最后一个调用。

2．bindService 方式启动服务的生命周期

除了使用 startService()方法来启动服务，还可以通过 bindService()方法来启动，也就是绑定服务。“绑定”即将启动组件和服务绑定在一起。前面讲的通过 startService 方式启动的服务是与组件相独立的，即使启动服务的组件被销毁了，服务仍然在后台运行不受干扰。但是通过 bindSerivce 方式绑定的服务就不一样了，它与绑定组件的生命周期是有关的。

多个（或一个）组件可以绑定到同一个服务上。如果只有一个组件绑定服务，当绑定的组件被销毁时，服务也就会停止了。如果是多个组件绑定到一个服务上，只有当绑定到该服务的所有组件都被销毁时，服务才会停止。

bindService（绑定服务）和 startService（启动服务）的生命周期是不一样的。bindService 的生命周期：onCreate()→onBind()（只一次，不可多次绑定）→onUnbind()→onDestroy()。其中 onBind()方法和 onUnbind()方法最为重要。

（1）onBind()：当其他组件想通过 bindService 与服务绑定时，系统将回调这个方法。在实现中，必须要返回一个 IBinder 接口，供客户端与服务进行通信。此方法必须要实现，它是 Service 的一个抽象方法。但是如果不允许绑定组件的话，返回 null 就可以了。

（2）onUnbind()：当所有与服务绑定的组件都被解除时，就会调用此方法。

3．startService 示例

【示例】 下面是一个音乐播放的应用，使用 startService 来启动本地服务。运行主界面如图 5-2 所示。启动 Android Studio，在 Ch5 工程中创建 StartServiceDemo 模块，并在模块中创建 MainActivity.java、activity_main.xml 和 MusicService.java 等文件。在 StartServiceDemo 模块的资源目录（ch5\StartServiceDemo\src\main\res）下创建 raw 目录，再把 lake.mp3 复制到 raw 目录下。

图 5-2　运行 StartServiceDemo 模块的主界面

StartServiceDemo 模块的 MainActivity.java 主要代码如下：

```
    public  class  MainActivity  extends  AppCompatActivity  implements  View.
OnClickListener  {
      private Intent intent;
      @Override
```

```
    protected void onCreate(Bundle savedInstanceState) {
        super.onCreate(savedInstanceState);
        setContentView(R.layout.activity_main);
        Button playBtn = (Button) findViewById(R.id.play);
        Button stopBtn = (Button) findViewById(R.id.stop);
        Button pauseBtn = (Button) findViewById(R.id.pause);
        Button exitBtn = (Button) findViewById(R.id.exit);
        playBtn.setOnClickListener(this);
        stopBtn.setOnClickListener(this);
        pauseBtn.setOnClickListener(this);
        exitBtn.setOnClickListener(this);
    }
    @Override
    public void onClick(View v) {
        int num = -1;
        intent = new Intent("com.mialab.startservicedemo.musicService");
        intent.setPackage(this.getPackageName());
        switch (v.getId()) {
            case R.id.play:
                Toast.makeText(this, "play music…", Toast.LENGTH_SHORT).show();
                num = 1;
                break;
            case R.id.stop:
                Toast.makeText(this, "stop music…", Toast.LENGTH_SHORT).show();
                num = 2;
                break;
            case R.id.pause:
                Toast.makeText(this, "pause music…", Toast.LENGTH_SHORT).show();
                num = 3;
                break;
            case R.id.exit:
                Toast.makeText(this, "退出…", Toast.LENGTH_SHORT);
                num = 4;
                stopService(intent);
                this.finish();
                return;
        }
        Bundle bundle = new Bundle();
        bundle.putInt("music", num);
        intent.putExtras(bundle);
        startService(intent);
    }
    @Override
    public void onDestroy(){
        super.onDestroy();
        if(intent != null){
            stopService(intent);
        }
    }
}
```

MainActivity中通过重写OnClickListener接口的onClick()方法实现对播放音乐的控制，利用Bundle绑定数字num后，再使用startService(intent)来启动服务MusicService。

StartServiceDemo 模块的 MusicService.java 主要代码如下：

```
public class MusicService extends Service {
    private static final String TAG = "MusicService";
    private MediaPlayer mediaPlayer;
    private boolean reset = false;
    @Override
    public IBinder onBind(Intent intent) {
        return null;
    }
    @Override
    public void onCreate() {
        Log.v(TAG, "onCreate");
        if (mediaPlayer == null) {
            mediaPlayer = MediaPlayer.create(this, R.raw.lake);
            mediaPlayer.setLooping(false);
        }
    }
    @Override
    public void onDestroy() {
        Log.v(TAG, "onDestroy");
        if (mediaPlayer != null) {
            mediaPlayer.stop();
            mediaPlayer.release();
        }
    }
    @Override
    public void onStart(Intent intent, int startId) {
        Log.v(TAG, "onStart");
        if (intent != null) {
            Bundle bundle = intent.getExtras();
            if (bundle != null) {
                int num = bundle.getInt("music");
                switch (num) {
                    case 1:
                        play();
                        break;
                    case 2:
                        stop();
                        break;
                    case 3:
                        pause();
                        break;
                }
            }
        }
    }
    public void play() {
        Log.v(TAG, "-----------" + reset + "----------");
        if (mediaPlayer == null)
            return;
        if (reset == true)
            mediaPlayer.seekTo(0);
        if (!mediaPlayer.isPlaying()) {
            mediaPlayer.start();
```

```
            }
        }
        public void pause() {
            reset = false;
            if (mediaPlayer != null && mediaPlayer.isPlaying()) {
                mediaPlayer.pause();
            }
        }
        public void stop() {
            reset = true;
            if (mediaPlayer != null) {
                mediaPlayer.stop();
                try {
                    mediaPlayer.prepare(); //在调用stop后如果需要再次通过start进行播放，
                                           //则需要在播放前调用prepare()函数
                } catch (IOException ex) {
                    ex.printStackTrace();
                }
            }
        }
    }
```

MusicService 启动过程中将回调：onCreate→onStartCommand（可多次调用）→onDestroy。

其中，

- onCreate()：创建 mediaPlayer。
- onStartCommand ()：通过获取“Bundle bundle = intent.getExtras();”，提取“int num = bundle.getInt ("music");”，然后执行相应的音乐播放操作。
- onDestroy()：停止并释放 mediaPlayer 音乐资源，当执行 context.stopService()时调用此方法。

此外，还要在 AndroidManifest.xml 中注册 MusicService。

```
<service
    android:name="com.mialab.startservicedemo.MusicService"
    android:enabled="true">
    <intent-filter>
        <action android:name="com.mialab.startservicedemo.musicService" />
    </intent-filter>
</service>
```

4．bindService 示例

【示例】 对上面音乐播放的应用，还可以改用 bindService 来绑定服务。启动 Android Studio，在 Ch5 工程中创建 BindServiceDemo 模块，并创建 MainActivity.java、activity_main.xml 和 BindMusicService.java 等文件。在 BindServiceDemo 模块的资源目录（ch5\BindServiceDemo\src\main\res）下创建 raw 目录，再把 lake.mp3 复制到 raw 目录下。

BindServiceDemo 模块的 MainActivity.java 主要代码如下：

```
    public class MainActivity extends AppCompatActivity implements View.
OnClickListener {
        private BindMusicService musicService = null;
        @Override
        protected void onCreate(Bundle savedInstanceState) {
            super.onCreate(savedInstanceState);
            setContentView(R.layout.activity_main);
            Button playBtn = (Button) findViewById(R.id.play);
            Button stopBtn = (Button) findViewById(R.id.stop);
```

```
        Button pauseBtn = (Button) findViewById(R.id.pause);
        Button exitBtn = (Button) findViewById(R.id.exit);
        playBtn.setOnClickListener(this);
        stopBtn.setOnClickListener(this);
        pauseBtn.setOnClickListener(this);
        exitBtn.setOnClickListener(this);
        Intent intent = new Intent("com.mialab.bindService");
        intent.setPackage(this.getPackageName());
        bindService(intent, sc, Context.BIND_AUTO_CREATE);
        Toast.makeText(this, "music comes…", Toast.LENGTH_SHORT).show();
    }

    @Override
    public void onClick(View view) {
        switch (view.getId()) {
            case R.id.play:
                Toast.makeText(this, "play music…", Toast.LENGTH_SHORT).show();
                musicService.play();
                break;
            case R.id.stop:
                Toast.makeText(this, "stop music…", Toast.LENGTH_SHORT).show();
                if (musicService != null) {
                    musicService.stop();
                } else {
                    Toast.makeText(this, "no service…", Toast.LENGTH_SHORT). show();
                }
                break;
            case R.id.pause:
                Toast.makeText(this, "pause music…", Toast.LENGTH_SHORT). show();
                if (musicService != null) {
                    musicService.pause();
                } else {
                    Toast.makeText(this, "no service…", Toast.LENGTH_SHORT). show();
                }
                break;
            case R.id.exit:
                Toast.makeText(this, "退出…", Toast.LENGTH_SHORT).show();
                this.finish();
                break;
        }
    }

    private ServiceConnection sc = new ServiceConnection() {
        @Override
        public void onServiceConnected(ComponentName name, IBinder service) {
            musicService = ((BindMusicService.MyBinder) (service)).getService();
            if (musicService != null) {
                musicService.play();
            }
        }

        @Override
        public void onServiceDisconnected(ComponentName arg0) {
            musicService = null;
```

```
        }
    };

    @Override
    public void onDestroy(){
        super.onDestroy();
        if(sc != null){
            unbindService(sc);
        }
    }
}
```

BindServiceDemo 模块的 BindMusicService.java 主要代码如下：

```
public class BindMusicService extends Service {
    private static final String TAG = "BindMusicService";
    private MediaPlayer mediaPlayer;
    private boolean reset = false;
    private final IBinder binder = new MyBinder();

    public class MyBinder extends Binder {
        BindMusicService getService() {
            return BindMusicService.this;
        }
    }

    @Override
    public IBinder onBind(Intent intent) {
        Log.v(TAG, "onBind…");
        return binder;
    }
    @Override
    public void onCreate() {
        super.onCreate();
        Log.v(TAG, "onCreate");
    }
    @Override
    public boolean onUnbind(Intent intent) {
        Log.v(TAG, "onUnbind…");
        return super.onUnbind(intent);
    }
    @Override
    public void onDestroy() {
        super.onDestroy();
        Log.v(TAG, "onDestroy…");
        if (mediaPlayer != null) {
            mediaPlayer.stop();
            mediaPlayer.release();
        }
    }
    public void play() {
        if (mediaPlayer == null) {
            mediaPlayer = MediaPlayer.create(this, R.raw.lake);
            mediaPlayer.setLooping(false);
        }
        if (reset == true)
```

```
            mediaPlayer.seekTo(0);
        if (!mediaPlayer.isPlaying()) {
            mediaPlayer.start();
        }
    }
    public void pause() {
        reset = false;
        if (mediaPlayer != null && mediaPlayer.isPlaying()) {
            mediaPlayer.pause();
        }
    }
    public void stop() {
        if (mediaPlayer != null) {
            mediaPlayer.stop();
            try {
                reset = true;
                mediaPlayer.prepare(); //在调用 stop 后如果需要再次通过 start 进行播放,
                                       //则需要在播放前调用 prepare 函数
            } catch (Exception ex) {
                ex.printStackTrace();
            }
        }
    }
}
```

与前面例子不同的是：

（1）MainActivity 通过“Intent intent = new Intent("com.mialab.bindService");”构建一个 Service 的 Action，然后使用“bindService(intent, sc, Context.BIND_AUTO_CREATE);”绑定服务。

（2）MainActivity 通过“private ServiceConnection sc = new ServiceConnection();”建立一个 Service 连接，回调 onServiceConnected()以获取 Service 实例，再回调 onServiceDisconnected()以释放连接。

（3）BindMusicService 中，通过重载 onBind(Intent intent)方法，返回 BindMusicService 实例给 MainActivity，然后再执行 onCreate()方法。这里要注意：如果调用 bindService 就不会调用 onStartCommand()方法。

（4）MainActivity 通过返回的 BindMusicService 实例，执行音乐播放控制的操作（如 play、pause、stop 等）。

5.1.3 跨进程调用 Service

1. AIDL 机制

通常系统中每个应用程序都会运行在自己的进程空间中，并且可以通过应用程序 UI 运行另一个服务进程，而且经常会在不同的进程间传递对象。在 Android 平台下，一个进程通常不能访问另一个进程的内存空间，所以要完成“对话”，就需要将对象分解成操作系统可以理解的基本单元，并且有序地通过进程边界。Android 提供了 AIDL 工具来处理这项工作。

AIDL（Android Interface Definition Language，接口描述语言）是一种 IDL 语言，用于生成可以在 Android 设备的两个进程之间进行“进程通信”（InterProcess Communication，IPC）的代码。如果在一个进程中（如 Activity）要调用另一个进程中的对象（如 Service）操作，就可以使用 AIDL 生成可序列化的参数。

AIDL IPC 机制是面向接口的，它使用代理类在客户端和实现端传递数据。

AIDL 是一种接口定义语言，用于约束两个进程间的通信规则，供编译器生成代码，实现 Android 设备上的两个进程间通信。AIDL IPC 机制和 EJB 所采用的 CORBA 很类似，进程之间的通信信息，首先会被转换成 AIDL 协议消息，然后发送给对方，对方收到 AIDL 协议消息后再转换成相应的对象。由于进程之间的通信信息需要双向转换，所以 Android 采用代理类在背后实现了信息的双向转换，代理类由 Android 编译器生成，对开发人员来说是透明的。

小贴士：在 Android 中，每个应用程序都有自己的进程，当需要在不同的进程之间传递对象时，该如何实现呢？显然，在 Java 中是不支持跨进程内存共享的。因此要传递对象，就需要把对象解析成操作系统能够理解的数据格式，以达到跨界对象访问的目的。在 Java EE 中，采用 RMI 通过序列化传递对象。在 Android 中，则采用 AIDL 方式实现。所以在 Android 中如果需要在不同进程间实现通信，就需要用 AIDL 去完成。AIDL 是一种接口定义语言，编译器通过*.aidl 文件的描述信息生成符合通信协议的 Java 代码，无须自己去写这段繁杂的代码，只要在需要时调用即可，通过这种方式就可以完成进程间的通信工作。

2. 使用 AIDL 实现进程间通信的示例

【示例】 演示一个操作 AIDL 的最基本流程。

（1）创建 AIDL 服务端。

启动 Android Studio，在 Ch5 工程中创建 AIDLService 模块，并在模块中创建 IPerson.aidl 和 AIDLService.java，AIDLService 模块目录如图 5-3 所示。

在 IPerson.aidl 中定义了一个 hello 方法，其代码如下：

```
package com.mialab.aidlservice;
interface IPerson {
    String hello(String someone);
}
```

构建后，Android Studio 会自动生成 AIDLService\build\generated\aidl_source_output_dir\debug\compileDebugAidl\out\com\mialab\aidlservice\IPerson.java 文件，选中 IPerson.java 文件，点击 View→Tool Windows→Structure，IPerson.java 的 Structure 如图 5-4 所示。

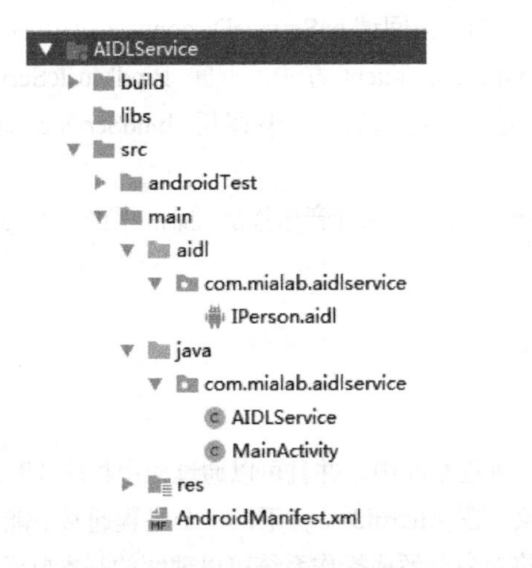

图 5-3 AIDLService 模块目录

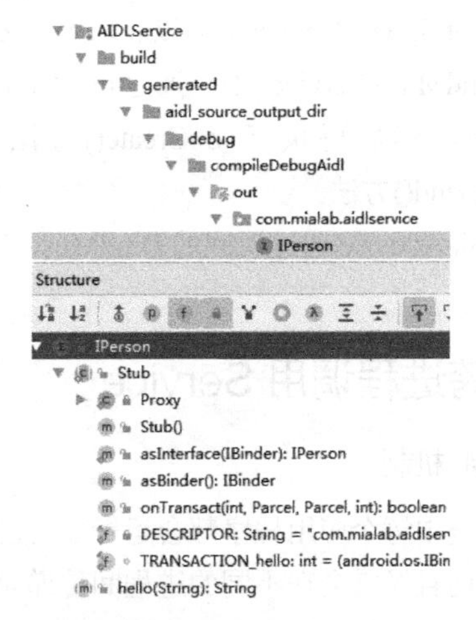

图 5-4 IPerson.java 的 Structure

IPerson 接口中的抽象内部类 Stub 继承 android.os.Binder 类并实现 IPerson 接口，其中比较重要的方法是 asInterface(android.os.IBinder obj)，该方法可将 IBinder 类型的对象转换成 IPerson 类型，必要时可生成一个代理对象返回结果。

AIDLService 模块的 AIDLService.java 主要代码如下：

```
public class AIDLService extends Service {
    private static final String TAG = "AIDLService";
    IPerson.Stub stub = new IPerson.Stub() {
```

```
        @Override
        public String hello(String someone) throws RemoteException {
            Log.d(TAG, "hello() called");
            return "Hello, " + someone;
        }
    };
    @Override
    public IBinder onBind(Intent intent) {
        Log.d(TAG, "onBind() called");
        return stub;
    }
    @Override
    public boolean onUnbind(Intent intent) {
        Log.d(TAG, "onUnbind() called");
        return true;
    }
    @Override
    public void onDestroy() {
        super.onDestroy();
        Log.d(TAG, "onDestroy() called");
    }
}
```

先实现 IPerson.Stub 这个抽象类的 hello 方法，然后再使用 onBind(Intent)方法返回 stub，获取的 IPerson.Stub 就是这个实例，hello 方法也会按照期望的执行。

要想让 Service 生效，还需要在 AndroidManifest.xml 中对 Service 进行配置。

```
<service android:name=".AIDLService" android:enabled="true" android: exported=
"true">
    <intent-filter>
        <action android:name="android.intent.action.AIDLService" />
    </intent-filter>
</service>
```

（2）创建 AIDL 客户端。

在 Ch5 工程中创建 AIDLService_Client 模块，AIDLService_Client 模块目录如图 5-5 所示，只需要把 IPerson.aidl 文件复制到相应的目录中即可，编译器同样会生成相对应的 IPerson.java 文件，这部分和服务端没什么区别。这样一来，服务端和客户端就在通信协议上达成了统一，其主要工作将在 MainActivity 中完成。

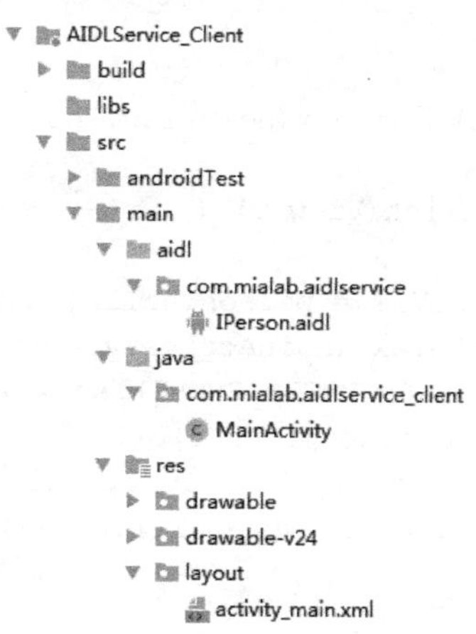

图 5-5　AIDLService_Client 模块目录

AIDLService_Client 模块的 MainActivity.java 主要代码如下：

```
public class MainActivity extends AppCompatActivity {
    private static String TAG = "MainActivity";
    private IPerson person;
    private ServiceConnection conn = new ServiceConnection() {
        @Override
        public void onServiceConnected(ComponentName name, IBinder service) {
            Log.d("ServiceConnection", "onServiceConnected() called");
            person = IPerson.Stub.asInterface(service);
            Log.d(TAG, "------------onServiceConnected()------------" );
            Log.d(TAG, "------------" + person + "-----------");
        }
        @Override
        public void onServiceDisconnected(ComponentName name) {
            Log.d("ServiceConnection", "onServiceDisconnected() called");
        }
    };

    @Override
    protected void onCreate(Bundle savedInstanceState) {
        super.onCreate(savedInstanceState);
        setContentView(R.layout.activity_main);
        final Button bindBtn = (Button) findViewById(R.id.bindBtn);
        final Button greetBtn = (Button) findViewById(R.id.greetBtn);
        final Button unbindBtn = (Button) findViewById(R.id.unbindBtn);
        bindBtn.setEnabled(true);
        greetBtn.setEnabled(false);
        unbindBtn.setEnabled(false);
        bindBtn.setOnClickListener(new View.OnClickListener() {
            @Override
            public void onClick(View v) {
                final Intent intent = new Intent();
                intent.setAction("android.intent.action.AIDLService");
                intent.setPackage("com.mialab.aidlservice");
                bindService(intent, conn, Context.BIND_AUTO_CREATE);
                bindBtn.setEnabled(false);
                greetBtn.setEnabled(true);
                unbindBtn.setEnabled(true);
            }
        });
        greetBtn.setOnClickListener(new View.OnClickListener() {
            @Override
            public void onClick(View v) {
                try {
                    String retVal = person.hello("Marry");
                    Toast.makeText(MainActivity.this, retVal,
                            Toast.LENGTH_SHORT).show();
                } catch (RemoteException e) {
                    Toast.makeText(MainActivity.this, "error",
                            Toast.LENGTH_SHORT).show();
                }
            }
        });
        unbindBtn.setOnClickListener(new View.OnClickListener() {
```

```
            @Override
            public void onClick(View v) {
                unbindService(conn);
                bindBtn.setEnabled(true);
                greetBtn.setEnabled(false);
                unbindBtn.setEnabled(false);
            }
        });
    }
}
```

从代码中可以看到，需要先重写 ServiceConnection 中的 onServiceConnected 方法将 IBinder 类型的对象转换成 IPerson 类型。然后再通过服务端 Service 定义的“android.intent.action.AIDL Service”这个标识符来绑定所需要的服务，这样客户端和服务端就实现了通信的连接，也就可以调用 IPerson 中的 hello 方法了。

（3）测试。

先运行 AIDLService（为保证服务端 AIDLService 可安装到手机模拟器或手机真实设备上），再运行 AIDLService_Client 模块。运行 AIDLService_Client 的界面如图 5-6 所示。

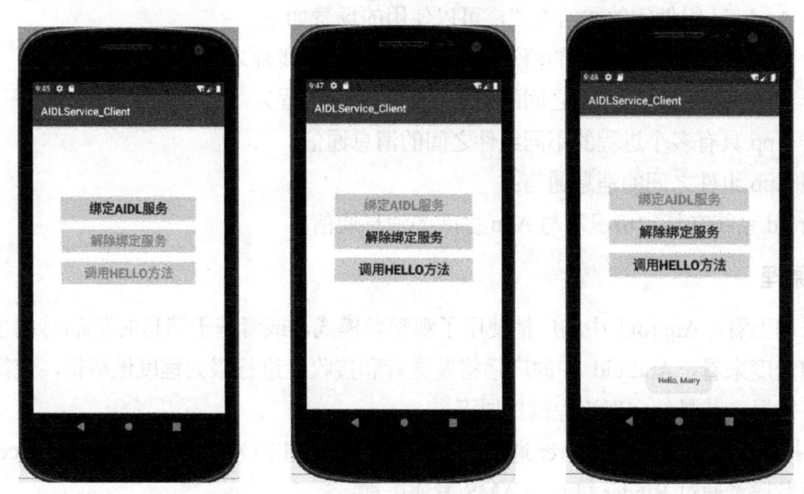

图 5-6　运行 AIDLService_Client 模块的界面

5.2　接收广播消息

5.2.1　简介

1. 广播发送者和广播接收者

Android 广播分为两个部分：广播发送者和广播接收者。通常情况下，BroadcastReceiver 指的就是广播接收者（广播接收器），用于异步接收广播 Intent。广播 Intent 是通过调用 Context.sendBroadcast()方法实现发送、BroadcastReceiver()方法实现接收的。广播 Intent 的发送是通过调用 Context.sendBroadcast()、Context.sendOrderedBroadcast()、Context.sendSticky-Broadcast()方法来实现的。通常，一个广播 Intent 可以被订阅的多个广播接收者所接收。

广播接收器只能接收广播，对广播的通知做出反应，很多广播都产生于系统代码，如开机启动、电池电量不足，或者时区发生改变等。

广播接收器没有用户界面，但是它可以为接收到的信息启动一个 Activity 或者使用 NotificationManager 来通知用户。

BroadcastReceiver 接收广播方式一般有以下两种。

（1）Normal Broadcasts（正常广播）：调用 Context.sendBroadcast()方法发送是完全异步的，它们都运行在一个未定义的顺序中，通常是在同一时间。

（2）Ordered Broadcasts（有序广播）：调用 Context.sendOrderedBroadcast()方法每次将广播发送到一个 Receiver。所谓有序，就是指每个 Receiver 执行后都可以传播到下一个 Receiver，也可以完全中止传播，即不传播给其他的 Receiver。Receiver 运行的顺序可以通过 matched intent-filter 里面的 android:priority 来控制，当 priority 优先级相同时，Receiver 可以按任意顺序运行。

经常说的“发送广播”和“接收”，表面上看广播是作为 Android 广播机制中的实体，实际上这个实体本身并不是以所谓的“广播”对象存在的，而是以“意图”（Intent）去表示的。广播的定义过程，实际上就是相应广播“意图”的定义过程，然后通过广播发送者将此“意图”发送出去。被相应的 BroadcastReceiver 接收后将会回调给 onReceive()函数。

2．使用场景

广播作为 Android 组件间的通信方式，可以使用的场景如下：

（1）同一 App 内部的同一组件的消息通信（单个或多个线程之间）；

（2）同一 App 内部的不同组件之间的消息通信（单个进程）；

（3）同一 App 具有多个进程的不同组件之间的消息通信；

（4）不同 App 组件之间的消息通信；

（5）Android 系统在特定情况下与 App 之间的消息通信。

3．实现原理

从实现原理上看，Android 中的广播使用了观察者模式，或者基于消息的发布/订阅事件模型。因此，从实现的角度来看，Android 中的广播将发送者和接收者进行极大程度地解耦，使得系统能够方便集成和更易扩展，其具体实现流程概括如下：

（1）广播接收者 BroadcastReceiver 通过 Binder 机制向 AMS（Activity Manager Service）进行注册；

（2）广播发送者通过 Binder 机制向 AMS 发送广播；

（3）AMS 查找符合相应条件（IntentFilter/Permission 等）的 BroadcastReceiver，将广播发送到 BroadcastReceiver（一般情况下是 Activity）相应的消息循环队列中；

（4）消息循环执行此广播，回调 BroadcastReceiver 中的 onReceive()方法。

对于不同的广播类型，以及不同的 BroadcastReceiver 注册方式，虽在具体实现上有所不同，但总体流程大致相同。

由此看来，广播发送者和广播接收者分别属于观察者模式中消息的发布和订阅，AMS 属于中间的处理中心。广播发送者和广播接收者的执行是异步的，发出去的广播不会关心有无接收者接收，也不确定接收者到底是何时才能接收到。

4．实际应用中的适用性

在上文列举的广播机制具体使用的场景中，分析实际应用的适用性。

第一种情形：同一 App 内部的同一组件的消息通信（单个或多个线程之间）。实际应用中肯定是不会用到广播机制的（虽然可以用），无论是使用扩展变量作用域、基于接口的回调还是 Handler-post/Handler-Message 等方式，都可以直接处理此类问题，若使用广播机制，显然有些“杀鸡焉用牛刀”的感觉。

第二种情形：同一 App 内部的不同组件之间的消息通信（单个进程）。对于此类需求，在有些较复杂的情况下，如果单纯地依靠基于接口的回调等方式不好处理，可以直接使用 EventBus（Android 开源库）等，相对而言，由于 EventBus 是针对统一进程的，因此用于处理此类需求非常适合，且轻松解耦。

其他情形：由于涉及不同进程间的消息通信，此时可根据实际业务情况，使用广播机制会显得非常适宜。

5.2.2 发送广播

1. 自定义 BroadcastReceiver

自定义广播接收器需要继承基类 BroadcastReceivre，并实现抽象方法 onReceive(context,intent)。广播接收器接收到相应广播后，会自动回调 onReceive()方法。默认情况下，广播接收器也运行在 UI 线程中，因此，onReceive()方法不能执行太耗时的操作，否则会造成 ANR（Application Not Responding），应用没有响应。

一般情况下，根据实际业务需求，onReceive()方法中都会涉及与其他组件之间的交互，如发送 Notification、启动 Service 等。

下面代码片段是一个广播接收器的自定义：

```
public class MyBroadcastReceiver extends BroadcastReceiver {
   public static final String TAG = "MyBroadcastReceiver";
   @Override
   public void onReceive(Context context, Intent intent) {
      Log.d(TAG, "intent: " + intent);
      String name = intent.getStringExtra("name");
      …
   }
}
```

一个 BroadcastReceiver 对象只有在被调用 onReceive(Context, Intent)时才有效，当从该函数返回后，该对象就无效了，从而结束生命周期。

因此从这个特征可以看出，在所调用的 onReceive(Context,Intent)函数里，如有过于耗时的操作，则不能使用线程来执行。对于耗时的操作，应该在 startService 中来完成。因为当得到其他异步操作所返回的结果时，BroadcastReceiver 可能已经无效了。

2. BroadcastReceiver 注册类型

BroadcastReceiver 可以分为静态注册和动态注册两种类型。

（1）静态注册。

直接在 AndroidManifest.xml 文件中进行注册，其规则如下：

```
<receiver android:enabled = ["true" | "false"]
android:exported = ["true" | "false"]
android:icon = "drawable resource"
android:labe l= "string resource"
android:name = "string"
android:permission = "string"
android:process = "string" >
…
</receiver>
```

其中，需要注意的属性如下。

① android:exported。用于标识此 BroadcastReceiver 能否接收其他 App 发出的广播，这个属性的默认值是根据 Receiver 中有无 intent-filter 决定的，如果有 intent-filter，则默认值为 true，否则为 false。

同样，在 Activity/Service 中的属性默认值一样遵循此规则。

② android:name。BroadcastReceiver 的类名。

③ android:permission。只有设置具有相应权限的广播发送方发送的广播，才能被此 BroadcastReceiver 所接收。

④ android:process。BroadcastReceiver 运行，默认为 App 的进程，也可以指定独立的进程。Android 四大基本组件都可以通过此属性指定自己的独立进程。

常见的注册形式如下：

```
<receiver android:name=".MyBroadcastReceiver" >
    <intent-filter>
        <action android:name="android.net.conn.CONNECTIVITY_CHANGE" />
    </intent-filter>
    <intent-filter>
        <action android:name="android.intent.action.BOOT_COMPLETED" />
    </intent-filter>
</receiver>
```

其中，intent-filter 用于指定此广播接收器接收特定的广播类型。本示例中给出的是，用于接收网络状态改变或开机启动时系统自身所发出的广播。当此 App 首次启动时，系统会自动实例化 MyBroadcastReceiver，并注册到系统中。

（2）动态注册。

动态注册时，不用在 AndroidManifest 中注册<receiver/>组件。可直接在代码中通过调用 Context 的 registerReceiver()函数，动态注册 BroadcastReceiver。

registerReceiver 的定义形式如下：

```
registerReceiver(BroadcastReceiver receiver, IntentFilter filter)
registerReceiver(BroadcastReceiver receiver, IntentFilter filter, String
broadcastPermission, Handler scheduler)
```

其示例代码如下：

```
public class MainActivity extends Activity {
    public static final String BROADCAST_ACTION = "com.example.corn";
    private BroadcastReceiver mBroadcastReceiver;
    @Override
    protected void onCreate(Bundle savedInstanceState) {
        super.onCreate(savedInstanceState);
        setContentView(R.layout.activity_main);
        mBroadcastReceiver = new MyBroadcastReceiver();
        IntentFilter intentFilter = new IntentFilter();
        intentFilter.addAction(BROADCAST_ACTION);
        registerReceiver(mBroadcastReceiver, intentFilter);
    }
    @Override
    protected void onDestroy() {
        super.onDestroy();
        unregisterReceiver(mBroadcastReceiver);
    }
}
```

注意，在 Android 所有与观察者模式有关的设计中，一旦涉及 register，必定在相应的时机需要 unregister。

因此，上例在 onDestroy()方法中需要 unregisterReceiver(mBroadcastReceiver)。

当 Activity 实例化时，会动态将 MyBroadcastReceiver 注册到系统中。当此 Activity 销毁时，动态注册的 MyBroadcastReceiver 将不会再接收到相应的广播。

动态注册方式隐藏在代码中，是比较难发现的。需要特别注意的是，在退出程序前要记得调用 Context.unregisterReceiver()方法。一般在 Activity 的 onStart()方法中进行注册，在 onStop()方法中进行注销。官方提醒，如果在 Activity.onResume()方法中注册了，就必须在 Activity.onPause()方法中注销。

3．BroadcastReceiver 示例

【示例】 实现简单的音乐播放，运行界面方法同前。启动 Android Studio，在 Ch5 工程中创建 BroadcastReceiverDemo 模块，并在此模块中创建 MainActivity.java、MusicPlayReceiver.java、MusicPlayService.java 和 activity_main.xml 等文件。

BroadcastReceiverDemo 模块的 MainActivity.java 主要代码如下：

```
    public class MainActivity extends AppCompatActivity implements View.
OnClickListener {
        private Intent intent;
        @Override
        protected void onCreate(Bundle savedInstanceState) {
            super.onCreate(savedInstanceState);
            setContentView(R.layout.activity_main);
            Button playBtn = (Button) findViewById(R.id.play);
            Button stopBtn = (Button) findViewById(R.id.stop);
            Button pauseBtn = (Button) findViewById(R.id.pause);
            Button exitBtn = (Button) findViewById(R.id.exit);
            playBtn.setOnClickListener(this);
            stopBtn.setOnClickListener(this);
            pauseBtn.setOnClickListener(this);
            exitBtn.setOnClickListener(this);
        }
        @Override
        public void onClick(View v) {
            int num = -1;
            intent = new Intent("com.mialab.broadcast.musicPlayReceiver");
            intent.setPackage(this.getPackageName());
            switch (v.getId()) {
                case R.id.play:
                    Toast.makeText(this, "play music…", Toast.LENGTH_SHORT).show();
                    num = 1;
                    break;
                case R.id.stop:
                    Toast.makeText(this, "stop music…", Toast.LENGTH_SHORT).show();
                    num = 2;
                    break;
                case R.id.pause:
                    Toast.makeText(this, "pause music…", Toast.LENGTH_SHORT). show();
                    num = 3;
                    break;
                case R.id.exit:
                    Toast.makeText(this, "退出…", Toast.LENGTH_SHORT);
                    num = 4;
                    this.finish();
                    break;
            }
            Bundle bundle = new Bundle();
            bundle.putInt("music", num);
```

```
            intent.putExtras(bundle);
            sendBroadcast(intent);
        }
        …
    }
```

BroadcastReceiverDemo 模块的 MusicPlayReceiver.java 主要代码如下：

```
public class MusicPlayReceiver extends BroadcastReceiver {
    private static String TAG = "MusicPlayReceiver";
    @Override
    public void onReceive(Context context, Intent intent) {
        Log.d(TAG, "--------------------onReceive-------------------");
        if(intent != null){
            Bundle bundle = intent.getExtras();
            Intent it = new Intent(context, MusicPlayService.class);
            it.putExtras(bundle);
            if(bundle != null){
                int op = bundle.getInt("music");
                Log.d(TAG, "" + op);
                if(op == 4){
                    context.stopService(it);         //stopService
                }else{
                    context.startService(it);        //startService
                }
            }
        }
    }
}
```

MusicPlayReceiver 在其 onReceive()方法中启动或停止后台服务 MusicPlayService。

MusicPlayReceiver 和 MusicPlayService 应在 AndroidManifest.xml 中注册。

```
    <service android:name="com.mialab.broadcastreceiverdemo.MusicPlayService" android:
enabled="true" >
        <intent-filter>
            <action android:name="com.mialab.broadcast.musicPlayService" />
        </intent-filter>
    </service>
    <receiver android:name="com.mialab.broadcastreceiverdemo.MusicPlayReceiver" >
        <intent-filter>
            <action android:name="com.mialab.broadcast.musicPlayReceiver" />
        </intent-filter>
    </receiver>
```

主要流程：

（1）MainActivity 通过重写 OnClickListener 接口 onClick()方法实现对播放音乐的控制。先在 onClick()方法中构造一个 Intent：“intent = new Intent("com.mialab.broadcast.musicPlay Receiver ");”。

（2）利用 Bundle 绑定数字 num 后，通过“sendBroadcast(intent);”广播传出去。

（3）MusicPlayReceiver 中，会处理 MainActivity 启动的“sendBroadcast(intent);”广播，通过重写 onReceive()方法，解析 MainActivity 中 Intent 的 Bundle 数据，然后通过“Intent it = new Intent(context, MusicPlayService.class);”初始化一个启动 Service 服务的 Intent。

（4）根据解析 Bundle 的 num 数值决定启动服务“context.startService(it);”或关闭服务“context. stopService(it);”。

5.2.3 有序广播

有序广播中的“有序”是针对广播接收者而言的，指的是发送出去的广播被 BroadcastReceiver 按照先后顺序接收。有序广播的定义过程与普通广播无异，只是其主要发送方式变为了 sendOrderedBroadcast(intent, receiverPermission, ...)。

对于有序广播，其主要特点如下。

（1）多个当前已经注册且有效的 BroadcastReceiver 接收有序广播时，是按照先后顺序接收的，先后顺序判定标准应遵循：将当前系统中动态注册和静态注册的 BroadcastReceiver 按照 priority 属性值从大到小排序，对于具有相同 priority 的动态广播和静态广播，动态广播会排在前面。

（2）先接收的 BroadcastReceiver 可以对此有序广播进行截断，使后面的 BroadcastReceiver 不再接收到此广播，但可以对广播进行修改，使后面的 BroadcastReceiver 接收到广播后解析得到错误的参数值。当然，一般情况下，不建议对有序广播进行此类操作，尤其是针对系统中的有序广播。

5.2.4 接收系统广播消息

Android 系统中内置了多个系统广播，只要涉及手机的基本操作，基本上都会发出相应的系统广播。如开启启动、网络状态改变、拍照、屏幕关闭与开启、点亮不足等。每个系统广播都具有特定的 intent-filter，其中主要包括系统广播发出后，将被相应的 BroadcastReceiver 接收；系统广播在系统内部当特定事件发生时，由系统自动发出。

5.3 本章小结

本章主要阐述了 Service 和 BroadcastReceiver 的使用。

（1）Service 用来在后台处理一些比较复杂的操作，最典型的例子就是音乐播放器的后台播放。虽然 Activity 中也能完成这样的功能，但当系统内存紧张时会先把 Activity“杀掉”，却很少有 Service 会“死掉”。另外，一般 Activity 会在 onDestroy 时释放一些内容，所以，当界面被覆盖时，Activity 就不安全了。

（2）Service 不是一个单独的进程，除非单独声明，否则它不会运行在单独的进程中，而是和启动它的程序运行在同一个进程中。Service 也不是线程，这意味着它将在主线程里运行。

（3）Service 与 Thread 的区别在于：Thread 是程序执行的最小单元，可以用 Thread 来执行一些异步的操作。Service 是 Android 的一种机制，当它运行时如果是 Local Service，那么对应的 Service 就会运行在主进程的 main 线程中。如果是 Remote Service，那么对应的 Service 则应运行在独立进程的 main 线程中。

（4）Thread 的运行是独立的，也就是说，当一个 Activity 被 finish 之后，如果没有主动停止 Thread 或者 Thread 里的 run 方法没有执行完毕的话，Thread 也会一直执行。因此这里会出现一个问题：一方面，当 Activity 被结束之后，就不再持有该 Thread 的引用，也不能再控制该 Thread 了。另一方面，无法在不同的 Activity 中对同一个 Thread 进行控制。如果一个 Thread 需要每隔一段时间就要连接服务器校验数据时，该 Thread 就需要在后台一直运行。这时候如果创建该 Thread 的 Activity 被结束了而该 Thread 却没有停止，那么将没有办法再控制该 Thread，除非“杀掉”该程序的进程。这时候如果创建并启动一个 Service，在 Service 里面创建、运行并控制该 Thread，便解决了该问题（因为任何 Activity 都可以控制同一个 Service，而系统也只会创建一个对应 Service 的实例）。

（5）可以把 Service 想象成一种消息服务，在任何有 Context 的地方都可以调用 Context.startService、

Context.stopService、Context.bindService、Context.unbindService 来控制它，也可以在 Service 里注册 BroadcastReceiver，通过发送 Broadcast 来达到控制的目的，这些都是 Thread 做不到的。

（6）启动 Service 有两种方法：Context.startService()和 Context.bindService()。对于 Context.startService()方法，调用者（Client）与服务端之间没有关联，即使调用者退出，服务仍可运行。对于 Context.bindService()方法，调用者与服务端绑定在一起，可以是多个调用者绑定一个服务端，当所有的调用者退出后，服务端也就会终止。

（7）有一点需要强调，如果有耗时操作在 Service 里，就必须开启一个单独的线程来处理。

（8）IntentService 是继承自 Service 的。IntentService 相对于 Service 来说，其优点在于：使用队列的方式将请求的 Intent 加入，然后开启一个 worker thread（线程）来处理队列中的 Intent；对于异步的 startService 请求，IntentService 会处理完成一个之后再处理第二个，每一个请求都会在一个单独的 worker thread 中处理，不会阻塞应用程序的主线程。因此，如果有耗时的操作与其在 Service 里面开启新线程，可以使用 IntentService 来处理耗时操作。

（9）广播接收者用于接收广播 Intent，广播 Intent 的发送是通过调用 Context.sendBroadcast()、Context.sendOrderedBroadcast()来实现的。通常一个广播 Intent 可以被多个广播接收者接收。

（10）BroadcastReceiver 自身并不实现图形用户界面，但是当它收到某个通知后，BroadcastReceiver 可以启动 Activity 作为响应，并通过 NotificationMananger 提醒用户，启动 Service 等。

（11）在 Android 中有各种各样的广播，如电池的使用状态、电话的接收和短信的接收都会产生一个广播，应用程序开发者也可以监听这些广播并做出相应的处理。

（12）BroadcastReceiver 生命周期只有 10s 左右，如果在 onReceive()内做超过 10s 的事情就会报错。每次广播到来时都会重新创建 BroadcastReceiver 对象，并且调用 onReceive()方法，执行完以后，该对象即被销毁。当 onReceive()方法在 10s 内没有执行完毕，Android 就会认为该程序无响应。所以在 BroadcastReceiver 里不能做一些比较耗时的操作，否则会弹出 ANR 的对话框。

（13）如果需要完成一项比较耗时的工作，应该通过发送 Intent 给 Service，由 Service 来完成。在此不能使用子线程来解决，因为 BroadcastReceiver 的生命周期很短，子线程可能还没有结束。如果 BroadcastReceiver 先结束，此时 BroadcastReceiver 的所在进程很容易在系统需要内存时被优先“销毁”，因为它属于空进程（没有任何活动组件的进程）。如果它的宿主进程被“销毁”，那么正在工作的子线程也会被“销毁”，所以采用子线程来解决是不可靠的。

（14）接收短信举例。

第一种方式：

```
public class MyBroadcastReceiver extends BroadcastReceiver {
  String SMS_RECEIVED = "android.provider.Telephony.SMS_RECEIVED" ;
  public void onReceive(Context context, Intent intent) {
    if (intent.getAction().equals( SMS_RECEIVED )) {  //action 名称
        //相关处理：地域变换、电量不足、来电来信
    }
  }
}
```

系统注册（在 AndroidManifest.xml 中）：

```
< receiver android:name = ".MyBroadcastReceiver" >
  < intent-filter android:priority = "1000" >
 < action android:name = " android.provider.Telephony.SMS_RECEIVED" />
  </ intent-filter >
</ receiver >
```

需要以下权限：

```
< uses-permission android:name = "android.permission.RECEIVE_SMS" />
< uses-permission android:name = "android.permission.SEND_SMS" />
```

第二种方式：

```
//广播接收者：广播的接收
private BroadcastReceiver myBroadcastReceiver = new BroadcastReceiver() {
    @Override
    public void onReceive(Context context, Intent intent) {
        //相关处理，如收短信、监听电量变化信息
    }
};
```

在代码中须注册：

```
IntentFilter intentFilter = new IntentFilter( "android.provider.Telephony.SMS_RECEIVED" );
registerReceiver( mBatteryInfoReceiver , intentFilter);
```

习 题 5

1．Service 的启动有哪两种方法？它们有何不同？
2．什么是进程内服务？试编程加以说明。
3．什么是跨进程服务？试编程加以说明。
4．如何发送广播？试编程加以说明。
5．如何接收系统广播消息？试编程加以说明。
6．什么是有序广播？试编程加以说明。
7．Service 和 Thread 的区别是什么？为什么要使用 Service？注意事项是什么？
8．什么是 IntentService？IntentService 的特点是什么？如何使用？
9．ANR 是什么？怎样避免和解决 ANR？
10．在什么场合下，适合使用 Android 开源库 EventBus？如何使用？试编程加以说明。

第6章　数据存储与访问

任何一个应用程序都少不了与数据进行交互，如 QQ、微信、手机银行等。通常情况下，可将这些数据存放在远程服务器的数据库中，通过接口进行读取。但是对于少量的数据，如程序配置信息（是否推送等），如果也采用存放远程数据库的方法，不仅运行效率低，还会占用数据库资源。因此，对于这种情况，Android 提供了三种解决办法：使用 SharedPreferences 存储数据、Android 内置轻量级数据库 SQLite 存储数据、借助 Java 的 I/O 体系实现文件存储。本章主要内容有：（1）SharedPreferences 存储；（2）SQLite 数据库使用（包括手动建库和代码建库）；（3）文件存储（包括内部存储和外部存储）；（4）Android 外部存储的公有存储空间和私有存储空间；（5）ContentProvider 的使用（包括创建数据提供者和使用数据提供者）。

6.1　SharedPreferences 存储

当仅有少量数据需要保存，且数据结构比较简单时，如一个简单的字符串，用数据库保存这类数据不仅运行效率低，而且还浪费数据库资源，类似这种情况，可以将这些数据保存于本地文件。Android 提供了 SharedPreferences 可进行保存。

SharedPreferences 是 Android 系统提供的一种轻量级数据保存方式，它的数据以键值对（Key-Value）的形式存放在 XML 文件中，直到应用程序被删除。

开发人员通过 SharedPreferences 将 Key-Value 格式的数据保存于 Android 的文件系统中，由于 SharedPreferences 屏蔽了对文件的操作过程，使得开发人员不用关注如何操作文件，仅需了解 SharedPreferences 中的操作函数，利用相关函数即可实现对数据的读/写。

6.1.1　将数据存储到 SharedPreferences

SharedPreferences 数据文件的存储路径是唯一的，以 XML 文件的形式保存于/data/data/<应用程序包名>/shared_prefs 目录下。将数据存储到 SharedPreferences 的基本步骤如下：

（1）调用 Context 提供的 getSharedPreferences()方法来获取 SharedPreferences 实例。

```
    SharedPreferences mySharedPreferences= getSharedPreferences("book_info", MODE_
PRIVATE);
```

getSharedPreferences(String name, int mode)方法涉及两个参数，第一个参数 name 表示要访问的文件，由于 SharedPreferences 文件存储的路径是唯一的，因此，此处 name 只需描述文件名即可。若为写入操作，当系统没有 name 文件时，则会自动创建一个名为 name 的文件。第二个参数 mode 表示操作模式，如 MODE_PRIVATE，则表示可以被本应用程序读/写。

（2）获得 Editor 对象来实现对数据的写入。

SharedPreferences 本身并不具备写入数据的能力，它是通过调用自身的 edit()方法来获取对应 Editor 对象的，由 Editor 对象来实现对数据的写入，其示例代码如下：

```
    Editor editor = mySharedPreferences.edit();
    editor.putString("bookname" , "Android 应用开发实践教程");
```

（3）调用 apply()或 editor()方法提交修改。

其示例代码如下：

```
editor.apply();      //editor.commit();也可以
```

apply()和 commit() 这两个方法的区别在于：

① apply()方法是将修改数据源提交到内存，而后再异步真正提交到硬件磁盘。commit()方法是同步提交到硬件磁盘。apply()方法会覆写之前内存中的值，异步写入磁盘的值只是最后的值，而 commit()方法每次都要写入磁盘，但磁盘的写入相对很低效，所以 apply()方法在频繁调用时要比 commit()方法的效率高很多。

② apply()方法没有返回值，commit()方法则要返回 boolean 类型的值，表明提交修改是否成功。apply()方法不会有任何失败的提示。

6.1.2　从 SharedPreferences 中读取数据

SharedPreferences 接口提供了以下方法来读取应用程序的数据。

- public abstract String getString (String key, String defValue)：获取 key 所指向的 value 值。若该 key 不存在，则返回默认值 defValue。
- public abstract Map<String, ?> getAll ()：获取所有的 SharedPreferences 数据，该函数返回值是一个以 Key-Value 形式描述的 Map 对象。
- public abstract boolean contains (String key)：判断是否包含 key 指向的数据。

其示例代码如下：

```
String bookName = mySharedPreferences.GetString("bookname", "none");
//读取 key 值为"bookname" 指向的数据，若无则返回"none"
```

6.1.3　SharedPreferences 举例

【示例】 将注册信息写入 SharedPreferences 并读取。

启动 Android Studio，在 Ch6 工程中创建 SharedPreferencesDemo 模块，并在 SharedPreferencesDemo 模块中创建 MainActivity.java、activity_main.xml、RegisterActivity.java、activity_register.xml、InfoActivity.java、activity_info.xml，以及 Spinner 控件列表项适配的 item_dropdown.xml、item_select.xml 等文件。

MainActivity 界面如图 6-1（a）所示，RegisterActivity 界面如图 6-1（b）所示，InfoActivity 界面如图 6-1（c）所示。InfoActivity 用来读取（显示）写入 SharedPreferences 中的注册信息。

图 6-1　运行 SharedPreferencesDemo 模块的界面

SharedPreferencesDemo 模块的 MainActivity.java 主要代码如下：

```
public class MainActivity extends AppCompatActivity {
    @Override
    protected void onCreate(Bundle savedInstanceState) {
        super.onCreate(savedInstanceState);
        setContentView(R.layout.activity_main);
        Button btn_1 = (Button) findViewById(R.id.btn1);
        Button btn_2 = (Button) findViewById(R.id.btn2);
        btn_1.setOnClickListener(new View.OnClickListener() {
            @Override
            public void onClick(View arg0) {
                Intent intent = new Intent(MainActivity.this, RegisterActivity.
class);
                startActivity(intent);
            }
        });
        btn_2.setOnClickListener(new View.OnClickListener() {
            @Override
            public void onClick(View arg0) {
                Intent intent = new Intent(MainActivity.this, InfoActivity.class);
                startActivity(intent);
            }
        });
    }
}
```

SharedPreferencesDemo 模块的 RegisterActivity.java 主要代码如下：

```
public class RegisterActivity extends AppCompatActivity implements View.
OnClickListener {
    private SharedPreferences mShared; //声明一个 SharedPreferences 对象
    private EditText mName;
    private EditText mPasswd;
    @Override
    protected void onCreate(Bundle savedInstanceState) {
        super.onCreate(savedInstanceState);
        setContentView(R.layout.activity_register);
        mName = (EditText) findViewById(R.id.name);
        mPasswd = (EditText) findViewById(R.id.passwd);
        findViewById(R.id.register_btn).setOnClickListener(this);
        initSpinner();
        //从 person_info.xml 中获取共享参数对象
        mShared = getSharedPreferences("person_info", MODE_PRIVATE);
    }
    private void initSpinner() {
        String[] eduArray = {"大学本科","大专","高中及以下","研究生"};
        Spinner sp = ((Spinner) findViewById(R.id.spinner));
        ArrayAdapter<String> adapter = new ArrayAdapter<String>(this,
                R.layout.item_select, eduArray);
        adapter.setDropDownViewResource(R.layout.item_dropdown);
        sp.setPrompt("请选择学历：");
        sp.setAdapter(adapter);
        sp.setSelection(0);
    }
    @Override
```

```
    public void onClick(View v) {
        if (v.getId() == R.id.register_btn) {
            String name = mName.getText().toString();
            String passwd = mPasswd.getText().toString();
            if (TextUtils.isEmpty(name)) {
                showToast("请先输入账号");
                return;
            } else if (TextUtils.isEmpty(passwd)) {
                showToast("请先输入密码");
                return;
            }
            SharedPreferences.Editor editor = mShared.edit();  //获得编辑器对象
            editor.putString("name", name);          //添加一个名叫name的字符串参数
            editor.putString("passwd", passwd);      //添加一个名叫passwd的字符串参数
            editor.apply();                          //提交编辑器中的修改
            showToast("数据已写入SharedPreferences");
        }
    }
    private void showToast(String desc) {
        Toast.makeText(this, desc, Toast.LENGTH_SHORT).show();
    }
}
```

SharedPreferencesDemo模块的InfoActivity.java主要代码如下：

```
public class InfoActivity extends AppCompatActivity {
    private SharedPreferences mShared_2;
    private TextView mName;
    private TextView mPasswd;
    @Override
    protected void onCreate(Bundle savedInstanceState) {
        super.onCreate(savedInstanceState);
        setContentView(R.layout.activity_info);
        mName = findViewById(R.id.name);
        mPasswd = findViewById(R.id.passwd);
        mShared_2 = getSharedPreferences("person_info", MODE_PRIVATE);
        String nameStr = mShared_2.getString("name","");
        String passwdStr = mShared_2.getString("passwd","");
        mName.setText("姓名为：" + nameStr);
        mPasswd.setText("密码为：" + passwdStr);
    }
}
```

其余文件详见Ch6\SharedPreferencesDemo。

小贴士：SharedPreferences存储数据的特点包括轻量且孤立的数据、文本形式的数据、需要持久化存储的数据、具体有应用程序的配置信息、游戏的玩家积分等。

为了查看SharedPreferencesDemo模块的SharedPreferences文件，可以通过点击View→Tool Windows→Device File Explorer，打开“Device File Explorer”窗口，在“/data/data/com.mialab.sharedpreferencesdemo/shared_prefs/”路径下找到该模块SharedPreferences文件的“person_info.xml”，并通过选择“Save As”选项导出进行查看。

6.2 SQLite 数据库

SQLite 数据库是 Android 系统内置的一款轻量级开源嵌入式数据库，D.RichardHip 于 2000 年发布 Alpha 版本的 SQLite，至今已经升级到 SQLite3。SQLite3 支持绝大部分 SQL92 语法，并且可以在所有主流的操作系统上运行。SQLite3 内置 SQL 数据库引擎，一般只要确保 SQLite3 的可执行文件存在即可开始创建、连接和使用数据库。

小贴士： 从本质上看，SQLite 的操作方式只是一种更为便捷的文件操作。当应用程序创建或打开一个 SQLite 数据库时，其实只是打开一个文件准备读/写。

6.2.1 手动建库

虽然应用程序完全可以在代码中动态建立 SQLite 数据库，但使用命令行手工建立和管理数据库仍然是非常重要的内容，对于调试使用数据库的应用程序非常有用。

Android SDK 的 platform-tools 目录里有 SQLite3 工具，同时，该工具也被集成在 Android 系统中。下面的内容将介绍如何连接到模拟器中的 Linux 系统，并在 Linux 系统中启动 SQLite3 工具，以及在 Android 程序目录中建立数据库和数据表，并使用命令在数据表中添加、修改和删除数据。

1. 使用 adb shell 命令连接到模拟器的 Linux 系统

在“开始菜单”左下角的搜索框中，输入“CMD”并按回车键，进入命令提示符窗口。adb（Android Debug Bridge）工具存放在 Android SDK 的 platform-tools 目录下，在命令提示符方式下进入该目录，输入命令“adb shell”，如图 6-2 所示，证明已连接到模拟器（或真实设备）中的 Linux 系统，应保证这时至少有正在运行的手机模拟器或者有打开的手机真实设备通过 USB 与 IDE 相连。

```
管理员: C:\windows\system32\cmd.exe - adb shell
Microsoft Windows [版本 6.1.7601]
版权所有 (c) 2009 Microsoft Corporation。保留所有权利。

C:\Users\Administrator>d:

D:\>cd D:\AndroidSDK\platform-tools

D:\AndroidSDK\platform-tools>adb shell
generic_x86:/ $
```

图 6-2 使用 adb shell 连接手机模拟器

2. 启动 SQLite3

在 Linux 命令提示符下输入“sqlite3”，启动 SQLite3 后会显示 SQLite 的版本信息，其显示内容如图 6-3 所示。

```
D:\AndroidSDK\platform-tools>adb shell
generic_x86:/ $ sqlite3
SQLite version 3.22.0 2018-01-22 18:45:57
Enter ".help" for usage hints.
Connected to a ←[1mtransient in-memory database←[0m.
Use ".open FILENAME" to reopen on a persistent database.
sqlite>
```

图 6-3 启动 SQLite3

启动 SQLite3 后，提示符变为“sqlite>”，表示用户进入了 SQLite 数据库的交互模式，此时可以输入 SQL 命令，建立、删除或修改数据库的内容。

正确退出 SQLite3 工具的方法是使用.exit 命令，如图 6-4 所示。

由于通过 adb shell 登录高版本的 Android 模拟器后，默认用户（shell）并不是 root，操作权限受限。通过“su”命令可将登录用户切换为 root，并通过“whoami”命令进行验证，如图 6-5 所示。由于 root 是 Linux 系统中的超级管理员用户账户，该账户拥有整个系统的最高权限，因此对 Android 系统下的目录和文件进行操作可更为方便。

```
sqlite> .exit
generic_x86:/ $ ls -l
ls: ./init.rc: Permission denied
ls: ./metadata: Permission denied
ls: ./init.environ.rc: Permission denied
ls: ./init.usb.configfs.rc: Permission denied
ls: ./adb_keys: Permission denied
ls: ./init.usb.rc: Permission denied
ls: ./init: Permission denied
ls: ./ueventd.rc: Permission denied
ls: ./init.zygote32.rc: Permission denied
total 52
```

图 6-4　shell 用户权限受限

```
1|generic_x86:/ $ whoami
shell
generic_x86:/ $ su
generic_x86:/ # whoami
root
generic_x86:/ #
```

图 6-5　登录用户切换为 root

3．创建数据库目录

在 Android 系统中，SQLite 数据库文件被存放在“/data/data/<应用程序包名>/databases”目录下，所有数据库都是私有的，仅允许本应用程序使用，其他应用程序访问都需通过 ContentProvider。但如果使用手工方式建立数据库，则必须手工建立数据库目录。

这里应用的包名是 com.mialab.ch5，它实际上标识的 Android 应用是创建 Ch5 工程时默认创建的 App 模块。将路径切换为/data/data/com.mialab.ch5，并在此路径下通过“mkdir databases”命令创建目录 databases，然后再通过“cd databases”命令将当前路径切换至/data/data/com.mialab.ch5 /databases 路径下，如图 6-6 所示。

```
generic_x86:/data/data # cd com.mialab.ch5
generic_x86:/data/data/com.mialab.ch5 # mkdir databases
generic_x86:/data/data/com.mialab.ch5 # ls -l
total 12
drwxrws--x 2 u0_a85 u0_a85_cache 4096 2019-04-22 07:31 cache
drwxrws--x 2 u0_a85 u0_a85_cache 4096 2019-04-22 07:31 code_cache
drwxrwxrwx 2 root   root         4096 2019-04-22 07:33 databases
generic_x86:/data/data/com.mialab.ch5 # cd databases
generic_x86:/data/data/com.mialab.ch5/databases #
```

图 6-6　创建数据库目录 databases

4．创建数据库

使用“sqlite3+文件名”的方式打开数据库文件，如果指定的文件不存在，SQLite3 工具将自动创建新文件。在/data/data/com.mialab.ch5/databases 路径下，通过“sqlite3 employee.db”命令创建 SQLite 数据库文件——employee.db，如图 6-7 所示。

```
generic_x86:/data/data/com.mialab.ch5/databases # sqlite3 employee.db
SQLite version 3.22.0 2018-01-22 18:45:57
Enter ".help" for usage hints.
sqlite>
```

图 6-7　创建 SQLite 数据库 employee.db

5．创建数据库表格

利用 create table 命令创建数据表格，由于 SQLite3 工具创建数据表格遵循的是 SQL 语法，因此，同样支持 insert、update、delete、select 等 SQL 命令。

在 SQLite 数据库 employee.db 中创建数据表，创建数据表 employee 的 SQL 语句如图 6-8 所示。其中，“.tables”命令用于显示数据库中的表，“.schema”命令用于查看创建表时使用的 SQL 语句，如图 6-9 所示。

```
sqlite> create table employee
   ...> (_id integer primary key autoincrement,
   ...> name text not null,
   ...> age integer,
   ...> height float);
sqlite> .tables
employee
sqlite>
```

图 6-8　创建 employee 数据表

```
sqlite> .tables
employee
sqlite> .schema
CREATE TABLE employee
(_id integer primary key autoincrement,
name text not null,
age integer,
height float);
CREATE TABLE sqlite_sequence(name,seq);
sqlite>
```

图 6-9　查看创建 employee 表的 SQL 语句

6．表记录的增加、删除、更改和查询

在 employee 表中插入表记录的 SQL 命令，如图 6-10 所示。

```
sqlite> insert into employee values(null,'Marry',46,1.72);
sqlite> insert into employee values(null,'LiBai',99,1.82);
sqlite> insert into employee values(null,'DuFu',96,1.86);
sqlite> select * from employee;
1|Marry|46|1.72
2|LiBai|99|1.82
3|DuFu|96|1.86
sqlite>
```

图 6-10　插入 employee 表记录并查询

可以使用“.mode MODE(输出格式)”命令更改结果的输出格式，如可以使用“column”格式显示 employee 数据表中的数据信息，如图 6-11 所示。

```
sqlite> .mode column
sqlite> select * from employee;
1          Marry      46         1.72
2          LiBai      99         1.82
3          DuFu       96         1.86
sqlite>
```

图 6-11　使用 column 格式查询表

在 employee 表中更新表记录的 SQL 命令，如图 6-12 所示。

```
sqlite> update employee set age=112 where name='LiBai';
sqlite> select * from employee;
1          Marry      46         1.72
2          LiBai      112        1.82
3          DuFu       96         1.86
sqlite>
```

图 6-12　在 employee 表中更新表记录

在 employee 表中删除（_id=3）记录的 SQL 命令，如图 6-13 所示。

```
sqlite> delete from employee where _id=3;
sqlite> select * from employee;
1          Marry      46         1.72
2          LiBai      112        1.82
sqlite>
```

图 6-13　在 employee 表中删除表记录

6.2.2 SQLiteDatabase

在 android.database.sqlite 包中，Android 提供了 SQLiteDatabase 类，该类封装了一些操作数据库的 API，使用该类可以完成对数据库中的数据增加、查询、更新和删除的操作。

由于在 Context 类中定义了 openOrCreateDatabase()方法，因此可以在活动页面的代码中，或任何能取到 Context 的地方获取 SQLiteDatabase 实例。

SQLiteDatabase 类的一些常用方法，如表 6-1 所示。

表 6-1 SQLiteDatabase 类的一些常用方法

返回类型	方法名	分类	描述
long	insert(String table, String nullColumnHack, ContentValues values)	数据处理类，用于数据表层面的操作	增加一行表记录到数据库
int	delete(String table, String whereClause, String[] whereArgs)		删除数据库表记录
int	update(String table, ContentValues values, String whereClause, String[] whereArgs)		更新数据库表记录
Cursor	query(String table, String[] columns, String selection, String[] selectionArgs, String groupBy, String having, String orderBy)		查询数据表
Cursor	rawQuery(String sql, String[] selectionArgs)		执行拼接好的 SQL 查询语句
void	execSQL(String sql)		执行 SQL 语句
void	execSQL(String sql, Object[] bindArgs)		执行带占位符的 SQL 语句
static SQLiteDatabase	openDatabase(String path, SQLiteDatabase.CursorFactory factory, int flags)	管理类，用于数据库层面的操作	打开指定路径下的数据库
static SQLiteDatabase	openOrCreateDatabase(String path, SQLiteDatabase.CursorFactoryfactory)		打开或创建指定路径下的数据库
boolean	isOpen()		判断数据库是否已打开
int	getVersion()		获取数据库的版本号
void	close()		关闭数据库
void	beginTransaction()	事务类，用于事务层面的操作	开始事务
void	endTransaction()		结束事务
void	setTransactionSuccessful()		设置事务成功标志

6.2.3 SQLiteOpenHelper

SQLiteOpenHelper 是 Android 提供的一个管理数据库的工具类，用于帮助创建或者打开和管理一个 SQLite 数据库。通常的用法是创建 SQLiteOpenHelper 的子类，IDE 会提示重写 onCreate()方法和 onUpgrade()方法。

（1）onCreate(SQLiteDatabase db)方法：当调用 SQLiteOpenHelper 的 getWritableDatabase()方法或 getReadableDatabase()方法获取 SQLiteDatabase 实例时，如果数据库不存在，系统会自动生成一个数据库，然后调用 onCreate()方法。onCreate()方法在初次生成数据库时才会被调用，通常会在这里处理一些创建表的逻辑。

（2）onUpgrade(SQLiteDatabase db, int oldVersion, int newVersion)方法：用于升级软件时更新数据库

的表结构。只要某次创建 SQLiteOpenHelper 时指定的数据库版本号高于之前指定的版本号，系统就会自动触发 onUpgrade()方法。

假定创建了 SQLiteOpenHelper 的子类——DemoOpenHelper.java，在 DemoOpenHelper.java 中将会重写 onCreate()方法和 onUpgrade()方法，其示例代码如下。

```
public class DemoOpenHelper extends SQLiteOpenHelper {
   public DemoOpenHelper(Context context, String name, int version) {
    super(context, name,null,version);
   }
   @Override
   public void onCreate(SQLiteDatabase db) {
      //创建数据表，初始化数据
   }
   @Override
   public void onUpgrade(SQLiteDatabase arg0, int arg1, int arg2) {
      //版本升级时维护数据表
   }
}
```

SQLiteOpenHelper 类提供的常用方法如表 6-2 所示。

表 6-2　SQLiteOpenHelper 类中的常用方法

返回类型	方法名	描述
	SQLiteOpenHelper(Context context, String name, SQLiteDatabase.CursorFactory factory, int version)	创建一个 SQLiteOpenHelper 对象来创建、打开及管理一个 SQLite 数据库
void	onCreate(SQLiteDatabase db)	数据库第一次创建时被调用
void	onUpgrade(SQLiteDatabase db, int oldVersion, int newVersion)	数据库需要升级时被调用
SQLiteDatabase	getReadableDatabase()	创建或者打开一个数据库
SQLiteDatabase	getWritableDatabase()	创建或者打开一个可读/写的数据库
void	onOpen(SQLiteDatabase db)	数据库打开时被调用
void	close()	关闭任何打开的数据库实例

SQLiteOpenHelper 提供的 getReadableDatabase()方法和 getWritableDatabase()方法，都可以打开（或创建）SQLite 数据库（如果数据库存在则直接打开，否则创建一个新的数据库），并返回一个可进行读/写操作的 SQLite 数据库对象。通常情况下这两种方法的实现效果是一样的。不同的是，当数据库不可写入的时候（如磁盘空间已满），getReadableDatabase()方法将返回一个“只能进行读操作不能进行写操作”的 SQLite 数据库对象，而 getWritableDatabase()方法将出现异常。

6.2.4　Cursor 和 ContentValues

1．Cursor

SQLiteDatabase 对象的 query()方法是返回一个 Cursor 对象，可将其视作查询结果的指针。Cursor 类支持在查询结果中以多种方式移动。Cursor 类中一些常用方法如表 6-3 所示。

表 6-3　Cursor 类中的一些常用方法

返回类型	方法名	描述
boolean	moveToFirst()	将指针移动到第一行
boolean	moveToNext()	将指针移动到下一行

续表

返 回 类 型	方 法 名	描 述
boolean	moveToPrevious()	将指针移动到上一行
boolean	moveToPosition(int position)	将指针移动到指定位置的数据上
int	getCount()	获取查询结果集中的条目个数
String	getColumnName(int columnIndex)	返回指定列号的属性名称
int	getColumnIndex(String columnName)	根据属性名称返回列号，如果不存在则返回-1
int	getColumnIndexOrThrow(String columnName)	返回指定属性名称的列号（从0开始），如果不存在则产生异常
int	getPosition()	返回当前的指针位置.

2．ContentValues

ContentValues 的使用方式类似于 SharedPreferences 类中提供的键值对数据存储。ContentValues 类提供的一些常用方法如表 6-4 所示。

表 6-4　ContentValues 类中的一些常用方法

返 回 类 型	方 法 名	描 述
void	put(String key, Integer value)	添加值到指定的集合中
Integer	getAsInteger(String key)	得到一个值并转换为 Integer 类型
int	size()	获取集合中有多少个 value
void	clear()	清空集合中的数据
boolean	containsKey(String key)	判断是否包含该字段

6.2.5　代码建库

【示例】 实现数据库中表记录的增加、删除、修改和查询。

启动 Android Studio，在 Ch6 工程中创建 SQLiteDemo 模块，并在 SQLiteDemo 模块中创建 MainActivity.java、InsertActivity.java、QueryActivity.java、UpdateActivity.java、DeleteActivity.java 及相应的布局文件 activity_main.xml、activity_insert.xml、activity_query.xml、activity_update.xml、activity_delete.xml，以及工具类（数据库适配器）DBAdapter.java、实体类 Book.java 等。

MainActivity 界面如图 6-14（a）所示，InsertActivity 界面如图 6-14（b）所示，QueryActivity 界面如图 6-14（c）和图 6-14（d）所示，UpdateActivity 界面如图 6-14（e）和图 6-14（f）所示，DeleteActivity 界面如图 6-14（g）所示。

（a）　（b）　（c）　（d）

图 6-14　运行并测试 SQLiteDemo 模块

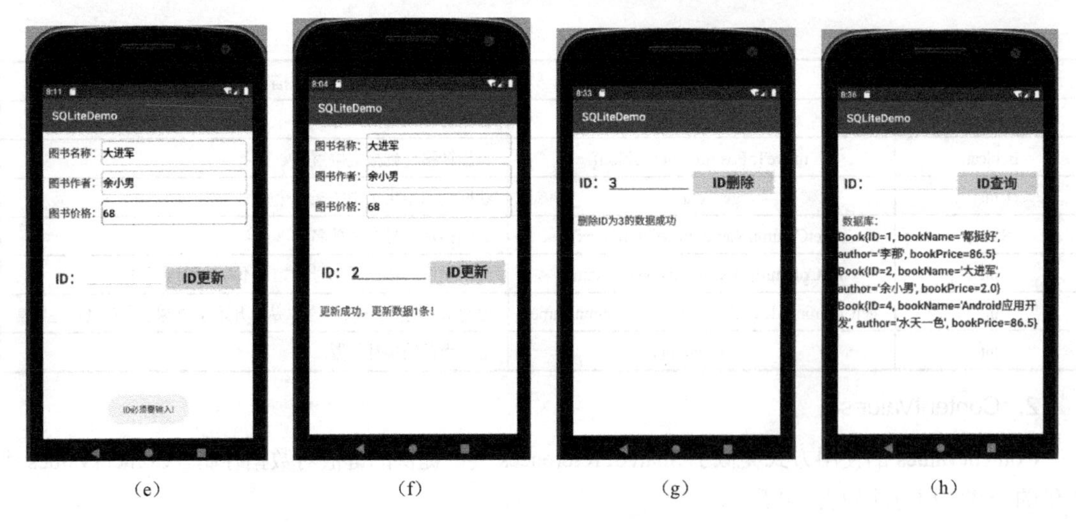

图 6-14　运行并测试 SQLiteDemo（续）

SQLiteDemo 模块的数据库工具类 DBAdapter.java，其主要代码如下：

```
public class DBAdapter {
   private static final String DB_NAME = "book.db";
   private static final String DB_TABLE = "bookinfo";
   private static final int DB_VERSION = 1;
   public static final String ID = "_id";
   public static final String BOOKNAME = "bookName";
   public static final String AUTHOR = "author";
   public static final String BOOKPRICE = "bookPrice";
   private SQLiteDatabase db;
   private final Context context;
   private BookDBHelper dbHelper;

   public DBAdapter(Context _context) {
      context = _context;
   }
   public void close() {
      if (db != null) {
         db.close();
         db = null;
      }
   }
   public void open() throws SQLiteException {
      dbHelper = new BookDBHelper(context, DB_NAME, null, DB_VERSION);
      try {
         db = dbHelper.getWritableDatabase();
      } catch (SQLiteException ex) {
         db = dbHelper.getReadableDatabase();
      }
   }
   public long insert(Book book) {
      ContentValues bookValues = new ContentValues();
      bookValues.put(BOOKNAME, book.getBookName());
      bookValues.put(AUTHOR, book.getAuthor());
      bookValues.put(BOOKPRICE, book.getBookPrice());
      return db.insert(DB_TABLE, null, bookValues);
   }
```

```
        public Book[] queryAll() {
            Cursor results = db.query(DB_TABLE, new String[] { ID, BOOKNAME, AUTHOR,
            BOOKPRICE}, null, null, null, null, null);
            return ConvertToBook(results);
        }
        public Book[] queryOne(long id) {
            Cursor results = db.query(DB_TABLE, new String[] { ID, BOOKNAME, AUTHOR,
            BOOKPRICE}, ID + "=" + id, null, null, null, null);
            return ConvertToBook(results);
        }
        private Book[] ConvertToBook(Cursor cursor){
            int resultCounts = cursor.getCount();
            if (resultCounts == 0 || !cursor.moveToFirst()){
                return null;
            }
            Book[] bookList = new Book[resultCounts];
            for (int i = 0 ; i<resultCounts; i++){
                bookList[i] = new Book();
                int newId = cursor.getInt(0);
                bookList[i].setID(newId);
                String newBookName = cursor.getString(cursor.getColumnIndex (BOOKNAME));
                bookList[i].setBookName(newBookName);
                String newAuthor = cursor.getString(cursor.getColumnIndex(AUTHOR));
                bookList[i].setAuthor(newAuthor);
                Float newBookPrice = cursor.getFloat(cursor.getColumnIndex (BOOKPRICE));
                bookList[i].setBookPrice(newBookPrice);
                cursor.moveToNext();
            }
            return bookList;
        }
        public long deleteAll() {
            return db.delete(DB_TABLE, null, null);
        }
        public long deleteOne(long id) {
            return db.delete(DB_TABLE, ID + "=" + id, null);
        }
        public long updateOne(long id , Book book){
            ContentValues bookValues = new ContentValues();
            bookValues.put(BOOKNAME, book.getBookName());
            bookValues.put(AUTHOR, book.getAuthor());
            bookValues.put(BOOKPRICE, book.getBookPrice());
            return db.update(DB_TABLE, bookValues, ID + "=" + id, null);
        }

        //SQLiteOpenHelper的子类，静态Helper类，用于建立、更新和打开数据库
        private static class BookDBHelper extends SQLiteOpenHelper {
            public BookDBHelper(Context context, String name, CursorFactory factory,
int version) {
                super(context, name, factory, version);
            }
            private static final String DB_CREATE = "create table " +
                    DB_TABLE + " (" + ID + " integer primary key autoincrement, " +
                    BOOKNAME + " text not null, " + AUTHOR + " text," + BOOKPRICE + "
float);";
```

```
        @Override
        public void onCreate(SQLiteDatabase _db) {
            _db.execSQL(DB_CREATE);
        }
        @Override
        public void onUpgrade(SQLiteDatabase _db, int _oldVersion, int _newVersion)
{
        }
    }
}
```

在数据库工具类DBAdapter.java中，定义了内部类BookDBHelper，它是SQLiteOpenHelper的子类，用于建立、更新和打开数据库。对数据库中表数据的操作，如增加、删除、修改和查询等都是由数据库工具类DBAdapter中封装的SQLiteDatabase对象应用（db）来完成的。

SQLiteDemo模块的MainActivity.java主要代码如下：

```
public class MainActivity extends AppCompatActivity implements View.
OnClickListener {
    @Override
    protected void onCreate(Bundle savedInstanceState) {
        super.onCreate(savedInstanceState);
        setContentView(R.layout.activity_main);
        findViewById(R.id.btn1).setOnClickListener(this);
        findViewById(R.id.btn2).setOnClickListener(this);
        findViewById(R.id.btn3).setOnClickListener(this);
        findViewById(R.id.btn4).setOnClickListener(this);
    }

    @Override
    public void onClick(View v) {
        if (v.getId() == R.id.btn1) {
            Intent intent = new Intent(MainActivity.this, InsertActivity.class);
            startActivity(intent);
        } else if (v.getId() == R.id.btn2) {
            Intent intent = new Intent(MainActivity.this, QueryActivity.class);
            startActivity(intent);
        } else if (v.getId() == R.id.btn3) {
            Intent intent = new Intent(MainActivity.this, UpdateActivity.class);
            startActivity(intent);
        } else if (v.getId() == R.id.btn4) {
            Intent intent = new Intent(MainActivity.this, DeleteActivity.class);
            startActivity(intent);
        }
    }
}
```

SQLiteDemo模块的InsertActivity.java主要代码如下：

```
public class InsertActivity extends AppCompatActivity {
    private DBAdapter dbAdapter;
    private EditText booknameText;
    private EditText authorText;
    private EditText bookpriceText;
    private TextView labelView;

    @Override
    protected void onCreate(Bundle savedInstanceState) {
```

```
        super.onCreate(savedInstanceState);
        setContentView(R.layout.activity_insert);
        booknameText = (EditText) findViewById(R.id.et_bookname);
        authorText = (EditText) findViewById(R.id.et_author);
        bookpriceText = (EditText) findViewById(R.id.et_bookprice);
        labelView = (TextView) findViewById(R.id.label);
        Button addButton = (Button) findViewById(R.id.btn_save);
        addButton.setOnClickListener(addButtonListener);
        dbAdapter = new DBAdapter(this);
        dbAdapter.open();
    }
    View.OnClickListener addButtonListener = new View.OnClickListener() {
        @Override
        public void onClick(View v) {
            String bookName = booknameText.getText().toString();
            String author = authorText.getText().toString();
            String priceStr = bookpriceText.getText().toString();
            if (TextUtils.isEmpty(bookName)) {
                showToast("请先填写图书名称");
                return;
            } else if (TextUtils.isEmpty(author)) {
                showToast("请先填写图书作者");
                return;
            } else if (TextUtils.isEmpty(priceStr)) {
                showToast("请先填写图书价格");
                return;
            }
            Book book = new Book();
            Float bookPrice = Float.parseFloat(bookpriceText.getText().toString());
            book.setBookName(bookName);
            book.setAuthor(author);
            book.setBookPrice(bookPrice);
            long num = dbAdapter.insert(book);
            if (num == -1) {
                labelView.setText("添加数据失败！");
            } else {
                labelView.setText("成功添加数据，ID：" + String.valueOf(num));
            }
        }
    };
    @Override
    protected void onStop() {
        super.onStop();
        dbAdapter.close();
    }
    private void showToast(String desc) {
        Toast.makeText(this, desc, Toast.LENGTH_SHORT).show();
    }
}
```

SQLiteDemo 模块的 QueryActivity.java 主要代码如下：

```
public class QueryActivity extends AppCompatActivity {
    private DBAdapter dbAdapter ;
    private EditText idQuery;
    private TextView labelView;
```

```
    private TextView displayView;
    @Override
    protected void onCreate(Bundle savedInstanceState) {
        super.onCreate(savedInstanceState);
        setContentView(R.layout.activity_query);
        idQuery = (EditText)findViewById(R.id.id_query);
        labelView = (TextView)findViewById(R.id.label);
        displayView = (TextView)findViewById(R.id.display);
        Button queryButton = (Button)findViewById(R.id.query);
        queryButton.setOnClickListener(queryButtonListener);
        dbAdapter = new DBAdapter(this);
        dbAdapter.open();
        displayAll();
    }
    private void displayAll() {
        Book[] books = dbAdapter.queryAll();
        if (books == null){
            labelView.setText("数据库中没有数据");
            return;
        }
        labelView.setText("数据库：");
        String msg = "";
        for (int i = 0 ; i<books.length; i++){
            msg += books[i].toString()+"\n";
        }
        displayView.setText(msg);
    }
    View.OnClickListener queryButtonListener = new View.OnClickListener() {
        @Override
        public void onClick(View v) {
            int id = Integer.parseInt(idQuery.getText().toString());
            Book[] books = dbAdapter.queryOne(id);
            if (books == null){
                labelView.setText("数据库中没有 ID 为"+String.valueOf(id)+"的数据");
                return;
            }
            labelView.setText("数据库：");
            displayView.setText(books[0].toString());
        }
    };
    @Override
    protected void onStop() {
        super.onStop();
        dbAdapter.close();
    }
}
```

SQLiteDemo 模块的 UpdateActivity.java 主要代码如下：

```
public class UpdateActivity extends AppCompatActivity {
    private DBAdapter dbAdapter;
    private EditText booknameText;
    private EditText authorText;
    private EditText bookpriceText;
    private EditText idUpdate;
    private TextView labelView;
```

```
    private TextView displayView;
    @Override
    protected void onCreate(Bundle savedInstanceState) {
        super.onCreate(savedInstanceState);
        setContentView(R.layout.activity_update);
        booknameText = (EditText) findViewById(R.id.et_bookname);
        authorText = (EditText) findViewById(R.id.et_author);
        bookpriceText = (EditText) findViewById(R.id.et_bookprice);
        idUpdate = (EditText) findViewById(R.id.id_update);
        labelView = (TextView) findViewById(R.id.label);
        displayView = (TextView) findViewById(R.id.display);
        Button updateButton = (Button) findViewById(R.id.btn_update);
        updateButton.setOnClickListener(updateButtonListener);
        dbAdapter = new DBAdapter(this);
        dbAdapter.open();
    }
    View.OnClickListener updateButtonListener = new View.OnClickListener() {
        @Override
        public void onClick(View v) {
            String bookName = booknameText.getText().toString();
            String author = authorText.getText().toString();
            String priceStr = bookpriceText.getText().toString();
            String idStr = idUpdate.getText().toString();
            if (TextUtils.isEmpty(bookName)) {
                showToast("请先填写图书名称");
                return;
            } else if (TextUtils.isEmpty(author)) {
                showToast("请先填写图书作者");
                return;
            } else if (TextUtils.isEmpty(priceStr)) {
                showToast("请先填写图书价格");
                return;
            }else if (TextUtils.isEmpty(idStr)) {
                showToast("ID必须要输入!");
                return;
            }
            Book book = new Book();
            Float bookPrice = Float.parseFloat(idStr);
            book.setBookName(bookName);
            book.setAuthor(author);
            book.setBookPrice(bookPrice);
            long id = Integer.parseInt(idUpdate.getText().toString());
            long count = dbAdapter.updateOne(id, book);
            if (count == -1) {
                labelView.setText("更新错误! ");
            } else if (count == 0) {
                labelView.setText("没有此ID，请重新输入! ");
            }  else{
                labelView.setText("更新成功，更新数据" + String.valueOf(count) +
"条! ");
            }
        }
    };
    @Override
```

```
    protected void onStop() {
        super.onStop();
        dbAdapter.close();
    }
    private void showToast(String desc) {
        Toast.makeText(this, desc, Toast.LENGTH_SHORT).show();
    }
}
```

SQLiteDemo 模块的 DeleteActivity.java 主要代码如下：

```
public class DeleteActivity extends AppCompatActivity {
    private DBAdapter dbAdapter;
    private EditText idDelete;
    private TextView labelView;
    private TextView displayView;
    @Override
    protected void onCreate(Bundle savedInstanceState) {
        super.onCreate(savedInstanceState);
        setContentView(R.layout.activity_delete);
        idDelete = (EditText) findViewById(R.id.id_delete);
        labelView = (TextView) findViewById(R.id.label);
        displayView = (TextView) findViewById(R.id.display);
        Button deleteButton = (Button) findViewById(R.id.btn_delete);
        deleteButton.setOnClickListener(deleteButtonListener);
        dbAdapter = new DBAdapter(this);
        dbAdapter.open();
    }
    View.OnClickListener deleteButtonListener = new View.OnClickListener() {
        @Override
        public void onClick(View v) {
            String idStr = idDelete.getText().toString();
            if (TextUtils.isEmpty(idStr)) {
                showToast("ID 必须要输入!");
                return;
            }
            long id = Integer.parseInt(idStr);
            long count = dbAdapter.deleteOne(id);
            if (count == 0) {
                labelView.setText("没有此 ID，请重新输入！删除数据" +
                        String.valueOf(count) + "条！");
            } else{
                labelView.setText("删除成功，删除数据" + String.valueOf(count) +
"条！");
            }
            String msg = "删除 ID 为" + idDelete.getText().toString() + "的数据" +
                    (count > 0 ? "成功" : "失败");
            labelView.setText(msg);
        }
    };
    @Override
    protected void onStop() {
        super.onStop();
        dbAdapter.close();
    }
    private void showToast(String desc) {
```

```
            Toast.makeText(this, desc, Toast.LENGTH_SHORT).show();
        }
    }
```

SQLiteDemo 模块的其余文件参见 Ch6\SQLiteDemo。

6.3 文件存储

手机的存储空间一般分为两块区域，一块用于内部存储，另一块用于外部存储（SD 卡）。早期的 SD 卡是可插拔式的存储芯片，后来越来越多的手机将 SD 卡固化到了手机内部，Android 仍然称之为外部存储。

内存（Memory/RAM）是一种临时的数据存储器，断电后数据就会消失，并且读取和写入的速度非常快，通常指手机的运行内存。内部存储不同于内存。

最初内部存储和外部存储是分开的，这种情况多发生于 Android4.4 以下系统，但是随着 Android 手机的发展，渐渐都做成了一体机，甚至将内部存储和外部存储集成在了一起，只是在逻辑上区分了内部存储和外部存储，到了现在很多厂商连外置 SD 卡的卡槽都不提供了。

Android 4.4 以下系统的内部存储和外部存储，如图 6-15 所示。

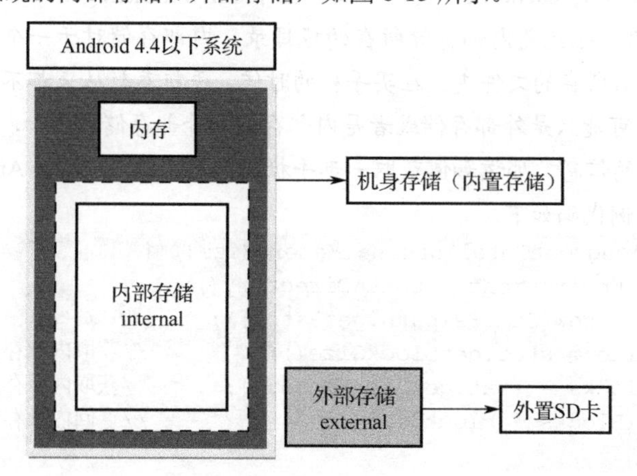

图 6-15　Android 4.4 以下系统的内部存储和外部存储

Android 4.4 及以上系统的内部存储和外部存储，如图 6-16 所示。

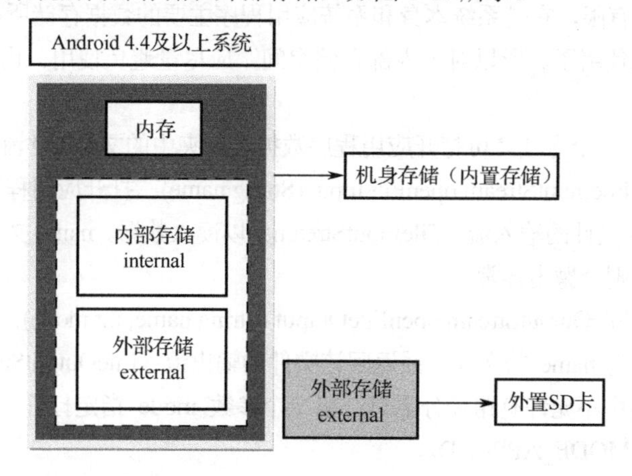

图 6-16　Android 4.4 及以上系统的内部存储和外部存储

虽然 SharedPreferences 能够实现轻量级数据的存储操作，但是对 Android 系统来说，文件的操作仍是必不可少的，Java 提供了一套完整的文件 I/O 工作机制实现对文件的操作与管理，借助于 Java 的文件管理体系，Android 同样可以实现对手机系统的文件操作。

6.3.1 内部存储

内部存储（Internal Storage）不是内存（Memory/RAM），它是手机里的一块存储区域，是系统本身和系统应用程序主要的数据存储所在地。手机的内部存储通常不会很大，一旦手机的内部存储容量被用完，可能会出现手机无法使用的情形，因此不宜存储视频等大文件，只适合存储一些小文件。如 SharedPreferences 和 SQLite 数据库都是存储在内部存储中的，不会占用太大的空间。Android 系统提供了的 API 如下。

```
Environment.getDataDirectory().getAbsolutePath(); //获取手机内部存储空间的绝对路径
getFilesDir().getAbsolutePath();      //获取当前应用包名文件夹下的 files 文件夹
getCacheDir().getAbsolutePath();      //获取当前应用包名文件夹下 cache 文件夹
getDir(String name, MODE_PRIVATE);  //在内部存储空间创建（或打开现有的）目录
```

小贴士：在手机中安装一个 App 就会在/data/data 目录下新增一个应用包名的文件夹，而 SQLiteDatabase 数据库、SharedPreferences 文件都是存放在这个文件夹中的。getFilesDir()是用来获取/data/data/<packagename>下的 files 文件夹路径（如果存在的话），getCacheDir()是用来获取/data/data/<packagename>下的 cache 文件夹路径的，至于 Environment.getDataDirectory()获取的路径，就是最外层的/data 目录，也就是内部存储所在的根目录。内部存储对于一个 App 来说，无非就是/data/data 目录下应用包名所在的文件夹。在买手机的时候，手机参数从来都不告诉内部存储有多大，通常所说的 64G、128G 可能只是外部存储或者是内部存储和外部存储的总和，无论怎样搜索都不会有关于一部手机内部存储的信息，那该如何获取一部手机的内部存储大小呢？Android 系统提供了相应的 API 可以得到，其示例代码如下。

```
public static long getTotalInternalMemorySize() {
    File path = Environment.getDataDirectory();
    StatFs stat = new StatFs(path.getPath());
    long blockSize = stat.getBlockSize();           //获取内部存储中每个区域的存储大小
    long totalBlocks = stat.getBlockCount();        //获取内部存储中所有的区块数量
    return totalBlocks * blockSize;                 //返回内部存储空间的总大小
}
```

内部存储位于系统中很特殊的一个位置，如果将文件存储于内部存储中，那么应用创建于内部存储的文件，是与这个应用关联起来的。当一个应用卸载之后，内部存储中的这些文件也会被删除。

内部存储空间十分有限，它是系统本身和系统应用程序主要的数据存储所在地，一旦内部存储空间耗尽，手机也就无法使用了。所以对于内部存储空间，应尽量避免使用。内部存储一般用 Context 来获取和操作。

Context 提供了如下两个方法，可打开应用程序数据文件夹中的文件 I/O 流。

（1）public abstract FileInputStream openFileInput (String name)：打开应用程序私有文件目录下名为 name 的文件，并返回该文件的输入流（FileInputStream）以读入数据。name 文件名中不允许有路径分隔符。若该文件不存在则会抛出异常。

（2）public abstract FileOutputStream openFileOutput (String name, int mode)：以 mode 方式打开应用程序私有文件目录下名为 name 的文件，并返回该文件的输出流（FileOutputStream）以写入数据。第一个参数 name 文件名中不允许有路径分隔符，第二个参数 mode 指定打开文件的模式。mode 取值 MODE_PRIVATE 或者 MODE_APPEND。

MODE_PRIVATE：默认模式，以读/写方式打开当前应用程序的文件，该模式意味着文件仅能被当前应用程序访问。

MODE_APPEND：表示追加模式，如果文件存在，则在文件的结尾处添加新数据，否则创建文件。

在 Android 平台下，可利用 openFileInput()和 openFileOutput()打开（内部存储中）文件的输入/输出流，再借助 Java 的 I/O 体系，实现对文件（/data/data/<packagename>/files/ <filename>）的读/写操作。

【示例】 把文本文件写入内部存储（/data/data/<packagename>）并读取。

启动 Android Studio，在 Ch6 工程中创建 FileDemo 模块，并在 FileDemo 模块中创建 MainActivity.java、TextWriteActivity.java、TextReadActivity.java、TestPathActivity.java，以及相应的布局文件 activity_main.xml、activity_text_write.xml、activity_text_read.xml、activity_test_path.xml。

MainActivity 界面如图 6-17（a）所示，TextWriteActivity 界面如图 6-17（b）所示，TextReadActivity 界面如图 6-17（c）所示，TestPathActivity 界面如图 6-17（d）所示。

图 6-17　运行 FilDemo 模块并测试

FileDemo 模块的 TextWriteActivity.java 主要代码如下：

```
public class TextWriteActivity extends AppCompatActivity {
    @Override
    protected void onCreate(Bundle savedInstanceState) {
        super.onCreate(savedInstanceState);
        setContentView(R.layout.activity_text_write);
        final EditText input = (EditText) findViewById(R.id.edText);
        Button btn_save = (Button) findViewById(R.id.btn_save);     //获取保存按钮
        Button btn_cancel = (Button) findViewById(R.id.btn_cancel);//获取取消按钮
        btn_save.setOnClickListener(new View.OnClickListener() {  //实现内部存储填
写的文本信息
            @Override
            public void onClick(View v) {
                FileOutputStream fos = null;                        //定义文件输出流
                String inputText = input.getText().toString();  //获取文本信息
                try {  //获得文件输出流，并指定文件保存的位置
                    fos = openFileOutput("boy.txt", MODE_PRIVATE);
                    fos.write(inputText.getBytes());                //保存文本信息
                    fos.flush();                                    //清除缓存
                } catch (IOException e) {
                    e.printStackTrace();
                } finally {
                    if (fos != null) {                              //输出流不为空时
                        try {
                            fos.close();                            //关闭文件输出流
```

```
                        Toast.makeText(TextWriteActivity.this, "文件写入成功！",
                        Toast.LENGTH_LONG).show();
                    } catch (IOException e) {
                        e.printStackTrace();
                    }
                }
            }
        }
    });
    btn_cancel.setOnClickListener(new View.OnClickListener() {
        @Override
        public void onClick(View v) {
            finish();
        }
    });
  }
}
```

注意，这里文本文件 boy.txt 的存储位置是/data/data/com.mialab.filedemo/files/boy.txt 。

为了查看 FileDemo 模块的写入文本文件 boy.txt，可以通过点击“View”→“Tool Windows”→“Device File Explorer”，打开“Device File Explorer”窗口，在“/data/data/ com.mialab.filedemo/files/”路径下找到文本文件 boy.txt，如图 6-18 所示，并可以通过选择“Save As”选项导出查看 boy.txt 的内容。

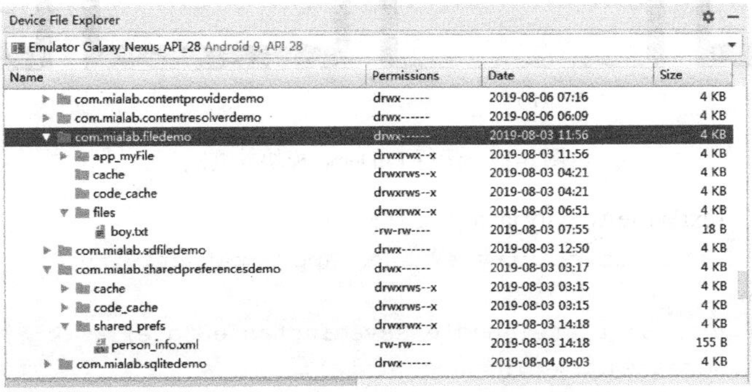

图 6-18 “Device File Explorer”中的/data/data/com.mialab.filedemo/files/boy.txt

FileDemo 模块的 TextReadActivity.java 主要代码如下：

```
public class TextReadActivity extends AppCompatActivity {
  @Override
  protected void onCreate(Bundle savedInstanceState) {
      super.onCreate(savedInstanceState);
      setContentView(R.layout.activity_text_read);
      final TextView display = (TextView) findViewById(R.id.diaplay);
      Button btnRead = (Button) findViewById(R.id.btn_read);      //获取读取按钮
      Button btnBack = (Button) findViewById(R.id.btn_back);      //获取返回按钮
      btnRead.setOnClickListener(new View.OnClickListener() {
          @Override
          public void onClick(View v) {
              FileInputStream fis = null;                        //定义文件输入流
              try {
                  fis = openFileInput("boy.txt");                //获得文件输入流
                  if(fis.available() == 0){
                      display.setText("对不起，该文件内容为空！");
```

```
                    return;
                }
                byte[] buffer = new byte[fis.available()];  //保存数据的数组
                while (fis.read(buffer) != -1);             //从输入流中读取数据
                String content = new String(buffer);
                display.setTextColor(Color.BLUE);
                display.setTextSize(32);
                display.setText(content);
            }catch (IOException e) {
                e.printStackTrace();
            } finally {
                if (fis != null) {                          //输入流不为空时
                    try {
                        fis.close();                        //关闭输入流
                    } catch (IOException e) {
                        e.printStackTrace();
                    }
                }
            }
        }
    });
    btnBack.setOnClickListener(new View.OnClickListener() {
        @Override
        public void onClick(View v) {
            finish();
        }
    });
  }
}
```

限于篇幅，其余文件代码参见教学资源包中的Ch6\FileDemo模块，在这里不再赘述。

6.3.2 外部存储（读/写SD卡上的文件）

当程序通过Context的openFileInput()和openFileOutput()打开文件输入流、输出流时，程序所打开的都是应用程序的数据文件夹中的文件，这样所存储的文件大小可能比较有限。

通常情况下，手机自身的存储空间会有一定的容量限制，一味地将文件存储在手机存储器上可能会出现空间不够或者空间不足引发的应用程序运行效率的问题，因此为了更好地利用存储器有限的空间，可以考虑将一部分大数据文件存放于外部存储。

对SD卡上的文件操作基本步骤如下：

（1）获取SD卡的访问权限。

为了能够实现读/写外部存储器的文件，首先要获取外部存储的读/写许可。在配置文件中为应用程序添加读/写SD卡的访问权限，其代码如下：

```
<uses-permission android:name="android.permission.WRITE_EXTERNAL_STORAGE"/>
<uses-permission android:name="android.permission.READ_EXTERNAL_STORAGE"/>
```

Android把外部存储分成了两块区域：公有存储空间和私有存储空间。从Android 7.0开始加强了SD卡的权限管理，即使App声明了完整的SD卡操作权限，系统仍然默认禁止该App访问外部存储；不过系统默认关闭存储只是关闭外部存储的公共空间，外部存储的私有空间依然可以正常读/写。如果要在外部存储的公有存储空间读/写文件，还需要请求“动态获取读/写SD卡”的权限，只有当用户授权读/写SD卡时才可执行读/写。

动态申请权限的示例代码如下：

```
private static final int  REQUEST_EXTERNAL_STORAGE = 1;
private static String[]  PERMISSIONS_STORAGE = {
      "android.permission.READ_EXTERNAL_STORAGE",
      "android.permission.WRITE_EXTERNAL_STORAGE" };
public static void  verifyStoragePermissions(Activity activity) {
   try {  //检测是否有写的权限
      int permission = ActivityCompat.checkSelfPermission(activity,
         "android.permission.WRITE_EXTERNAL_STORAGE");
      if (permission != PackageManager.PERMISSION_GRANTED) {
      //没有写的权限，去申请写的权限，会弹出对话框
      ActivityCompat.requestPermissions(activity,  PERMISSIONS_STORAGE,
         REQUEST_EXTERNAL_STORAGE);
      }
   } catch (Exception e) {
      e.printStackTrace();
   }
}
```

把 verifyStoragePermissions()方法放在 Activity 的 onCreate()方法中即可。

（2）查看 SD 卡是否可用。

调用 Environment 的 getExternalStorageState()方法来判断 SD 卡是否可用，以及应用程序具有对 SD 的读/写权限，其代码如下：

```
Environment.getExternalStorageState().equals(Environment.MEDIA_MOUNTED);
```

其中，getExternalStorageState()返回了外部存储的当前状态，MEDIA_MOUNTED 表示外部存储可以进行读/写访问。

（3）获取 SD 卡的路径。

调用 Environment 的 getExternalStorageDirectory()方法获取 SD 卡的路径。

获取外部存储公共空间的存储路径，调用的是 Environment.getExternalStorage PublicDirectory (String type)方法；获取外部存储应用私有空间的存储路径，调用的是 Context 的 getExternalFilesDir (String type)方法。

小贴士：Android 4.4 系统及以上的手机将机身存储（手机自身带的存储称为机身存储）在概念上分成了内部存储和外部存储这两部分。那么 Android 4.4 及以上系统的手机如果插了 SD 卡，SD 卡又是什么呢？如果 SD 卡也是外部存储的话，那又该如何区分机身存储的外部存储与 SD 卡的外部存储呢？是的，SD 卡也是外部存储。

在 Android 4.4 及以上的系统中，API 提供了这样一个方法来遍历手机的外部存储路径：

```
File[] files;
if (Build.VERSION.SDK_INT >= Build.VERSION_CODES.KITKAT) {
   files = getExternalFilesDirs(Environment.MEDIA_MOUNTED);
   for(File file:files){
       Log.e("main",file);
   }
}
```

其中 getExternalFilesDirs()可以获取多个外部存储空间，返回一个数组。这样就可以得到手机自身所带的外部存储和外接 SD 卡所定义的外部存储了。

外部存储常用的 API 如下。

```
//获取手机外部存储的路径：/storage/emulated/0
Environment.getExternalStorageDirectory().getAbsolutePath();
//获取某种特定内容（如视频、照片等）所在的路径（传入参数是内容类型）：/storage/emulated/0/DCIM
Environment.getExternalStoragePublicDirectory(Environment.DIRECTORY_DCIM).get
AbsolutePath();
```

```
    //获取某个应用在外部存储中的files路径:/storage/emulated/0/Android/data/demo.suning.
com.demo/files
    getExternalFilesDir(" ").getAbsolutePath();
    //获取某个应用在外部存储中的cache路径:/storage/emulated/0/Android/data/demo.suning.
com.demo/cache
    getExternalCacheDir().getAbsolutePath();
```

获取特定内容的类型有如下:

```
    DIRECTORY_DCIM                  //相机拍摄的图片和视频保存的位置
    DIRECTORY_DOWNLOADS             //下载文件保存的位置
    DIRECTORY_MOVIES                //电影保存的位置
    DIRECTORY_MUSIC                 //音乐保存的位置
    DIRECTORY_PICTURES              //下载的图片保存的位置
```

（4）操作 SD 卡里的文件。

借助于 Java 的 I/O 体系实现对 SD 卡文件的操作。

【示例】 把文本文件写入外部存储的私有存储空间（/storage/emulated/0/Android/data/ < package name>/）并读取。

启动 Android Studio，在 Ch6 工程中创建 SDFileDemo 模块，并在 SDFileDemo 模块中创建 MainActivity.java、TextWriteActivity.java、TextReadActivity.java、FilePathActivity.java，以及相应的布局文件 activity_main.xml、activity_text_write.xml、activity_text_read.xml、activity_file_path.xml。

MainActivity 界面如图 6-19（a）所示，TextWriteActivity 界面如图 6-19（b）所示，TextReadActivity 界面如图 6-19（c）所示，FilePathActivity 界面如图 6-19（d）所示。

（a）　　（b）　　（c）　　（d）

图 6-19　运行 SDFilDemo 模块并测试

SDFileDemo 模块的 TextWriteActivity.java 主要代码如下：

```
public class TextWriteActivity extends AppCompatActivity {
  private EditText edText;
  private String mPath;
  private TextView display;
  @Override
  protected void onCreate(Bundle savedInstanceState) {
     super.onCreate(savedInstanceState);
     setContentView(R.layout.activity_text_write);
     edText = findViewById(R.id.edText);
     display = findViewById(R.id.tv_display);
     Button btnSave = (Button) findViewById(R.id.btn_save);
```

```
        Button btnCancal = (Button) findViewById(R.id.btn_cancel);
        //获取当前 App 外部存储的私有存储目录
        mPath = getExternalFilesDir(Environment.DIRECTORY_DOWNLOADS).toString()
+ "/";
        btnSave.setOnClickListener(new View.OnClickListener() {
            @Override
            public void onClick(View arg0) {
                String input = edText.getText().toString();
                if (TextUtils.isEmpty(input)) {
                    showToast("输入文本不得为空！");
                    return;
                }
                if    (Environment.getExternalStorageState().equals(Environment.
MEDIA_MOUNTED)) {
                    String file_path = mPath + "a.txt";
                    saveText(file_path, input);   //把文本字符串写入文本文件
                    display.setText("亲，输入文本的保存路径为：\n" + file_path);
                    display.setTextColor(Color.BLUE);
                    showToast("数据已写入 SD 卡文件");
                } else {
                    showToast("未发现已挂载的 SD 卡，请检查");
                }
            }
        });
        btnCancal.setOnClickListener(new View.OnClickListener() {
            @Override
            public void onClick(View arg0) {
                finish();
            }
        });
    }
    private void saveText(String file_path, String input) {
        try {
            //根据指定文件路径构建文件输出流对象
            FileOutputStream fos = new FileOutputStream(file_path);
            fos.write(input.getBytes());        //把字符串写入文件输出流
            fos.close();                        //关闭文件输出流
        } catch (Exception e) {
            e.printStackTrace();
        }
    }
    private void showToast(String desc) {
        Toast.makeText(this, desc, Toast.LENGTH_SHORT).show();
    }
}
```

SDFileDemo 模块的 TextReadActivity.java 主要代码如下：

```
public class TextReadActivity extends AppCompatActivity {
    private String mPath;
    private TextView read;
    @Override
    protected void onCreate(Bundle savedInstanceState) {
        super.onCreate(savedInstanceState);
        setContentView(R.layout.activity_text_read);
        read = findViewById(R.id.tv_read);
```

```
        Button btnRead = (Button) findViewById(R.id.btn_read);
        Button btnBack = (Button) findViewById(R.id.btn_back);
        //获取当前 App 的私有存储目录
        mPath = getExternalFilesDir(Environment.DIRECTORY_DOWNLOADS).toString()
+ "/";
        btnRead.setOnClickListener(new View.OnClickListener() {
            @Override
            public void onClick(View arg0) {
                if (Environment.getExternalStorageState().equals(Environment.MEDIA_
MOUNTED)) {
                    showToast("SD 卡已挂载!");
                } else {
                    showToast("未发现已挂载的 SD 卡，请检查");
                }
                String file_path = mPath + "a.txt";  //打开并显示选中的文本文件内容
                String content = readText(file_path);
                read.setText("文件内容如下：\n" + content);
                read.setTextColor(Color.BLUE);
            }
        });
        btnBack.setOnClickListener(new View.OnClickListener() {
            @Override
            public void onClick(View arg0) {
                finish();
            }
        });
    }
    private String readText(String path) {
        String readStr = "";
        try {
            //根据指定文件路径构建文件输入流对象
            FileInputStream fis = new FileInputStream(path);
            byte[] b = new byte[fis.available()];
            fis.read(b);                    //从文件输入流读取字节数组
            readStr = new String(b);        //把字节数组转换为字符串
            fis.close();                    //关闭文件输入流
        } catch (Exception e) {
            e.printStackTrace();
        }
        return readStr;                     //返回文本文件中的文本字符串
    }
    private void showToast(String desc) {
        Toast.makeText(this, desc, Toast.LENGTH_LONG).show();
    }
}
```

限于篇幅，其余文件代码参见教学资源包中的 Ch6\SDFileDemo 模块，在这里不再赘述。

这里写入的文本文件 a.txt 的存储位置是/storage/emulated/0/Android/data/com.mialab.sdfiledemo/files/Download/a.txt 。

为了查看 SDFileDemo 模块的写入文本文件 boy.txt，可以通过点击“View”→“Tool Windows”→“Device File Explorer”，弹出“Device File Explorer”窗口，在“/storage/emulated/0/Android/data/com.mialab.sdfiledemo/files/Download/”路径下找到文本文件 a.txt，如图 6-20 所示，并通过选择“Save As”选项导出查看 a.txt 的内容。

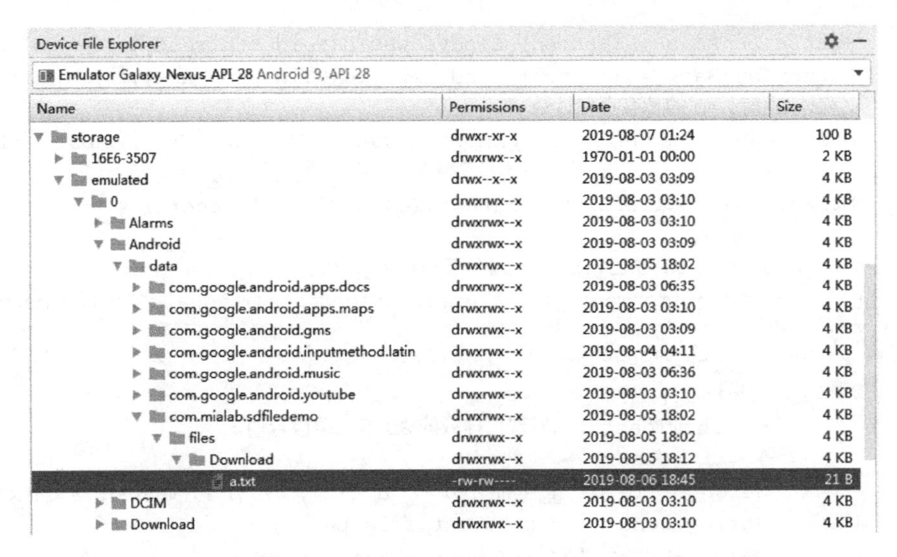

图 6-20　/storage/emulated/0/Android/data/com.mialab.sdfiledemo/files/Download/a.txt 路径

小贴士：/data 目录下的文件物理上存放在内部存储中，/storage 目录下的文件物理上存放在外部存储中。/system 用于存放系统文件，/cache 用于存放一些缓存文件，物理上它们也是存放在内部存储中的。

6.4　数据共享

6.4.1　ContentProvider

通常 Android 应用程序之间的数据不能够直接访问，为了实现不同应用程序之间的数据交互，Android 提供了一个接口 ContentProvider，它是不同应用程序之间进行数据交互的标准 API，一个应用程序把自己的数据通过 ContentProvider 暴露给其他应用程序，而其他应用程序通过 ContentResolver 来获取暴露的数据，如图 6-21 所示。

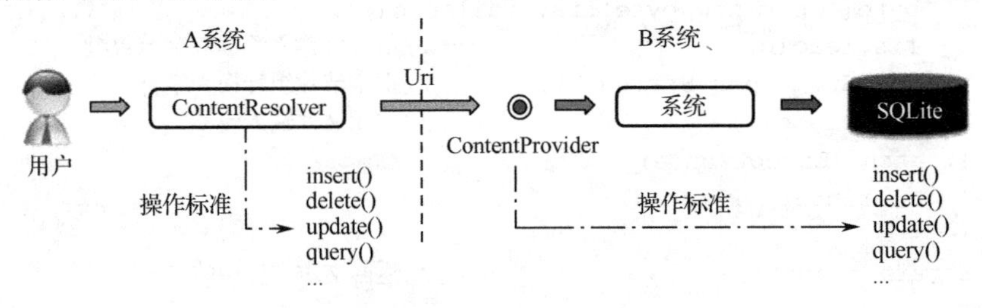

图 6-21　ContentProvider 的调用关系

应用程序之间交互的数据既可以是数据库中的数据，也可以是文件系统中的数据，还可以是网络数据，提供者实现 ContentProvider 类中对数据操作的接口函数，而调用者使用 ContentResolver 对象通过 URI 间接调用 ContentProvider。

ContentProvider 在程序的操作中所提供的是一个操作的标准，所以用户如果要想依靠此标准进行数据操作的时候，必须使用 android.content.ContentResolver 类来完成，而这个类中所给出的操作方法与 ContentProvider 是一一对应的，当用户调用了 ContentResolver 类的方法时，就相当于调用了 ContentProvider 类中的对应方法。

扩展 ContentProvider 类需要重写的方法，如表 6-5 所示。

表 6-5　扩展 ContentProvider 类需要重写的方法

返回类型	方法名	描　述
boolean	onCreate ()	该方法在 ContentProvider 创建后会被调用
Uri	insert (Uri uri, ContentValues values)	根据该 Uri 插入 values 对应的数据
Cursor	query (Uri uri, String[] projection, String selection, String[] selectionArgs, String sortOrder)	根据指定的 Uri 执行查询操作，所有的查询结果通过 Cursor 对象返回
int	update (Uri uri, ContentValues values, String selection, String[] selectionArgs)	根据指定的 Uri 执行记录的更新操作，并且返回更新记录的条目数量
int	delete (Uri uri, String selection, String[] selectionArgs)	根据 Uri 删除 selection 条件所匹配的全部记录
String	getType (Uri uri)	返回当前 Uri 所代表数据的 MIME 类型。如果该 Uri 对应的数据包括多条记录，那么 MIME 类型字符串应该以 vnd.android.cursor.dir/开头；如果该 Uri 对应的数据只包含一条记录，那么 MIME 类型字符串应该以 vnd.android.cursor.item/开头

小贴士：ContentProvider 为 App 存取内部数据提供统一的外部接口，让不同的应用之间得以共享数据。ContentProvider 操作的是本设备其他应用的内部数据，是一种中间层次的数据存储形式。ContentProvider 作为中间接口，本身并不直接保存数据，而是通过 SQLiteOpenHelper 与 SQLiteDatabase 间接操作底层的 SQLite 数据库。所以要想使用 ContentProvider，首先得实现 SQLite 数据库的帮助类，然后再由 ContentProvider 封装对外的接口。

6.4.2　Uri

Uri 作为提供者与调用者交互的重要桥梁，可以分为四个组成部分，其语法结构如下：

```
content: //<authorities>/<data_path>/<id>
```

Uri 各组成部分的解释如下。

- content://：通用前缀，表示该 Uri 用于 ContentProvider 定位资源，类似于网址前缀 http://。
- <authorities>：用于唯一标识一个 ContentProvider，外部调用者可以根据此标识访问该 ContentProvider。通常可将 authority 设置为包名和类名的全称，以保证唯一性。
- <data_path>：资源部分，可用来表示待操作的数据集。
- <id>：数据编号，用来唯一确定数据集中的一条记录，以匹配数据集中_ID 字段的值。这个部分是可以省略的。

通常利用字符串来定义 Uri，如“content://com.mialab.contentprovider/data”，为了将字符串转化为 Uri，可以调用 Uri 工具类的 parse()方法进行转换，其代码如下：

```
Uri uri = Uri.parse("content://com.mialab.contentprovider/data");
```

Uri 类中的常用操作方法如表 6-6 所示。

表 6-6　Uri 类中的常用操作方法

返回类型	方法名	描　述
Uri	parse(String uriString)	将字符串转换为 Uri 对象
Uri	withAppendedPath(Uri baseUri, String pathSegment)	baseUri 再添加 pathSegment 字符串，然后转换为 Uri 对象
Uri	fromFile(File file)	从指定的文件获取 Uri 对象

6.4.3 UriMatcher 和 ContentUris

1．UriMatcher

为了确定该 ContentProvider 实际能处理的 Uri，以及确定每个方法中 Uri 参数所操作的数据，Android 系统提供了 UriMatcher 工具类。UriMatcher 工具类提供了以下两种方法。

（1）void addUri(String authority, String path, int code)：用于向 UriMatcher 对象注册 Uri，其中 authority 和 path 组成一个 Uri，code 代表该 Uri 对应的匹配码。

（2）int match(Uri uri)：注册 Uri 与传入的 Uri 匹配，如果匹配成功，则返回相应的 code（匹配码）；如果匹配失败，则返回-1。

首先注册所有需要匹配的 Uri 路径：

```
UriMatcher uriMatcher = new UriMatcher(UriMatcher.NO_MATCH);
uriMatcher.addURI("com.mialab.provider.contentprovider","data_table",1);
uriMatcher.addURI("com.mialab.provider.contentprovider","data_table/#",2);
```

其中，常量 UriMatcher.NO_MATCH 表示不匹配任何路径的返回码（-1）。

注册完需要匹配的 Uri 后，就可以使用 uriMatcher.match(uri)方法对输入的 Uri 进行匹配，如果匹配成功则返回匹配码。

如果使用 match()方法匹配 content://com.mialab.provider.contentprovider/data_table 路径，则返回匹配码为 1。

“#”号为通配符，如果使用 match()方法匹配 content://com.mialab.provider.contentprovider/data_table/8 路径，则返回匹配码为 2。

2．ContentUris

除此以外，Android 还提供一个 ContentUris 工具类，用于操作 Uri 字符串，它提供了以下两种方法。

（1）static Uri withAppendedId (Uri uri, long id)：用于为路径加上 ID 部分。

```
Uri  uri = Uri.parse("content://com.mialab.contentprovider/data");
Uri  newUri = ContentUris.withAppendedId(uri,1);
```

此时 newUri 为 content://com.mialab.contentprovider/data/1。

（2）static long parseId (Uri uri)：用于从指定的 Uri 中解析出所包含的 ID 值。

```
Uri  uri = Uri.parse("content://com.mialab.provider.contentprovider/data/1");
long  dataId = ContentUris.parseId(uri);   //dataId 值为 1
```

6.4.4 ContentResolver

ContentResolver 提供的方法与 ContentProvider 是一一对应的，如 query、insert、update、delete、getType 等方法，连方法的参数类型都一模一样。

扩展 ContentResolver 类需要重写的方法如表 6-7 所示。

表 6-7　扩展 ContentProvider 类需要重写的方法

返回类型	方法名	描述
Uri	insert (Uri uri, ContentValues values)	调用指定 ContentProvider 中的 insert ()方法
Cursor	query (Uri uri, String[] projection, String selection, String[] selectionArgs, String sortOrder)	调用指定 ContentProvider 中的 query ()方法
int	update (Uri uri, ContentValues values, String selection, String[] selectionArgs)	调用指定 ContentProvider 中的 update ()方法

续表

返回类型	方法名	描述
int	delete (Uri uri, String selection, String[] selectionArgs)	调用指定 ContentProvider 中的 delete ()方法
String	getType (Uri uri)	调用指定 ContentProvider 中的 getType ()方法

要获取 ContentResolver 对象，可以使用 Context 提供的 getContentResolver()方法，其示例代码如下：

```
ContentResolvercr = getContentResolver();
```

6.4.5 创建数据提供者

通过继承 ContentProvider 类创建一个数据提供者，创建步骤可分为 3 步：

（1）创建一个继承 ContentProvider 的 Java 类，重载类中的 6 个函数：onCreate()、insert()、delete()、update()、qurey()、getType()。

（2）声明 CONTENT_URI，实现 UriMatcher。

（3）注册 ContentProvider。应用程序必须在配置文件 AndroidManifest.xml 中注册 Contentrovider 后方可使用。注册 ContentProvider 只需在<application/>元素下添加<provider/>子元素即可，其代码如下：

```
<provider  android:name=".BookProvider"
        android:authorities="com.mialab.bookprovider"
        android:enabled="true"
        android:exported="true">
</provider>
```

其中，name 属性值指定 ContentProvider 实现类的类名，authorities 指定 ContentProvider 对应的 Uri，android:exported 指定该 ContentProvider 是否允许其他应用调用。

【示例】 创建数据提供者。启动 Android Studio，在 Ch6 工程中创建 ContentProviderDemo 模块，并在 ContentProviderDemo 模块中创建 ContentProvider 的子类 BookProvider.java，以及用来存储常量的 BookPara.java。

ContentProviderDemo 模块的 BookProvider.java 主要代码如下：

```
public class BookProvider extends ContentProvider {
   private static final String DB_NAME = "book_test.db";
   private static final String DB_TABLE = "bookinfo_2";
   private static final int DB_VERSION = 1;
   private SQLiteDatabase db;
   private BookHelper bookHelper;
   private static final int MULTIPLE_BOOK = 1;
   private static final int SINGLE_BOOK = 2;
   private static final UriMatcher uriMatcher;
   static { //在新构造的ContentProvider类中，经常需要判断Uri是单条数据还是多条数据
            //最简单的方法就是构造一个UriMatcher
       uriMatcher = new UriMatcher(UriMatcher.NO_MATCH);
       uriMatcher.addURI(BookPara.AUTHORITY, BookPara.PATH_MULTIPLE, MULTIPLE_
BOOK);
       uriMatcher.addURI(BookPara.AUTHORITY, BookPara.PATH_SINGLE, SINGLE_BOOK);
   }
   @Override
   public boolean onCreate() {
       Context context = getContext();
       bookHelper = new BookHelper(context, DB_NAME, null, DB_VERSION);
```

```
            db = bookHelper.getWritableDatabase();
            if (db == null)
                return false;
            else
                return true;
        }

        @Override
        public Uri insert(Uri uri, ContentValues values) {
            long id = db.insert(DB_TABLE, null, values);
            if (id > 0) {
                Uri newUri = ContentUris.withAppendedId(BookPara.CONTENT_URI, id);
                getContext().getContentResolver().notifyChange(newUri, null);
                return newUri;
            }
            throw new SQLException("Failed to insert row into " + uri);
        }
        @Override
        public Cursor query(Uri uri, String[] projection, String selection,
                    String[] selectionArgs, String sortOrder) {
            SQLiteQueryBuilder qb = new SQLiteQueryBuilder();
            qb.setTables(DB_TABLE);
            switch (uriMatcher.match(uri)) {
                case SINGLE_BOOK:
                    qb.appendWhere(BookPara.ID + "=" + uri.getPathSegments().get(1));
                    break;
                default:
                    break;
            }
            Cursor cursor = qb.query(db, projection, selection, selectionArgs, null,
null, sortOrder);
            cursor.setNotificationUri(getContext().getContentResolver(), uri);
            return cursor;
        }
        @Override
        public int update(Uri uri, ContentValues values, String selection, String[]
selectionArgs) {
            int count;
            switch (uriMatcher.match(uri)) {
                case MULTIPLE_BOOK:
                    count = db.update(DB_TABLE, values, selection, selectionArgs);
                    break;
                case SINGLE_BOOK:
                    String segment = uri.getPathSegments().get(1);
                    count = db.update(DB_TABLE, values, BookPara.ID + "=" + segment,
selectionArgs);
                    break;
                default:
                    throw new IllegalArgumentException("Unknow URI:" + uri);
            }
            getContext().getContentResolver().notifyChange(uri, null);
            return count;
        }
        @Override
```

```
        public int delete(Uri uri, String selection, String[] selectionArgs) {
            int count = 0;
            switch (uriMatcher.match(uri)) {
                case MULTIPLE_BOOK:
                    count = db.delete(DB_TABLE, selection, selectionArgs);
                    break;
                case SINGLE_BOOK:
                    String segment = uri.getPathSegments().get(1);
                    count  =  db.delete(DB_TABLE,  BookPara.ID  +  "="  +  segment,
selectionArgs);
                    break;
                default:
                    throw new IllegalArgumentException("Unsupported URI:" + uri);
            }
            getContext().getContentResolver().notifyChange(uri, null);
            return count;
        }
        @Override
        public String getType(Uri uri) {
            switch (uriMatcher.match(uri)) {
                case MULTIPLE_BOOK:
                    return BookPara.MINE_TYPE_MULTIPLE;
                case SINGLE_BOOK:
                    return BookPara.MINE_TYPE_SINGLE;
                default:
                    throw new IllegalArgumentException("Unkown uri:" + uri);
            }
        }

        private static class BookHelper extends SQLiteOpenHelper {
            public BookHelper(Context context, String name, SQLiteDatabase.CursorFactory
factory,
                 int version) {
                super(context, name, factory, version);
            }
            private static final String DB_CREATE = "create table " +
                    DB_TABLE + " (" + BookPara.ID + " integer primary key autoincrement,"+
                    BookPara.BOOKNAME + " text not null, " + BookPara.AUTHOR + " text,"
                        + BookPara.BOOKPRICE + " float);";
            @Override
            public void onCreate(SQLiteDatabase _db) {
                _db.execSQL(DB_CREATE);
            }
            @Override
            public void onUpgrade(SQLiteDatabase _db, int _oldVersion, int _newVersion)
{
                _db.execSQL("DROP TABLE IF EXISTS " + DB_TABLE);
                onCreate(_db);
            }
        }
    }
```

暴露ContentProviderDemo模块（或者应用）数据接口的正是BookProvider。

ContentProviderDemo 模块的 BookProvider.java 主要代码如下：

```
public class BookPara {
    public static final String MIME_DIR_PREFIX = "vnd.android.cursor.dir";
    public static final String MIME_ITEM_PREFIX = "vnd.android.cursor.item";
    public static final String MINE_ITEM = "vnd.mialab.book";
    public static final String MINE_TYPE_SINGLE = MIME_ITEM_PREFIX + "/" + MINE_
ITEM;
    public static final String MINE_TYPE_MULTIPLE = MIME_DIR_PREFIX + "/" + MINE_
ITEM;
    public static final String AUTHORITY = "com.mialab.bookprovider";
    public static final String PATH_SINGLE = "book/#";
    public static final String PATH_MULTIPLE = "book";
    public static final String CONTENT_URI_STRING = "content://" + AUTHORITY +
"/" + PATH_MULTIPLE;
    public static final Uri CONTENT_URI = Uri.parse(CONTENT_URI_STRING);
    public static final String ID = "_id";
    public static final String BOOKNAME = "bookName";
    public static final String AUTHOR = "author";
    public static final String BOOKPRICE = "bookPrice";
}
```

值得注意的是，须在配置文件 AndroidManifest.xml 中注册 Contentrovider 后方可使用，注册 ContentProvider 的代码可参见 Ch6\ContentProviderDemo 模块。

6.4.6 使用数据提供者

当一个程序通过 ContentProvider 暴露数据之后，其他应用程序即可通过 ContentResolver 对象共享这些数据。ContentResolver 是通过 Uri 来查询 ContentProvider 中提供的数据的。程序开发者只要知道 Uri 和数据集的数据格式，即可对数据进行操作。获取共享数据的操作步骤如下。

（1）获取 ContentResolver 对象。

调用 getContentResolver()方法获取 ContentResolver 对象。

```
ContentResolver resolver = getContentResolver();
```

（2）操作数据。

与 ContentProvider 类中数据操作（增加、删除、修改、查询）方法对应，ContentResolver 对象也有 4 个方法，分别为 insert()、delete()、update()和 query()，这 4 个方法的参数列表与 ContentProvider 中对应的方法一致。调用任意一个方法都可触发 Uri 对应的 ContentProvider 方法，其示例代码如下：

```
Uri uri = Uri.parse("content://com.mialab.contentprovider/data");
String selection = "id=1";
int result = resolver.delete(uri,selection,null);  //删除 data 中 id 为 1 的数据
```

【示例】 使用数据提供者（这里是“Ch6\ContentProviderDemo 中暴露其数据接口的 BookProvider”），实现对 ContentProviderDemo 模块（应用）中数据的增加、删除、修改、查询。

启动 Android Studio，在 Ch6 工程中创建 ContentResolverDemo 模块，类似于前面所讲的 Ch6\SQLiteDemo 模块，同样在 ContentResolverDemo 模块中创建 MainActivity.java、InsertActivity.java、QueryActivity.java、UpdateActivity.java、DeleteActivity.java 和相应的布局文件 activity_main.xml、activity_insert.xml、activity_query.xml、activity_update.xml、activity_delete.xml，以及存储常量的 BookPara.java。

ContentResolverDemo 模块的运行界面，同 Ch6\SQLiteDemo，可参见图 6-14。

ContentResolverDemo 模块的 InsertActivity.java 主要代码如下：

```
public class InsertActivity extends AppCompatActivity {
    private EditText booknameText;
    private EditText authorText;
```

```
    private EditText bookpriceText;
    private TextView labelView;
    private ContentResolver resolver;
    @Override
    protected void onCreate(Bundle savedInstanceState) {
        super.onCreate(savedInstanceState);
        setContentView(R.layout.activity_insert);
        booknameText = (EditText) findViewById(R.id.et_bookname);
        authorText = (EditText) findViewById(R.id.et_author);
        bookpriceText = (EditText) findViewById(R.id.et_bookprice);
        labelView = (TextView) findViewById(R.id.label);
        Button addButton = (Button) findViewById(R.id.btn_save);
        addButton.setOnClickListener(addButtonListener);
        resolver = this.getContentResolver();
    }

    View.OnClickListener addButtonListener = new View.OnClickListener() {
        @Override
        public void onClick(View v) {
            String bookName = booknameText.getText().toString();
            String author = authorText.getText().toString();
            String priceStr = bookpriceText.getText().toString();
            if (TextUtils.isEmpty(bookName)) {
                showToast("请先填写图书名称");
                return;
            } else if (TextUtils.isEmpty(author)) {
                showToast("请先填写图书作者");
                return;
            } else if (TextUtils.isEmpty(priceStr)) {
                showToast("请先填写图书价格");
                return;
            }
            Float bookPrice = null;
            try {
                bookPrice = Float.parseFloat(bookpriceText.getText().toString());
            } catch (NumberFormatException e) {
                showToast("数据格式不对！");
                bookpriceText.setText("");
                e.printStackTrace();
                return;
            }
            ContentValues values = new ContentValues();
            values.put(BookPara.BOOKNAME, bookName);
            values.put(BookPara.AUTHOR, author);
            values.put(BookPara.BOOKPRICE, bookPrice);
            Uri newUri = null;
            try {
                newUri = resolver.insert(BookPara.CONTENT_URI, values);
                labelView.setText("成功添加数据，URI: " + newUri);
            } catch (Exception e) {
                labelView.setText("数据添加失败");
                e.printStackTrace();
            }
        }
```

```
    };
    private void showToast(String desc) {
        Toast.makeText(this, desc, Toast.LENGTH_LONG).show();
    }
}
```

ContentResolverDemo 模块的 QueryActivity.java 主要代码如下：

```
public class QueryActivity extends AppCompatActivity {
    private EditText idQuery;
    private TextView labelView;
    private TextView displayView;
    private ContentResolver resolver;

    @Override
    protected void onCreate(Bundle savedInstanceState) {
        super.onCreate(savedInstanceState);
        setContentView(R.layout.activity_query);
        idQuery = (EditText)findViewById(R.id.id_query);
        labelView = (TextView)findViewById(R.id.label);
        displayView = (TextView)findViewById(R.id.display);
        Button queryButton = (Button)findViewById(R.id.query);
        queryButton.setOnClickListener(queryButtonListener);
        resolver = this.getContentResolver();
        displayAll();
    }

    private void displayAll() {
        Cursor cursor = resolver.query(BookPara.CONTENT_URI,
                new String[] { BookPara.ID, BookPara.BOOKNAME, BookPara.AUTHOR,
                    BookPara.BOOKPRICE}, null, null, null);
        if (cursor == null){
            labelView.setText("数据库中没有数据");
            return;
        }
        labelView.setText("数据库：" + String.valueOf(cursor.getCount()) + "条记录");
        String msg = "";
        if (cursor.moveToFirst()){
            do{
                msg += "ID：" + cursor.getInt(cursor.getColumnIndex(BookPara.ID)) +
"，";
                msg += "图书名称：" +
                cursor.getString(cursor.getColumnIndex(BookPara.BOOKNAME))+ "，";
                msg += "作者：" + cursor.getString(cursor.getColumnIndex(BookPara.
AUTHOR)) + "， ";
                msg += "图书价格：" +
                cursor.getFloat(cursor.getColumnIndex(BookPara.BOOKPRICE)) + "\n";
            }while(cursor.moveToNext());
        }
        displayView.setText(msg);
        displayView.setTextColor(Color.BLUE);
    }

    View.OnClickListener queryButtonListener = new View.OnClickListener() {
        @Override
        public void onClick(View v) {
```

```
            String idStr = idQuery.getText().toString();
            try {
                int inputId = Integer.parseInt(idStr);
                if((inputId < 0)||(inputId == 0)){
                    showToast("输入 ID 须为大于 0 的整数！");
                    idQuery.setText("");
                    return;
                }
            } catch (NumberFormatException e) {
                showToast("输入 ID 数据格式不对！");
                idQuery.setText("");
                e.printStackTrace();
                return;
            }
            Uri uri = Uri.parse(BookPara.CONTENT_URI_STRING + "/" + idStr);
            Cursor cursor = resolver.query(uri, new String[] {BookPara.ID,
                BookPara.BOOKNAME, BookPara.AUTHOR, BookPara.BOOKPRICE},
            null, null, null);
            if (cursor == null){
                labelView.setText("数据库中没有数据");
                return;
            }
            String msg = "";
            if (cursor.moveToFirst()){
               msg += "ID: " + cursor.getInt(cursor.getColumnIndex(BookPara.ID)) +
", ";
               msg += "图书名称: " +
               cursor.getString(cursor.getColumnIndex(BookPara.BOOKNAME))+ ", ";
               msg += "作者: " + cursor.getString(cursor.getColumnIndex(BookPara.
AUTHOR)) + ", ";
               msg += "图书价格: " + cursor.getFloat(cursor.getColumnIndex(BookPara.
BOOKPRICE))
               + "\n";
            }
            labelView.setText("数据库: ");
            labelView.setTextColor(Color.RED);
            displayView.setText(msg);
            displayView.setTextColor(Color.BLUE);
        }
    };
    private void showToast(String desc) {
        Toast.makeText(this, desc, Toast.LENGTH_LONG).show();
    }
}
```

ContentResolverDemo 模块的 DeleteActivity.java 主要代码如下：

```
public class DeleteActivity extends AppCompatActivity {
    private EditText idDelete;
    private TextView labelView;
    private TextView displayView;
    private ContentResolver resolver;
    @Override
    protected void onCreate(Bundle savedInstanceState) {
        super.onCreate(savedInstanceState);
        setContentView(R.layout.activity_delete);
```

```
        idDelete = (EditText) findViewById(R.id.id_delete);
        labelView = (TextView) findViewById(R.id.label);
        displayView = (TextView) findViewById(R.id.display);
        Button deleteButton = (Button) findViewById(R.id.btn_delete);
        deleteButton.setOnClickListener(deleteButtonListener);
        resolver = this.getContentResolver();
    }
    View.OnClickListener deleteButtonListener = new View.OnClickListener() {
        @Override
        public void onClick(View v) {
            String idStr = idDelete.getText().toString();
            if (TextUtils.isEmpty(idStr)) {
                showToast("ID不能为空!");
                return;
            }
            try {
                int inputId = Integer.parseInt(idStr);
                if ((inputId < 0) || (inputId == 0)) {
                    showToast("输入ID须为大于0的整数! ");
                    idDelete.setText("");
                    return;
                }
            } catch (NumberFormatException e) {
                showToast("ID数据格式不对! ");
                idDelete.setText("");
                e.printStackTrace();
                return;
            }
            Uri uri = Uri.parse(BookPara.CONTENT_URI_STRING + "/" + idStr);
            int result = resolver.delete(uri, null, null);
            String msg = "删除 ID 为" + idStr + "的数据" + (result > 0 ? "成功" :
"失败");
            labelView.setText(msg);
            labelView.setTextColor(Color.BLUE);
        }
    };
    private void showToast(String desc) {
        Toast.makeText(this, desc, Toast.LENGTH_LONG).show();
    }
}
```

限于篇幅，其余文件代码参见教学资源包中的ContentResolverDemo模块，在这里不再赘述。

6.5 相关阅读：Android系统中内部存储和外部存储的若干疑问

以下是一些教学过程中提出的常见疑问。

疑问1：getFilesDir().getAbsolutePath()和getCacheDir().getAbsolutePath()有什么区别？

解答：getFilesDir()获取的是files目录，getCacheDir()获取的是cache目录，它们位于同一级目录，只是为了存放不同类型的数据。如图6-22所示，由文件名不难看出：cache中存放缓存数据，databases中存放使用SQLite存储的数据，files中存放普通数据（log数据、json型数据等），shared_prefs中存放使用SharedPreference的数据。这些文件夹都可由系统自动创建。假定Ch6\SharedPreferencesDemo

模块在代码中创建了数据库，并实现了读/写内部存储的文件，则在/data/data/<packagename>下会自动生成 shared_prefs 文件夹、databases 文件夹和文件夹目录，这里的<packagename>就是 com.mialab.sharedpreferencesdemo。

Device File Explorer

Emulator Galaxy_Nexus_API_28 Android 9, API 28

Name	Permissions	Date	Size
▼ com.mialab.sharedpreferencesdemo	drwx------	2019-08-08 16:29	4 KB
▶ cache	drwxrws--x	2019-08-03 03:15	4 KB
▶ code_cache	drwxrws--x	2019-08-03 03:15	4 KB
▶ databases	drwxrwxrwx	2019-08-08 16:23	4 KB
▶ files	drwxrwxrwx	2019-08-08 16:21	4 KB
▶ shared_prefs	drwxrwx--x	2019-08-03 14:18	4 KB
test.db	-rw-------	2019-08-08 16:29	16 KB

图 6-22 “/data/data/<packagename>”中的 shared_prefs 文件夹、databases 文件夹和文件夹目录

疑问 2：getFilesDir().getAbsolutePath()和 getExternalFilesDir().getAbsolutePath()有什么区别呢？

解答：先看它们的路径。

（1）getFilesDir().getAbsolutePath()的值为：/data/user/0/< packagename > /files。

（2）getExternalFilesDir().getAbsolutePath()值为：/storage/emulated/0/Android/data/packname/files。

很显然，前者是内部存储，后者是外部存储。

它们的共同点是路径都带有包名，表明是这个 App 的“专属文件”，这类文件应该是随着 App 卸载而一起被删除的，并且在 Android 系统的“设置”里清除该应用的数据时，这两个文件夹下的数据都会被清除。

疑问 3：什么是 App“专属文件”？

解答：在上面疑问 2 中，提到了“专属文件”，所谓“专属文件”就是指该文件是属于某个具体应用的，其文件路径都带有相应的包名，当 App 卸载时，它们会随应用一起删除。当在“设置”里面手动清除某个应用数据时（不是清除缓存），它们也会一起被清除掉。Android 系统使用这种“专属文件”的目的，一个是为了方便文件管理，避免文件随意存储；另一个是为了当应用被卸载时不会留下很多垃圾文件。

疑问 4：既然内部存储与外部存储都有 App“专属文件”，那么应该使用哪个呢？

解答：很显然应该使用外部存储。因为内部存储本身就比较小，而且已经存储了一些系统的文件。因此，内部存储尽量不要去使用。一般程序员会判断是否有外部存储，如果没有，再使用内部存储，其代码如下：

```
public static String getFilePath(Context context, String dir) {
    String  directoryPath = "";
    if (MEDIA_MOUNTED.equals(Environment.getExternalStorageState()) ) {//判断外部存储是否可用
        directoryPath  = context.getExternalFilesDir(dir).getAbsolutePath();
    }else{                         //没有外部存储就使用内部存储
        directoryPath=context.getFilesDir() + File.separator + dir;
    }
    File file = new File(directoryPath);
        if(!file.exists()){        //判断文件目录是否存在
        file.mkdirs();
    }
    return directoryPath;
}
```

6.6 本章小结

本章主要介绍了如何对轻量级数据实现存储与访问。根据数据量的规模和使用需求，介绍了三种存储方式：使用 SharedPreferences、使用文件和使用 SQLite 数据库。SharedPreferences 是一种轻量级 XML 文档，可以用于存储少量的数据，应用程序可以读取自身的 SharedPreferences 文件，还可以访问其他应用程序的 SharedPreferences 文件。借助于 Java 的 I/O 体系，应用程序可以读/写系统中的文件，也可以读/写外部存储器中的文件。SQLite 数据库是 Android 内置的数据库，用户可以通过 SQLiteDataBase 访问 SQLite 数据库，对数据进行增加、删除、修改和查询。考虑到有些应用需要访问其他应用程序的数据，本章还介绍了利用 ContentProvider 和 ContentResolver 实现数据共享。

这里要注意的是：Android 使用 getWritableDatabase()方法和 getReadableDatabase()方法都可以获取一个用于操作数据库的 SQLiteDatabase 实例。getWritableDatabase()方法取得的实例不仅具有写的功能，而且同时具有读和写的功能。同样，getReadableDatabase()方法取得的实例也具有对数据库进行读和写的功能。

两者的区别在于：getWritableDatabase()取得的实例是以读/写的方式打开数据库，如果打开的数据库磁盘满了，则只能读不能写，此时如果调用了 getWritableDatabase()的实例，那么将会发生错误（异常）。getReadableDatabase 取得的实例是先调用 getWritableDatabase 以读/写的方式打开数据库，如果数据库的磁盘满了，则返回打开失败，继而用 getReadableDatabase 的实例以只读的方式打开数据库。

ContentProvider 共享数据是通过定义一个对外开放的统一接口来实现的。然而，应用程序并不直接调用这些方法，而是使用一个 ContentResolver 对象调用它的方法作为替代。ContentResolver 可以与任意内容提供者进行会话，与其合作来对所有相关交互通信进行管理。当外部应用需要对 ContentProvider 中的数据进行增加、删除、修改和查询操作时，可以使用 ContentResolver 类来完成，要获取 ContentResolver 对象，可以使用 Context 提供的 getContentResolver()方法“ContentResolvercr = getContentResolver();”。

ContentProvider 可以向其他应用程序提供数据，与之对应的 ContentResolver 则负责获取 ContentProvider 提供的数据。ContentProvider 类提供了修改、添加、删除、更新数据等方法，ContentResolver 类也提供了与 ContentProvider 类相对应的方法。

习 题 6

1．Android 的数据存储方式有哪些？各有什么特点？

2．使用 SharedPreferences，实现登录页面的“记住密码”功能。

3．使用数据库 SQLite，实现登录页面的“记住密码”功能。

4．使用文件存储的“内部存储”方式，实现登录页面的“记住密码”功能。

5．使用文件存储的“外部存储”方式，实现登录页面的“记住密码”功能。

6．手动创建 SQLite 数据库 contact.db，在 contact.db 中创建数据表 userinfo，以及数据表 userinfo 中的字段，如表 6-8 所示。

表 6-8　userinfo 数据表

字 段 名	数 据 类 型	描　　述
_id	integer	主键，自动增长
username	varchar	用户名
userphone	varchar	用户手机号
address	varchar	用户地址
E_mail	varchar	电子邮箱

7．手动实现对 userinfo 数据表中记录的增加、删除、修改和查询。

8．开发一款“通讯录”App，编程实现：对 userinfo 数据表中记录的增加、删除、修改和查询，其中的“查询”页面能够显示所有联系人的手机号（用户名和用户手机号），输入用户名能够显示该用户名的手机号。

9．在上述 8 的基础上，编程实现“拨打电话”和“发短信”的功能。

10．使用 ContentProvider 和 ContentResolver 是如何实现数据共享的？

11．如何访问系统 ContentProvider？试编程举例加以说明。

12．利用本章所学的知识，尝试开发一款 App，实现“购物车”功能。

13．Android 的 ORM 数据库框架有哪些？可上网搜寻相关资料。

14．推荐一个你喜欢的 Android 数据库框架，并编程举例说明它是如何使用的？如 GreenDao 或者 LitePal 等，其他的数据库框架也可以。

第二部分

提　高　篇

第7章 Java并发编程

本章导读

Android 的应用程序支持多线程，多线程编程为充分利用系统资源提供了便利，同时也为设计复杂 UI 和耗时操作提供了途径，提升了安卓用户的使用体验。线程池的基本思想是开辟一块内存空间，里面存放了众多（未销毁）的线程，池中线程执行调度由池管理器来处理。当有线程任务时，从池中取一个，执行完就后线程对象归池，这样可以避免反复创建线程对象所带来的性能开销，节省了系统的资源。本章主要知识点有：（1）Java 线程池简介；（2）Executor 与 ExecutorService;（3）ThreadPoolExecutor;（4）Future 模式；（5）使用线程池实现网络多图片下载。

7.1 Java线程池简介

如何理解线程池呢？简单地说（可能并不十分准确，但能够帮助理解），线程池是一个装着“线程”的池子，它包含了很多已经启动且处于睡眠状态的线程。当用户有请求时，就会直接使用池子里面的线程而不用去创建。对于请求数量较少的情况没有多少作用，但是当系统请求数量很多的时候就会为系统节约大量的资源，避免系统总是忙于线程的创建和销毁，而能够让系统更好地完成其他功能。

根据系统环境情况，可以自动或手动设置线程数量，达到运行的最佳效果。使用线程池控制线程数量，其他线程需要排队等候。当一个任务执行完毕，再从队列中取出最前面的任务开始执行。一个新任务需要运行时，如果线程池中有等待的工作线程，就可以开始运行了，否则进入等待队列。

通过使用线程池，可以减少创建和销毁线程的次数，使每个工作线程都可以被重复利用，以执行多个任务。有时也需要根据系统的承受能力，调整线程池中工作线程的数目，以防止内存消耗过多。

线程的使用在 Java 应用中占有极其重要的地位，在 JDK1.4 及其之前的 JDK 版本中，关于线程池的使用，需要用户自己动手实现。JDK1.5 出现后，并发线程这块发生了根本的变化，最重要的莫过于新的启动、调度、管理线程等一大堆 API 了。在 JDK1.5 以后，通过 Executor 来启动线程比用 Thread 的 start()更好，用户既可以控制线程的启动、执行和关闭过程，还可以使用线程池的特性，这归结于 JDK1.5 之后加入了 java.util.concurrent 包。该包主要介绍 Java 线程及线程池的使用，为用户在开发中处理线程问题提供了极大的帮助。

软件包 java.util.concurrent 包含了并发编程中常用的实用工具类。Executor 是一个简单的标准化接口，用于定义类似于线程的自定义子系统，包括线程池、异步 IO 和轻量级任务框架。ExecutorService 提供了多个完整的异步任务执行框架，它用于管理任务的排队和安排，并允许受控制地关闭。ScheduledExecutorService 子接口及相关的接口添加了对延迟和定期任务执行的支持。ExecutorService 提供了安排异步执行的方法，可执行由 Callable 表示的任何函数，类似于 Runnable。Future 模式能够返回函数的结果，允许确定执行是否完成，并提供取消执行的方法。RunnableFuture 是拥有 run 方法的 Future，run 方法执行时将设置其结果。

类 ThreadPoolExecutor 和 ScheduledThreadPoolExecutor 提供了可调的、灵活的线程池。Executors 类提供了大多数 Executor 的常见类型和配置的工厂方法，以及使用它们的几种实用工具方法。其他基于 Executor 的实用工具还包括具体类 FutureTask，它提供 Future 的常见可扩展实现，以及 ExecutorCompletionService，可有助于协调对异步任务组的处理。

7.2 Executor 与 ExecutorService

7.2.1 Executor

Java 中线程池的顶级接口是 Executor，但是严格意义上讲 Executor 并不是一个线程池，而只是一个执行线程的工具。真正的线程池接口是 ExecutorService。

java.util.concurrent 包中接口 Executor 源码如下。

```
public interface Executor{
    void execute(Runnable command);
}
```

Executor 用来执行已提交的 Runnable 任务对象。通常使用 Executor 而不是显式地创建线程。不过，Executor 接口并没有严格地要求其执行是异步的。在最简单的情况下，执行程序可以在调用者的线程中立即运行已提交的任务：

```
class DirectExecutor implements Executor {
    public void execute(Runnable r) {
        r.run();
    }
}
```

更常见的是，任务是在某个不是调用者线程的线程中执行的。以下执行程序将为每个任务生成一个新线程。

```
class ThreadPerTaskExecutor implements Executor {
    public void execute(Runnable r) {
        new Thread(r).start();
    }
}
```

7.2.2 ExecutorService

ExecutorService 继承了 Executor，ExecuteService 接口介绍如下：

```
public interface ExecutorService extends Executor {
   void shutdown();
   List<Runnable> shutdownNow();
   boolean isShutdown();
   boolean isTerminated();
   boolean awaitTermination(long timeout, TimeUnit unit) throws InterruptedException;
   <T> Future<T> submit(Callable<T> task);
   <T> Future<T> submit(Runnable task, T result);
   Future<?> submit(Runnable task);
   <T> List<Future<T>> invokeAll(Collection<? extends Callable<T>> tasks)
       throws InterruptedException;
   …
}
```

ExecutorService 提供了管理终止的方法，以及可为跟踪一个或多个异步任务执行状况而生成 Future 的方法。如果关闭 ExecutorService，可导致其拒绝新任务。

ExecutorService 提供两个方法来关闭线程池（ExecutorService 的实现类）。

shutdown()方法在终止前允许执行以前提交的任务，而 shutdownNow()方法则阻止等待任务的启动，并试图停止当前正在执行的任务。在终止后，执行程序没有任务在执行，也没有任务在等待执行，并且无法提交新任务。程序员应该关闭未使用的 ExecutorService，以允许回收其资源。

通过创建并返回一个可用于取消执行（或等待完成）的 Future，方法 submit 扩展了基本方法 Executor.execute(java.lang.Runnable)。

7.2.3 常用线程池

Executors 类为创建 ExecutorService 提供了便捷的工厂方法。

要配置一个线程池是比较复杂的，尤其是在对于线程池的原理不是很清楚的情况下，配置的线程池很有可能不是较优的，因此在 Executors 类中提供了一些静态工厂方法，可生成一些常用的线程池。

（1）newCachedThreadPool()：创建一个可缓存的线程池。如果线程池的大小超过了处理任务所需要的线程，那么就会回收部分空闲（60s 不执行任务）的线程，当任务数增加时，此线程池又可以智能地添加新线程来处理任务。此线程池不会对线程池大小进行限制，线程池大小完全依赖于操作系统（或者说 JVM）能够创建的最大线程大小。

（2）newFixedThreadPool()：创建固定大小的线程池。每次提交一个任务就会创建一个线程，直到线程达到线程池的最大容量。线程池的大小一旦达到最大值就会保持不变，如果某个线程因为执行异常而结束，那么线程池就会补充一个新线程。

（3）newSingleThreadExecutor()：创建一个单线程的线程池。这个线程池只有一个线程在工作，也就是相当于单线程串行执行所有任务。如果这个唯一的线程因为异常结束，那么就会有一个新的线程来替代它。此线程池保证了所有任务都能按照提交的顺序执行。

（4）newScheduledThreadPool：创建一个定长线程池，此线程池支持定时，以及周期性执行任务的需求。

以下是常用线程池的举例。

（1）newCachedThreadPool。

CachedThreadPool 模式首先会按照需要创建足够多的线程来执行任务（Task）。随着程序执行的过程，当有的线程执行完了任务，可以被重新循环使用时，才不再创建新的线程来执行任务。相关代码详见教学资源包中第 7 章的 CachedThreadPoolDemo 工程，这是使用 Eclipse 创建的。

首先，任务定义如下（实现了 Runnable 接口，并且重写了 run 方法）：

```
public class LiftOff implements Runnable {
    protected int countDown = 10;      //Default
    private static int taskCount = 0;
    private final int id = taskCount++;
    public LiftOff() {
    }
    public LiftOff(int countDown) {
        this.countDown = countDown;
    }
    public String status() {
        return "#" + id + "(" + (countDown > 0 ? countDown : "LiftOff!") + ") ";
    }
    @Override
    public void run() {
        while (countDown-- > 0) {
            System.out.println(status());
            Thread.yield();
        }
```

```
    }
}
```

采用 CachedThreadPool 方式执行编写的客户端程序如下。

```
public class CachedThreadPool {
    public static void main(String[] args) {
        ExecutorService exec = Executors.newCachedThreadPool();
        for(int i = 0; i < 10; i++) {
            exec.execute(new LiftOff());
        }
        exec.shutdown();
    }
}
```

上述程序中共有 10 个任务，采用了 CachedThreadPool 模式，exec 每遇到一个 LiftOff 的对象（Task），就会创建一个线程来处理任务。现在假设遇到第 4 个任务时，之前用于处理第 1 个任务的线程已经执行完任务，那么此时将不会创建新的线程来处理任务，而是使用之前处理第 1 个任务的线程来处理第 4 个任务。接着如果遇到第 5 个任务时，前面那些任务都还没有执行完，那么就会创建新的线程来执行第 5 个任务。否则，使用之前执行完任务的线程来处理新的任务。

（2）newFixedThreadPool。

FixedThreadPool 模式用于创建一个定长线程池，可控制线程最大并发数，超出最大并发数的线程会在队列中等待。规定数目的线程处理所有任务，一旦有线程处理完了任务就会被用来处理新的任务（如果有的话）。这种模式与上述的 CachedThreadPool 是不同的，CachedThreadPool 模式下处理一定数量任务的线程数目是不确定的，而 FixedThreadPool 模式下处理最多的线程数目是一定的，其示例代码如下。

相关代码参见教学资源包中的 FixedThreadPoolDemo 工程，这是使用 Eclipse 创建的。

```
public class ThreadPoolExecutorTest {
    public static void main(String[] args) {
        ExecutorService fixedThreadPool = Executors.newFixedThreadPool(3);
        for (int i = 0; i < 18; i++) {
            final int index = i;
            fixedThreadPool.execute(new Runnable() {
                public void run() {
                    try {
                        System.out.println(index);
                        Thread.sleep(2000);
                    } catch (InterruptedException e) {
                        e.printStackTrace();
                    }
                }
            });
        }
        fixedThreadPool.shutdown();
    }
}
```

因为线程池大小为 3，每个任务输出 index 后延迟 2s，所以每 2s 打印 3 个数字。定长线程池大小最好根据系统资源进行设置，如 Runtime.getRuntime().availableProcessors()。

（3）newSingleThreadExecutor。

SingleThreadExecutor 模式只会创建一个线程。它与 FixedThreadPool 类似，不过线程数是一个。如果多个任务被提交给 SingleThreadExecutor，那么这些任务就会被保存在一个队列中，并且会按照任务提交的顺序，一个任务先执行完成再执行另外一个任务。

SingleThreadExecutor 模式可以保证只有一个任务会被执行。

SingleThreadExecutor 模式编写的客户端程序如下。

相关代码参见教学资源包中第 7 章的 SingleThreadExecutorDemo 工程，这是使用 Eclipse 创建的。

```
public class SingleThreadExecutor {
    public static void main(String[] args) {
        ExecutorService exec = Executors.newSingleThreadExecutor();
        for (int i = 0; i < 2; i++) {
            exec.execute(new LiftOff());
        }
        exec.shutdown();
    }
}
```

第 1 个任务执行完成之后才开始执行第 2 个任务。

（4）newScheduledThreadPool。

ScheduledThreadPool 模式创建一个定长线程池，支持定时及周期性任务执行。

延迟执行示例代码如下。

相关代码参见教学资源包中 ScheduledThreadPoolDemo 工程，这是使用 Eclipse 创建的。

```
public class ScheduledThreadPool {
    public static void main(String[] args) {
        ScheduledExecutorService            scheduledThreadPool       =       Executors.
newScheduledThreadPool(5);
        scheduledThreadPool.scheduleAtFixedRate(new Runnable() {
            public void run() {
                System.out.println("delay 1 seconds, and excute every 3 seconds");
            }
        }, 1, 3, TimeUnit.SECONDS);
    }
}
```

上述代码表示延迟 1s 后每 3s 执行一次。

7.3 ThreadPoolExecutor

ThreadPoolExecutor 是 ExecutorService 的一个实现类，它执行池线程提交的任务。

线程池可以解决两个不同的问题：由于减少了每个任务调用的开销，它们通常可以在执行大量异步任务时提供增强的性能，并且还可以提供绑定和管理资源（包括执行任务集时使用的线程）的方法。

强烈建议程序员使用较为方便的 Excutors 工厂方法 Executors.newCachedThreadPool()（无界线程池，可以进行自动线程回收）、Executors.newFixedThreadPool(int)（固定大小线程池）和 Executors.newSingleThreadExecutor()（单个后台线程），它们均为大多数使用场景预定义了设置。

7.3.1 ThreadPoolExecutor 的构造方法

ThreadPoolExecutor 构造方法摘要如表 7-1 所示。

表 7-1　ThreadPoolExecutor 构造方法摘要

构造方法摘要
ThreadPoolExecutor(int corePoolSize, int maximumPoolSize, long keepAliveTime, TimeUnit unit, BlockingQueue<Runnable> workQueue) 说明：用给定的初始参数和默认的线程工厂及被拒绝的执行处理程序创建新的 ThreadPoolExecutor
ThreadPoolExecutor(int corePoolSize, int maximumPoolSize, long keepAliveTime, TimeUnit unit, BlockingQueue<Runnable> workQueue, RejectedExecutionHandler handler) 说明：用给定的初始参数和默认的线程工厂创建新的 ThreadPoolExecutor
ThreadPoolExecutor(int corePoolSize, int maximumPoolSize, long keepAliveTime, TimeUnit unit, BlockingQueue<Runnable> workQueue, ThreadFactory threadFactory) 说明：用给定的初始参数和默认被拒绝的执行处理程序创建新的 ThreadPoolExecutor
ThreadPoolExecutor(int corePoolSize, int maximumPoolSize, long keepAliveTime, TimeUnit unit, BlockingQueue<Runnable> workQueue, ThreadFactory threadFactory, RejectedExecutionHandler handler) 说明：用给定的初始参数创建新的 ThreadPoolExecutor

在 JDK 帮助文档中对 ThreadPoolExecutor 构造方法的参数是这样说明的。

- corePoolSize：线程池中所保存的线程数，包括空闲线程。
- maximumPoolSize：线程池中允许的最大线程数。
- keepAliveTime：当线程数大于线程池中核心数目（corePoolSize）时，此为终止前多余的空闲线程等待新任务的最长时间。
- unit：参数的时间单位。
- workQueue：执行前用于保持任务的队列。此队列仅保持由 Execute 方法提交的 Runnable 任务。
- threadFactory：执行程序创建新线程时使用的工厂。
- Handler：由于超出线程范围和队列容量使执行被阻塞时，所使用的处理程序。

关于 Executors 类提供的几种常用线程池的创建，从以下源码可以看到 ThreadPoolExecutor 是 Executors 类的底层实现。

```
    public class Executors {
      public static ExecutorService newFixedThreadPool(int nThreads) {
          return new ThreadPoolExecutor(nThreads, nThreads, 0L, TimeUnit. MILLISECONDS,
                      new LinkedBlockingQueue<Runnable>());
      }
      public static ExecutorService newSingleThreadExecutor() {
          return new FinalizableDelegatedExecutorService(new ThreadPoolExecutor(1,
1,  0L,
                     TimeUnit.MILLISECONDS, new LinkedBlockingQueue<Runnable>()));
      }
       public static ExecutorService newCachedThreadPool() {
          return new ThreadPoolExecutor(0, Integer.MAX_VALUE, 60L, TimeUnit. SECONDS,
                     new SynchronousQueue<Runnable>());
        }
        …
    }
```

对上述代码的说明如下：

（1）关于核心和最大池的大小：ThreadPoolExecutor 将根据 corePoolSize 设置的边界自动调整池的大小。当新任务在方法 execute(java.lang.Runnable)中提交时，如果运行的线程少于 corePoolSize，则会创建新线程来处理请求，即使其他辅助线程是空闲的。如果运行的线程多于 corePoolSize 而少于 maximumPoolSize，则仅当队列满时才创建新线程。如果设置的 corePoolSize 和 maximumPoolSize 相同，则可创建固定大小的线程池。如果将 maximumPoolSize 设置为基本的无界值（如 Integer.MAX_VALUE），则允许线程池适应任意数量的并发任务。在大多数情况下，核心和最大池大小仅基于 ThreadPoolExecutor 的构造函数来设置，不过也可以使用 setCorePoolSize(int)和 setMaximumPoolSize(int) 进行动态更改。

（2）关于保持活动时间：如果线程池中当前有多于 corePoolSize 的线程，则这些多出的线程在空闲时间超过 keepAliveTime 时将会终止，这提供了当线程池处于非活动状态时，可减少资源消耗的方法。

（3）关于排队：所有 BlockingQueue 都可用于传输和保持提交的任务，使用此队列与线程池的大小进行交互。如果运行的线程少于 corePoolSize，则 Executor 始终首选添加新的线程，而不进行排队。如果运行的线程等于或多于 corePoolSize，则 Executor 始终首选将请求加入队列，而不添加新的线程。如果无法将请求加入队列，则创建新的线程，除非创建此线程已超出 maximumPoolSize，在这种情况下，任务将被拒绝。

对于排队有 3 种通用策略。

① 直接提交。

工作队列的默认选项是 SynchronousQueue，它将任务直接提交给线程而不用保持。在此，如果不存在可用于立即运行任务的线程，而试图把任务加入队列则会导致失败，因此要构造一个新的线程。但是由于该 Queue 本身的特性，在某次添加元素后必须等待其他线程取走后才能继续添加。在这里不是核心线程便是新创建的线程。当直接提交时通常要求无界 maximumPoolSizes 以避免拒绝新提交的任务。

② 无界队列。

无界队列即 LinkedBlockingQueue。如 newFixedThreadPool，根据前文提到的规则：当运行的线程少于 corePoolSize 时，Executor 会始终首选添加新的线程，而不进行排队。那么当任务继续增加时，会发生什么情况呢？如果运行的线程等于或多于 corePoolSize，则 Executor 始终首选将请求加入队列，而不添加新的线程。这样，任务就加入队列之中了，那么什么时候才会添加新线程呢？

如果无法将请求加入队列，就会创建新的线程，除非创建此线程超出 maximumPoolSize，在这种情况下，任务将被拒绝。这就很有意思了，可能会出现无法加入队列的情况吗？它不像 SynchronousQueue 那样有其自身的特点，对于无界队列来说，总是可以将请求加入队列的（资源耗尽要另当别论）。换句说，永远也不会触发产生新的线程，corePoolSize 会一直运行，执行完当前的，就会从队列中取出任务开始运行。因此要防止任务疯长，如任务运行的时间比较长，而添加任务的速度远远超过处理任务的时间，而且还不断增加，不久就会“爆”了。

使用无界队列将导致在所有 corePoolSize 线程都忙时，新任务在队列中等待。这样，创建的线程就不会超过 corePoolSize（因此 maximumPoolSize 的值也就无效了）。当每个任务完全独立于其他任务时，适合使用无界队列。

③ 有界队列。

当使用有限的 maximumPoolSizes 时，有界队列（如 ArrayBlockingQueue）有助于防止资源耗尽，但是可能较难调整和控制。队列大小和最大线程池的大小可能需要相互折中：使用大型队列和小型线程池可以最大限度地降低 CPU 使用率、操作系统资源和上下文切换的开销，但是可能导致人工降低吞吐量。如果任务频繁阻塞（如 I/O 边界），则系统可能为超过用户许可的更多线程安排时间。使用小型队列通常要求较大的线程池，CPU 使用率较高，但是可能遇到不可接受的调度开销，这样也会降低吞吐量。该使用方法较复杂，JDK 不推荐使用。

（4）关于被拒绝的任务：当 Executor 已经关闭，并且 Executor 将有界用于最大线程和工作队列容量，且已经饱和时，在方法 execute(java.lang.Runnable)中提交的新任务将被拒绝。在以上两种情况下，Execute 方法都将调用其 RejectedExecutionHandler 的 RejectedExecutionHandler.rejectedExecution (java.lang.Runnable, java.util.concurrent.ThreadPool Executor) 方法。

7.3.2 编制 ThreadPoolExecutor

在同等数量级的操作下，使用线程池的效率要远远高于单线程。线程池可以降低创建线程带来的开销，而线程池中的线程结束后进行的是回收操作并不会真正将线程销毁。在这个过程中，线程池带来的内存消耗肯定会大于单线程。在使用线程池的时候要慎重对待这个问题。

下面是一个测试，以证明使用线程池的效率要远远高于单线程。

相关代码参见教学资源包中第 7 章的 ThreadpoolDemo1 工程，这是使用 Eclipse 创建的。

```
public class ThreadpoolDemo1 {
    public static void main(String[] args) {
        useThreadPool(20000);
        useOneThread(20000);
    }
    static void useThreadPool(int count) {
        final List<Integer> list = new LinkedList<Integer>();      //定义存储集合
        long startTime = System.currentTimeMillis();
        ThreadPoolExecutor tpe = new ThreadPoolExecutor(1, 1, 60, TimeUnit. SECONDS,
                new LinkedBlockingQueue<Runnable>(count));
        final Random random = new Random();                              //产生随机数
        for (int i = 0; i < count; i++) {
            tpe.execute(new Runnable() {
                @Override
                public void run() {
                    list.add(random.nextInt());
                }
            });
        }
        tpe.shutdown();
        try {
            tpe.awaitTermination(1, TimeUnit.DAYS);          //设置最长等待时间为 1 天
        //用于等待子线程结束，再继续执行下面的代码。
        } catch (InterruptedException e) {
            e.printStackTrace();
        }
        System.out.println(System.currentTimeMillis() - startTime);
        System.out.println(list.size());
    }
    static void useOneThread(int count) {
        final List<Integer> list = new LinkedList<Integer>();
        long startTime = System.currentTimeMillis();
        final Random random = new Random();                        //产生随机数
        for (int i = 0; i < count; i++) {
            Thread thread = new Thread() {
                public void run() {
                    list.add(random.nextInt());
                }
            };
            thread.start();
            try {
                thread.join();
            } catch (InterruptedException e) {
                e.printStackTrace();
```

```
            }
        }
        System.out.println(System.currentTimeMillis() - startTime);
        System.out.println(list.size());
    }
}
```

其执行结果如下。

```
38
20000
2514
20000
```

可以看到在 count 参数为 20 000 时，使用线程池会比单独创建线程的速度快很多倍。

此处再举一个使用线程池提交任务被拒绝的例子，ThreadpoolDemo2 工程（源码）目录如图 7-1 所示。

相关代码详见教学资源包第 7 章的 ThreadpoolDemo2 工程，这是使用 Eclipse 创建的。

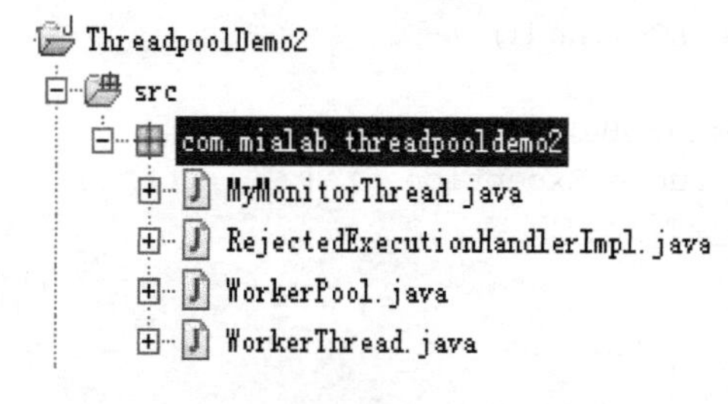

图 7-1　ThreadpoolDemo2 工程（源码）目录

程序的入口是 WorkerPool 的 main 方法。在初始化 ThreadPoolExecutor 时，保持初始池大小为 2，最大池大小为 4，而工作队列大小为 2。因此如果已经有 4 个正在执行的任务而此时又分配了更多任务的话，工作队列将仅仅保留新任务中的两个，其他的将会被 RejectedExecutionHandlerImpl 处理。

WorkerPool.java 的主要代码如下：

```
public class WorkerPool {
    public static void main(String args[]) throws InterruptedException {
        //创建 RejectedExecutionHandler 接口对象
        //RejectedExecutionHandlerImpl 类是用户自定义 RejectedExecutionHandler 接口的
实现
        RejectedExecutionHandlerImpl          rejectionHandler      =      new
RejectedExecutionHandlerImpl();
        ThreadFactory threadFactory = Executors.defaultThreadFactory();
        //创建线程池
       ThreadPoolExecutor executorPool = new ThreadPoolExecutor(2, 4, 10, TimeUnit.
SECONDS,
            new ArrayBlockingQueue<Runnable>(2), threadFactory, rejectionHandler);
        MyMonitorThread monitor = new MyMonitorThread(executorPool, 3);
        //创建监测线程
        Thread monitorThread = new Thread(monitor);
        monitorThread.start();              //启动监测线程
        for (int i = 0; i < 10; i++) {      //提交任务给线程池执行
            executorPool.execute(new WorkerThread("cmd" + i));
        }
        Thread.sleep(30000);
        executorPool.shutdown();            //关闭线程池
        Thread.sleep(5000);
        monitor.shutdown();                 //关闭监测线程
```

```
        }
    }
```

WorkerThread 是实现 Runnable 接口将被执行的任务。

WorkerThread.java 的主要代码如下：

```
public class WorkerThread implements Runnable {
    private String command;
    public WorkerThread(String s) {
            this.command = s;
        }
    @Override
    public void run() {
        System.out.println(Thread.currentThread().getName() + " Start. Command =
" + command);
        processCommand();
        System.out.println(Thread.currentThread().getName() + " End.");
    }
    private void processCommand() {
        try {
            Thread.sleep(5000);
        } catch (InterruptedException e) {
            e.printStackTrace();
        }
    }
    @Override
    public String toString() {
        return this.command;
    }
}
```

ThreadPoolExecutor 提供了一些方法，可以使用这些方法来查询线程池的当前状态、线程池的大小，以及线程池中活动线程的数量、任务数量。

在这里提供了监控线程 MyMonitorThread 对执行任务的线程池进行监控，在特定的时间间隔内打印线程池的相关信息。MyMonitorThread.java 的主要代码如下：

```
public class MyMonitorThread implements Runnable {
    private ThreadPoolExecutor executor;
    private int seconds;
    private boolean run = true;
    public MyMonitorThread(ThreadPoolExecutor executor, int delay) {
        this.executor = executor;
        this.seconds = delay;
    }
    public void shutdown() {
        this.run = false;
    }
    @Override
    public void run() {
        while (run) {
            System.out.println(String.format("[monitor] [%d/%d] Active: %d,
                    Completed: %d, Task: %d, isShutdown: %s, isTerminated: %s",
                                        this.executor.getPoolSize(),
                                        this.executor.getCorePoolSize(),
                                        this.executor.getActiveCount(),
                                        this.executor.getCompletedTaskCount(),
```

```
                                        this.executor.getTaskCount(),
                                        this.executor.isShutdown(),
                                        this.executor.isTerminated()));
            try {
                Thread.sleep(seconds * 1000);
            } catch (InterruptedException e) {
                e.printStackTrace();
            }
        }
    }
}
```

RejectedExecutionHandlerImpl.java 是用户自定义 RejectedExecutionHandler 接口的实现。RejectedExecutionHandlerImpl.java 的主要代码如下：

```
public class RejectedExecutionHandlerImpl implements RejectedExecutionHandler {
    @Override
    public void rejectedExecution(Runnable r, ThreadPoolExecutor executor) {
        System.out.println(r.toString() + " is rejected");
    }
}
```

其执行结果如下。

```
pool-1-thread-1 Start. Command = cmd0
pool-1-thread-3 Start. Command = cmd4
cmd6 is rejected
cmd7 is rejected
cmd8 is rejected
cmd9 is rejected
pool-1-thread-2 Start. Command = cmd1
pool-1-thread-4 Start. Command = cmd5
[monitor] [0/2] Active: 0, Completed: 0, Task: 6, isShutdown: false, isTerminated: false
[monitor] [4/2] Active: 4, Completed: 0, Task: 6, isShutdown: false, isTerminated: false
pool-1-thread-1 End.
pool-1-thread-1 Start. Command = cmd2
pool-1-thread-3 End.
pool-1-thread-3 Start. Command = cmd3
pool-1-thread-2 End.
pool-1-thread-4 End.
[monitor] [4/2] Active: 2, Completed: 4, Task: 6, isShutdown: false, isTerminated: false
[monitor] [4/2] Active: 2, Completed: 4, Task: 6, isShutdown: false, isTerminated: false
pool-1-thread-1 End.
pool-1-thread-3 End.
[monitor] [4/2] Active: 0, Completed: 6, Task: 6, isShutdown: false, isTerminated: false
[monitor] [2/2] Active: 0, Completed: 6, Task: 6, isShutdown: false, isTerminated: false
[monitor] [2/2] Active: 0, Completed: 6, Task: 6, isShutdown: false, isTerminated: false
[monitor] [2/2] Active: 0, Completed: 6, Task: 6, isShutdown: false, isTerminated: false
```

```
    [monitor] [2/2] Active: 0, Completed: 6, Task: 6, isShutdown: false, isTerminated: 
false
    [monitor] [2/2] Active: 0, Completed: 6, Task: 6, isShutdown: false, isTerminated: 
false
    [monitor] [0/2] Active: 0, Completed: 6, Task: 6, isShutdown: true, isTerminated: 
true
    [monitor] [0/2] Active: 0, Completed: 6, Task: 6, isShutdown: true, isTerminated: 
true
```

7.4　Future 模式

使用过 Java 并发包的读者或许对 Future（interface）已经比较熟悉了，其实 Future 本身就是一种被广泛运用的并发设计模式，可在很大程度上简化需要数据流同步的并发应用开发。

Future 模型是将异步请求和代理模式联合的模型产物，类似商品的订单模型。

客户端发送一个长时间的请求，服务端不需等待该数据处理完成，便立即返回一个伪造的代理数据（相当于商品订单，不是商品本身），用户也无须等待，先去执行其他的若干操作后，再去调用服务器已经完成组装的真实数据。该模型充分利用了等待的时间片段。

本节讲述两个例子：futuredemo_1 工程和 futuredemo_2 工程。

futuredemo_1 工程和 futuredemo_2 工程的目录结构如图 7-2 所示。

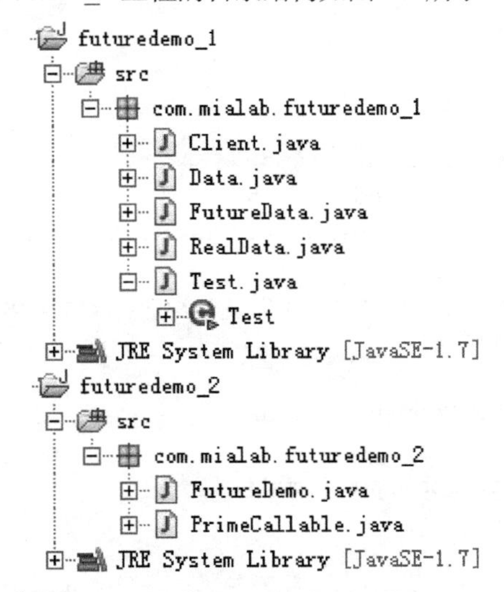

图 7-2　futuredemo_1 工程和 futuredemo_2 工程的目录结构

1．Future 模式应用举例

- Test 类的主要作用：启动系统，调用 Client 发出请求。
- Client 类的主要作用：返回 Data 对象，并开启 Client 线程装配 RealData。
- 接口 Data 的主要作用：返回数据的接口。它提供了 getResult()方法。
- FutureData 类的主要作用：表示 Future 数据，构造速度很快，但它是一个虚拟的数据，需要装配 RealData。
- RealData 类的主要作用：表示真实数据，构造比较慢。

相关代码参见教学资源包中第 7 章的 futuredemo_1 工程，这是使用 Eclipse 创建的。

（1）Test.java 的主要代码如下：

```
public class Test {
    public static void main(String[] args) {
        Client client = new Client();
        Data data = client.request("name");                    //返回一个 FutureData
        System.out.println("请求完毕！");
        try {
            //处理其他业务
            //这个过程中，真实数据 RealData 组装完成，重复利用等待时间
            Thread.sleep(2000);
        } catch (Exception e) {
        }
        System.out.println("数据 = " + data.getResult()); //真实数据
    }
}
```

（2）Client.java 的主要代码如下：

```
public class Client {
    public Data request(final String queryStr) {
        final FutureData future = new FutureData();
        new Thread() {  //开启一个新的线程来构造真实数据
            public void run() {
                RealData realData = new RealData(queryStr);
                future.setRealData(realData);
            }
        }.start();
        return future;
    }
}
```

（3）Data.java 的主要代码如下：

```
public interface Data {
    public  String getResult();
}
```

（4）FutureData.java 的主要代码如下：

```
public class FutureData implements Data {
    protected RealData realData = null;
    protected boolean isReady = false;
    public synchronized void setRealData(RealData realData) {
        if (isReady) {
            return;
        }
        this.realData = realData;
        isReady = true;
        notifyAll();  //唤醒所有正在等待该对象的线程
    }
    @Override
    public synchronized String getResult() {
        while (!isReady) {
            try {
                wait();
            } catch (Exception e) {
            }
        }
        return realData.result;
```

```
    }
}
```

（5）RealData.java 的主要代码如下：

```
public class RealData implements Data {
    protected String result;
    public RealData(String para) {   //构造比较慢
        StringBuffer sb = new StringBuffer();
        for (int i = 0; i < 10; i++) {
            sb.append(para);
            try {
                Thread.sleep(1000);
            } catch (Exception e) {
            }
            result = sb.toString();
        }
    }
    @Override
    public String getResult() {
        return result;
    }
}
```

其执行结果如下：

```
请求完毕!
数据 = namenamenamenamenamenamenamenamenamename
```

说明：FutureData 是对 RealData 的包装，是对真实数据的一个代理，封装了获取真实数据的等待过程。客户端在调用的方法中，通过单独启用一个线程来完成真实数据的组织，这对于调用客户端的 main 方法是封闭的。因为 FutureData 中的 notifyAll 函数和 wait 函数，主程序会等待组装完成后才会继续主线程，也就是说，如果没有组装完成，main 函数会一直等待。

2. 以 java.util.concurrent.Future 为例

这里就以 java.util.concurrent.Future 为例简单介绍 Future 的具体工作方式。

Future 对象本身可以看成是一个显式的引用，一个对异步处理结果的引用。由于其异步性质，在创建之初，它所引用的对象可能还并不可用（如在运算中、网络传输中或等待中）。这时，得到 Future 的程序流程如果并不急于使用 Future 所引用的对象，那么它可以做其他任何想做的事，当流程进行到需要 Future 背后引用的对象时，可能有以下两种情况：

（1）希望能看到这个对象可用，并完成一些相关的后续流程。如果实在不可用，也可以进入其他的分支流程。

（2）若没有耐心一直等下去，可以限定时间，即设置一个超时时间。

对于前一种情况，可以通过调用 Future.isDone()判断引用的对象是否就绪，并采取不同的处理；后一种情况只需调用 get()或 get(long timeout, TimeUnit unit)通过同步阻塞方式就可等待对象就绪。实际运行期是阻塞还是立即返回就取决于 get()的调用时机和对象就绪的先后顺序。

简单而言，Future 模式可以在连续流程中满足数据驱动的并发需求，既获得了并发执行的性能提升，又不失连续流程的简洁优雅。

小贴士：在 Java 中一般通过继承 Thread 类或者实现 Runnable 接口这两种方式来创建多线程，但是这两种方式都有个缺陷，就是不能在执行完成后获取执行的结果。因此在 Java 1.5 之后提供了 Callable 接口和 Future 接口，通过它们就可以在任务执行完毕后得到结果。

java.util.concurrent.Callable 类与 java.util.concurrent.Future 类可以协助用户完成 Future 模式。Future

模式在请求发生时，会先产生一个 Future 对象给发出请求的客户。它的作用类似于代理对象，而同时所代理的真正目标对象的生成是由一个新的线程持续进行的。真正的目标对象生成后，可将其设置到 Future 中，而当客户端真正需要目标对象时，目标对象也已经准备好，可以供客户提取使用。

Callable 是一个接口，与 Runnable 类似，包含一个必须实现的方法，可以启动为让另一个线程来执行。不过 Callable 工作完成后，可以传回结果对象。Callable 接口的定义如下：

```
public interface Callable<V> {
    Vcall() throws Exception;
}
```

可以使用 Callable 完成某个费时的工作，等工作结束后传回结果对象，如求质数。这里是在 Eclipse JavaEE 中建立 futuredemo_2 工程，是 Java Project。

相关代码参见教学资源包第 7 章的 futuredemo_2 工程。

FutureDemo.java 的主要代码如下：

```
public class FutureDemo {
    public static void main(String[] args) {
        Callable<int[]> primeCallable = new PrimeCallable(1000);
        FutureTask<int[]> primeTask = new FutureTask<int[]>(primeCallable);
        Thread t = new Thread(primeTask);
        t.start();
        try {
            Thread.sleep(5000);                    //假设现在执行其他事情
            if (primeTask.isDone()) {              //回来检查质数是否已经查找完毕
                int[] primes = primeTask.get();
                for (int prime : primes) {
                    System.out.print(prime + " ");
                }
                System.out.println();
            }
        } catch (InterruptedException e) {
            e.printStackTrace();
        } catch (ExecutionException e) {
            e.printStackTrace();
        }
    }
}
```

PrimeCallable.java 的主要代码如下：

```
public class PrimeCallable implements Callable<int[]> {
    private int max;
    public PrimeCallable(int max) {
        this.max = max;
    }
    @Override
    public int[] call() throws Exception {
        int[] prime = new int[max + 1];
        List<Integer> list = new ArrayList<Integer>();
        for (int i = 2; i <= max; i++)
            prime[i] = 1;
        for (int i = 2; i * i <= max; i++) {      //这里可以改进
            if (prime[i] == 1) {
                for (int j = 2 * i; j <= max; j++) {
                    if (j % i == 0)
                        prime[j] = 0;
```

```
                }
            }
        }
        for (int i = 2; i < max; i++) {
            if (prime[i] == 1) {
                list.add(i);
            }
        }
        int[] p = new int[list.size()];
        for (int i = 0; i < p.length; i++) {
            p[i] = list.get(i).intValue();
        }
        return p;
    }
}
```

程序中的求质数方法是很简单的，但效率不好，这里只是为了示范方便。

假设现在求质数的需求是在启动 PrimeCallable 的几秒之后，则可以使用 Future 来获得 Callable 执行的结果，从而在未来的时间点获得结果。

java.util.concurrent.FutureTask 是一个代理，真正执行找质数功能的是 Callable 对象。使用另一个线程启动 FutureTask 后，就可以做其他的事情了。等到某个时间点，用 isDone()观察任务是否完成，如果完成了，就可以获得结果，其执行结果如下，显示了所有找到的质数。

```
2 3 5 7 11 13 17 19 23 29 … 941 947 953 967 971 977 983 991 997
```

7.5 项目实战："移动商城"（二）

7.5.1 任务说明

【示例】 使用线程池异步加载网络多张图片在手机上显示，且滑动手机屏幕时，能实现图片显示快速流畅的效果。

7.5.2 项目讲解

1. 总体框架

【示例】 启动 Android Studio，在 Ch7 工程中创建 MobileMall_B 模块。Ch7\MobileMall_B（源码）目录层次结构如图 7-3 所示。

Ch7\MobileMall_B（源码）src\main\java 路径下各个包的主要功能如下。

- com.mialab.mobilemall 包：包含主界面类 MainActivity.java。
- com.mialab. mobilemall.base 包：基础包。它主要包含基础类 BaseActivity.java。
- com.mialab. mobilemall.beans 包：此包中放置的是实体类。
- com.mialab. mobilemall.common 包：放置配置类、通用类的包。它主要包含 Constants.java。
- com.mialab.mobilemall.adapter 包：此包中放置的是自定义适配器类。
- com.mialab. mobilemall.tools 包：工具类包。它主要包含 ImageWorker.java、ImageCache.java、OnHandleCacheListener.java、DiskCache.java、FileUtils.java、FunctionUtil.java。
- com.mialab. mobilemall.ui 包：界面类包。它主要包含宽带列表界面 BroadbandActivity.java、"历史"（曾购买记录、浏览记录等）界面 HistoryActivity.java。

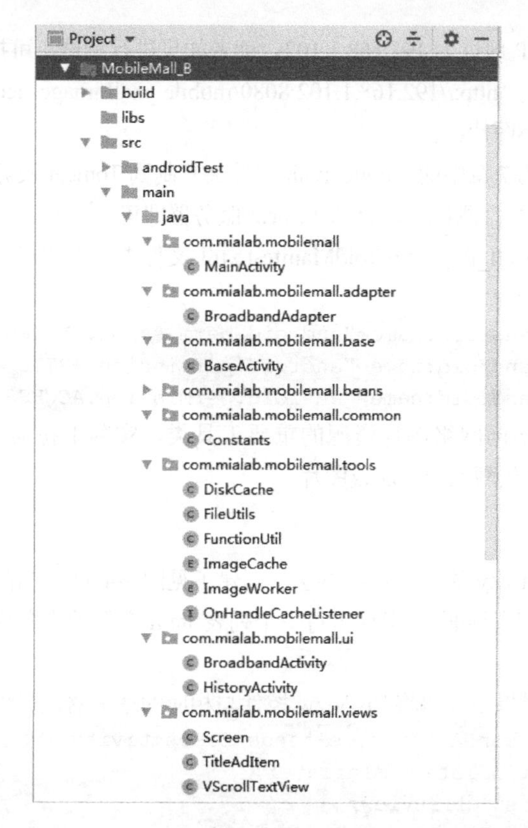

图 7-3　Ch7\MobileMall_B（源码）目录层次结构

- com.mialab. mobilemall.views 包：视图(控件)类包。它主要包含 Screen.java、VScrollTextView.java 等。

该应用（Ch7\MobileMall_B）的基本功能和运行界面，同第 3 章的“移动商城（一）”（Ch3\MobileMall），只不过这里是从网络异步加载图片的。

MobileMall_B 主要运行界面如图 7-4 所示。

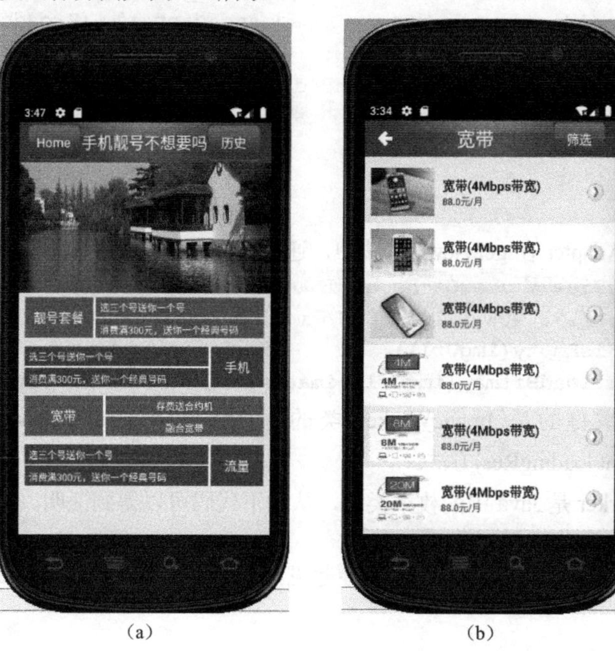

（a）　　　　　　　　（b）

图 7-4　MobileMall_B 运行界面

测试时编者计算机的IP地址为192.168.1.102。读者如果进行测试，可换为自己的IP地址。

网络图片的存储路径为“http://192.168.1.102:8080/mobile_mall/images/icons/”，在icons文件夹中提供了给手机客户端访问的小图片。

这里需要把教学资源第7章中的mobile_mall文件夹复制到Tomcat服务器的webapps目录下，作为网络访问资源（服务器端）。测试时，启动Tomcat服务器即可。

还需要在Ch7\MobileMall_B的AndroidManifest.xml文件中，添加网络访问的权限和访问外部存储的权限，其代码如下所示。

```
<uses-permission android:name="android.permission.INTERNET" />
<uses-permission android:name="android.permission.WRITE_EXTERNAL_STORAGE" />
<uses-permission android:name="android.permission.ACCESS_NETWORK_STATE" />
```

ImageWorker 是异步访问网络图片资源的重要工具类，实际上扮演了 ImageLoader 的角色。ImageWorker的主要功能是从网上异步加载图片。

2. 程序主要流程

程序的入口是MainActivity的onCreate方法，创建了见图7-4（b）的商品列表主界面，点击“宽带文本区域”进入“宽带列表界面”，但这里每一个列表item左侧的小图片都是异步从网络加载的，而且是并发访问。

BroadbandActivity通过以下代码将ListView控件与相应的适配器进行绑定。

```
public class BroadbandActivity extends BaseActivity {
    private LayoutInflater mInflater;
    private ListView listView;
    private BroadbandAdapter broadbandAdapter;
    …
    @Override
    protected void onCreate(Bundle savedInstanceState) {
        …
        listView = (ListView) this.findViewById(R.id.simple_list);
        mImageWorker =getImageWorker();
        mImageWorker.setOnScreen(Constants.APP_TAG, true);
        broadbandAdapter = new BroadbandAdapter (BroadbandActivity.this,
mImageWorker, listView);
        listView.setAdapter(broadbandAdapter);
        …
    }
    …
}
```

适配器BroadbandAdapter在getView()方法中，通过以下代码从网络异步加载图片。

```
ImageView imageView = viewCache.getImageView();
imgUrl = "http://192.168.1.102:8080/ mobile_mall /images/icons/phone.png";
if (!TextUtils.isEmpty(imgUrl))
    imageWorker.loadBitmap(imgUrl, imageView, R.drawable.broadband_default);
```

可以看到代码中关键的是 ImageWorker 类的 public void loadBitmap(final String path, final ImageView imageView,int loadingRes)方法。

实际上，ImageWorker是Java中的枚举类型，从如下代码可以得到证明。

```
public enum ImageWorker {…}
```

可以把枚举类型视作Java中的类。为了方便起见，在下文中不称其为ImageWorker枚举类型，而称为ImageWorker类。

7.5.3 典型代码及技术要点

在本程序中将对ImageWorker类进行分析，其主要代码如下：

```
public enum ImageWorker {
    INSTANCE;
    private ImageCache mImageCache;
    private int loadingResId = 0, errorResId = 0, bgResId = 0;
    private volatile boolean onScreen = true;
    public static int ScreenWeith = 480;
    private ExecutorService searchThreadPool;
    private HashMap<String, ImageCacheParams> params;
    private Handler mHandler;
    public static final int IO_BUFFER_SIZE = 8 * 1024;
    private OnHandleCacheListener mIHandleCache;
    public static ImageWorker newInstance() {
        return INSTANCE;
    }
    private ImageWorker() {
        mHandler = new Handler();
        mImageCache = ImageCache.createCache();
        searchThreadPool = Executors.newFixedThreadPool(8);
        mIHandleCache = new OnHandleCacheListener() {
            @Override
           public void onSetImage(final ImageView imageView, final Bitmap bitmap)
{
                mHandler.post(new Runnable() {
                    @Override
                    public void run() {
                        imageView.setImageBitmap(bitmap);
                        …
                    }
                });
            }
            @Override
            public void onError(final ImageView imageView) {
                mHandler.post(new Runnable() {
                    @Override
                    public void run() {
                        if (imageView != null) {
                            …
                        }
                    }
                });
            }
        };
    }
    protected boolean cancelWork(final ImageView view, final String path) {
        SearchTask task = getSearchTask(view);
        if (task != null) {
            final String taskPath = task.getPath();
            if (TextUtils.isEmpty(taskPath) || !taskPath.equals(path)) {
                task.cancelWork();
            } else {
                return false;
```

```
            }
        }
        return true;
    }
    private void restartThreadPool() {
        synchronized (searchThreadPool) {
            if    (searchThreadPool.isTerminated()    ||    searchThreadPool.
isShutdown()) {
                searchThreadPool = null;
                searchThreadPool = Executors.newFixedThreadPool(8);
            }
        }
    }
    private void shutdownThreadPool() {
        searchThreadPool.shutdown();
    }
    public class AsyncDrawable {…}
    public class SearchTask implements Runnable { …}
    …
}
```

可以看到，ImageWorker 类包含两个内部类 SearchTask 和 AsyncDrawable，还有一个监听器 OnHandleCacheListener。

1．内部类 SearchTask

ImageWorker 的内部类 SearchTask.java 的代码如下。

```
public class SearchTask implements Runnable {
        String path;
        volatile boolean stop = false;
        OnHandleCacheListener mIHandleCache;
        int reqW = 0;
        int reqH = 0;
        private WeakReference<ImageView> mImageViewReference;

        public void cancelWork() {  //停止任务
            stop = true;
        }
        public SearchTask(final String path, final ImageView imageView, int w,
                int h, final OnHandleCacheListener mIHandleCache) {
            this.path = path;
            reqW = w;
            reqH = h;
            mImageViewReference = new WeakReference<ImageView>(imageView);
            this.mIHandleCache = mIHandleCache;
        }

        @Override
        public void run() {…}
        …
}
```

注意：如果生成 SearchTask 对象，会含有弱引用 mImageViewReference 指向 ImageView 对象，以及 boolean 变量 stop 被声明为 volatile。

2. 内部类 AsyncDrawable

ImageWorker 内部类 AsyncDrawable.java 的主要代码如下：

```
public class AsyncDrawable {
    private final WeakReference<SearchTask> task;
    private int defaultId = 0;
    private int errorResId = 0;
    private int bgresId = 0;
    public AsyncDrawable(int loadingRes, int errResId, int bgResid, SearchTask
searchTask) {
        task = new WeakReference<SearchTask>(searchTask);
        defaultId = loadingRes;
        errorResId = errResId;
        bgresId = bgResid;
    }
    …
}
```

注意：如果生成 AsyncDrawable 对象，会含有弱引用 task 指向 SearchTask 对象。

3. 枚举类型 ImageCache

枚举类型 ImageCache.java 的主要代码如下：

```
public enum ImageCache {
    INSTANCE;
    //默认内存缓存大小
    private static final int DEFAULT_MEM_CACHE_SIZE = 1024 * 1024 * 4;  //4MB
    //是否使用内存缓存
    private static final boolean DEFAULT_MEM_CACHE_ENABLED = true;
    //是否使用 SD 卡缓存
    private static final boolean DEFAULT_DISK_CACHE_ENABLED = true;
    //是否在使用缓存前清理 SD 卡
    private static final boolean DEFAULT_CLEAR_DISK_CACHE_ON_START = false;
    private ImageCacheParams mImageCacheParams;
    private DiskCache mDiskCache;                              //SD 卡
    private LruCache<String, Bitmap> mMemoryCache;             //内存

    public static ImageCache createCache() {
        return INSTANCE;
    }
    private ImageCache() {
    }
    public void setCacheParams(ImageCacheParams cacheParams) {
        init(cacheParams);
    }
    private void init(ImageCacheParams cacheParams) {
        mImageCacheParams = cacheParams;
        //Set up disk cache
        if (cacheParams.diskCacheEnabled) {… }
        //Set up memory cache
        if (cacheParams.memoryCacheEnabled) {
            mMemoryCache = new LruCache<String, Bitmap>( cacheParams.
memCacheSize) {
                @Override
                protected int sizeOf(String key, Bitmap bitmap) {
                    return getBitmapSize(bitmap);
```

```
                }
            };
        }
    }
    public int getBitmapSize(Bitmap bitmap) {
        return bitmap.getRowBytes() * bitmap.getHeight();
    }
    public void addBitmapToCache(String data, Bitmap bitmap) { //将图片添加到缓存中
        if (data == null || bitmap == null) {
            return;
        }
        if (mMemoryCache != null && mMemoryCache.get(data) == null) {
            mMemoryCache.put(data, bitmap);           //Add to memory cache
        }
    }
    public Bitmap getBitmapFromMem(String path) {     //从内存中取得图片
        if (mMemoryCache != null) {
            final Bitmap memBitmap = mMemoryCache.get(path);
            if (memBitmap != null) {
                return memBitmap;
            }
        }
        return null;
    }
    public void clearCaches() {                       //清理缓存
        if (mMemoryCache != null) {
            mMemoryCache.evictAll();
        }
    }
    //A holder class that contains cache parameters.
    public static class ImageCacheParams {
        public boolean memoryCacheEnabled = DEFAULT_MEM_CACHE_ENABLED;
        public boolean diskCacheEnabled = DEFAULT_DISK_CACHE_ENABLED;
        public int memCacheSize = DEFAULT_MEM_CACHE_SIZE;
        public  boolean  clearDiskCacheOnStart  =  DEFAULT_CLEAR_DISK_CACHE_ON_
START;
    }
    …
}
```

注意：图片缓存是使用 Android 提供的 LruCache 类来创建的。

以下介绍一些预备知识来帮助理解 ImageWorker 异步下载网络图片的代码。

4. 预备知识一：强引用、软应用和弱引用

垃圾回收的机制主要是看对象是否有引用指向它。Java 对象的引用包括强引用、软引用、弱引用等。

强引用是指创建一个对象并把这个对象赋值给一个引用变量。当强引用不为 null 时，它指向的对象即使内存不足也永远不会被垃圾回收。当强引用被置为 null 时，该对象则被标记为可回收的，但是 GC 可能依旧没有回收它，这和 GC 的回收算法有关，同时该对象仍然占着内存。总之，不能保证可回收的对象都能被 GC 回收。

软引用通过 SoftReference 类来实现。软引用指向的对象，不用置 null 也可以被 GC 回收，对象是否被释放取决于 GC 算法，以及 GC 运行时可用的内存数量。通俗地讲，内存空间足够，GC 就不会回

收它；如果内存空间不足了，GC 就会回收这些对象的内存。

弱引用通过 WeakReference 类来创建，其示例代码如下。

```
public class Test {
        public static void main(String[] args) {
            String str = new String("JAVA 讲义");       //创建是在常量池中
            //创建一个弱引用 str 指向 String 对象
            WeakReference<String> wr = new WeakReference<String>(str);
            str = null;
            System.out.println(wr.get());              //输出 Java 讲义
            System.gc();                               //强制垃圾回收
            System.out.println(wr.get());              //输出 null
    }
}
```

GC 运行时如果碰到了弱引用对象，不管当前内存空间是否足够，都会回收它的内存，但是也有可能需要 GC 多次才能发现和释放弱引用的对象。

软引用和弱引用都可以与引用队列（ReferenceQueue ）关联，如果软引用或弱引用所引用的对象被 JVM 回收，这个软引用或弱引用就会被加入到与之关联的引用队列中。这样就可以知道软引用或者弱引用是否被回收。

```
MyObject obj = new MyObject();
ReferenceQueue rq = new ReferenceQueue();
WeakReference wr = new WeakReference(obj, rq);
obj = null;  //如果不置 null，则 obj 对象依旧为强引用，就没有了弱引用的效果
System.gc();
wr.get();
rq.poll(); .
```

如果没有被回收，则 wr.get()返回 obj 对象，rq.poll()返回 null。

如果被 GC 回收了，则 wr.get()返回 null，rq.poll()返回对象的弱引用 wr。

5. 预备知识二：使用 LruCache 缓存图片

在应用程序 UI 界面中加载一张图片是很简单的事情，但是当需要在界面上加载一大堆图片时，情况就变得复杂起来。在很多情况下（如使用 ListView、GridView 或者 ViewPager 这样的组件），屏幕上显示的图片会通过滑动屏幕等事件不断增加，最终导致 OOM（Out Of Memory，内存泄漏）。

为了保证内存的使用始终维持在一个合理的范围，通常会把被移除屏幕的图片进行回收处理。此时垃圾回收器也会认为用户不再持有这些图片的引用，从而对这些图片进行 GC 操作。

虽然使用这种思路来解决问题是非常好的，但为了能让程序快速加载运行，又必须要考虑到某些图片被回收后，用户又要将它重新滑入屏幕这种情况。这时重新加载一遍刚加载过的图片无疑是性能的瓶颈，因此用户需要想办法去避免这种情况的发生。

这时，使用内存缓存技术就可以很好地解决这个问题，它可以让组件快速地重新加载和处理图片。

下面来介绍如何使用内存缓存技术对图片进行缓存，从而使应用程序在加载很多图片时仍能提高响应速度和流畅性。

内存缓存技术对那些大量占用应用程序宝贵内存的图片提供了快速访问的方法，其中最核心的类是 LruCache（此类在 android-support-v4 的包中提供）。这个类非常适合用来缓存图片，它的主要算法原理是把最近使用的对象用强引用存储在 LinkedHashMap 中，并且把最近最少使用对象在缓存值达到预设定值之前从内存中移除。

为了能够选择一个合适的缓存大小给 LruCache，至少有以下因素需要考虑。

- 你的设备可以为每个应用程序分配多大的内存？

- 设备屏幕上一次最多能显示多少张图片？有多少张图片需要进行预加载？
- 设备屏幕的大小和分辨率分别是多少？一个超高分辨率的设备比一个较低分辨率的设备，在持有相同数量图片的时候，需要更大的缓存空间。
- 图片的尺寸和大小是多少？每张图片会占据多大的内存空间？
- 图片被访问的频率有多高？会不会有一些图片的访问频率比其他图片要高？如果有的话，应该让这些图片常驻在内存中，或者使用多个 LruCache 对象来区分不同组的图片。
- 你能维持好数量和质量之间的平衡吗？有些时候，存储多张低像素的图片，不如在后台去开线程加载高像素的图片更加有效。

实际上，并没有一个指定的缓存大小可以满足所有的应用程序，因为它是由用户决定的。用户应该去分析程序内存的使用情况，然后制订出一个合适的解决方案。一个太小的缓存空间，有可能会造成图片频繁地被释放和重新加载，这并没有好处。然而一个太大的缓存空间，也可能会引起 java.lang.OutOfMemory 的异常。

下面是一个使用 LruCache 来缓存图片的例子。

```
private LruCache<String, Bitmap> mMemoryCache;
@Override
protected void onCreate(Bundle savedInstanceState) {
    //获取到可用内存的最大值，使用内存超出这个值就会引起 OutOfMemory 异常
    //LruCache 通过构造函数传入缓存值，以 KB 为单位
    int maxMemory = (int) (Runtime.getRuntime().maxMemory() / 1024);
    //使用最大可用内存值的 1/8 作为缓存的大小
    int cacheSize = maxMemory / 8;
    mMemoryCache = new LruCache<String, Bitmap>(cacheSize) {
        @Override
        protected int sizeOf(String key, Bitmap bitmap) {
            //重写此方法来衡量每张图片的大小，默认返回图片数量
            return bitmap.getByteCount() / 1024;
        }
    };
}
public void addBitmapToMemoryCache(String key, Bitmap bitmap) {
    if (getBitmapFromMemCache(key) == null) {
        mMemoryCache.put(key, bitmap);
    }
}
public Bitmap getBitmapFromMemCache(String key) {
    return mMemoryCache.get(key);
}
```

在这个例子当中，使用了系统分配给应用程序的八分之一内存来作为缓存大小。假设是 4MB 的缓存空间，如果一个全屏幕的 GridView 使用 4 张 800×480 分辨率的图片来填充，则大概会占用 1.5MB 的空间（800*480*4=1536000）。因此，这个缓存大小可以存储 2.5 张图片。

当向 ImageView 中加载一张图片时，首先会在 LruCache 的缓存中进行检查。如果找到了相应的键值，则会立刻更新 ImageView ，否则会开启一个后台线程来加载这张图片。

```
public void loadBitmap(int resId, ImageView imageView) {
    final String imageKey = String.valueOf(resId);
    final Bitmap bitmap = getBitmapFromMemCache(imageKey);
    if (bitmap != null) {
        imageView.setImageBitmap(bitmap);
    } else {
        imageView.setImageResource(R.drawable.image_placeholder);
```

```
        BitmapWorkerTask task = new BitmapWorkerTask(imageView);
        task.execute(resId);
    }
}
```

BitmapWorkerTask 还要把新加载图片的键值对放到缓存中。

```
class BitmapWorkerTask extends AsyncTask<Integer, Void, Bitmap> {
    @Override
    protected Bitmap doInBackground(Integer… params) { //在后台加载图片
        final Bitmap bitmap = decodeSampledBitmapFromResource(
                getResources(), params[0], 100, 100);
        addBitmapToMemoryCache(String.valueOf(params[0]), bitmap);
        return bitmap;
    }
}
```

6. 预备知识三：Java 中的关键字 volatile

volatile 修饰符用来“告诉”JVM，访问该变量的线程必须使自己对该变量的私有副本与内存中的主副本保持一致。

Java 代码在编译后会变成 Java 字节码，字节码被类加载器加载到 JVM 里，由 JVM 执行字节码，最终需要转化为汇编指令在 CPU 上执行。

volatile 是轻量级的 synchronized（volatile 不会引起线程上下文的切换和调度），它在多处理器开发中保证了共享变量的“可见性”。“可见性”的意思是指当一个线程修改一个共享变量时，另外一个线程能读到这个修改的值。

由于内存访问速度远不及 CPU 的处理速度，为了提高处理速度，处理器并不直接和内存进行通信，而是先将系统内存的数据读到内部缓存后再进行操作，但操作完不知道何时会写到内存中。普通共享变量被修改之后，什么时候被写入主存是不确定的，当其他线程去读取时，此时内存中可能还是原来的旧值，因此无法保证可见性。如果对声明了 volatile 的变量进行写操作，JVM 就会向处理器发送一条 Lock 前缀的指令，表示将当前处理器缓存行的数据写回到系统内存中。

在 Java 内存模型中有 main memory，每个线程也有自己的 memory（如寄存器）。为了提高性能，一个线程会在自己的 memory 中保持要访问的变量副本。这样就会出现同一个变量在某个瞬间的一个线程 memory 的值，可能与另外一个线程 memory 的值或者 main memory 的值不一致的情况。

一个变量被声明为 volatile，就意味着这个变量随时会被其他线程修改，因此不能将它缓存在线程 memory 中。以下例子展现了 volatile 的作用：

```
public class StoppableTask extends Thread {
    private volatile boolean stop;
    public void run() {
        while (!stop) {
            //do some stuff…
        }
    }
    public void tellMeToStop() {
        stop = true;
    }
}
```

假如 stop 没有被声明为 volatile，线程执行 run 时检查的是自己的副本，就不能及时得知其他线程已经调用 tellMeToStop()修改了 pleaseStop 的值。

7．程序的进一步流程

① ImageWorker 的 loadBitmap(final String path, final ImageView imageView,int loadingRes)方法中又调用了 loadBitmap(final String path, final ImageView imageView, int w, int h, int loadingRes, int errRes, int bgRes)方法，其主要代码如下：

```
    public enum ImageWorker {
    …
    public  void  loadBitmap(final  String  path,  final  ImageView  imageView,  int 
loadingRes) {
          loadBitmap(path, imageView, ScreenWeith / 3, ScreenWeith / 3,loadingRes,
errorResId, bgResId);
    }

        public void loadBitmap(final String path, final ImageView imageView, int w,
int h, int loadingRes,
          int errRes, int bgRes) {
              Bitmap result = mImageCache.getBitmapFromMem(path);
              imageView.setImageResource(0);
              imageView.setBackgroundResource(bgRes > 0 ? bgRes : 0);
              if (result != null && !result.isRecycled()) {
                mIHandleCache.onSetImage(imageView, result);
              } else if (cancelWork(imageView, path)) {
                final  SearchTask  task  =  new  SearchTask(path,  imageView,  w,  h,
mIHandleCache);
                final  AsyncDrawable  asyncDrawable  =  new  AsyncDrawable(loadingRes,
errRes, bgRes, task);
                imageView.setTag(asyncDrawable);
                imageView.setBackgroundResource(asyncDrawable.getDefaultResID());
                if    (!searchThreadPool.isTerminated()    &&    !searchThreadPool.
isShutdown()) {
                    searchThreadPool.execute(task);
                }
              }
        }
    }
```

上述的 loadBitmap()方法中，先从缓存中获取图片，如没有，再执行 ImageWorker 的 cancelWork(imageView, path)方法，并判断其返回值。

② cancelWork(imageView, path)方法的代码参见相关 ImageWorker 的代码。

在 cancelWork(imageView, path)方法中，先看能否得到传进去 imageView 关联的任务 task，刚开始是没有的，cancelWork(imageView, path)方法的返回值为 true。

③ 创建 SearchTask 对象 task，准确地说，是创建了堆上的 SearchTask 对象及栈上的指向 SearchTask 对象的引用 task。

④ 创建 AsyncDrawable 对象 asyncDrawable，准确地说，是创建了堆上的 AsyncDrawable 对象及栈上的指向 AsyncDrawable 对象的引用 asyncDrawable。

⑤ 为传进来的 imageView 设置 Tag，对于 Tag 的类型，Android API 要求应为 Object 类型，此处为 asyncDrawable。AsyncDrawable 对象中是含有弱引用 task 的（此处的 task 和前面的 task 是不同的引用），task 指向 SearchTask 对象，这也就是 imageView 关联的任务。imageView 通过设置 Tag 和弱引用 task 发生了关联。

这里应把 imageView 视为图片像框，而不是图片。

⑥ 线程池没有关闭，便会执行 searchThreadPool.execute(task)方法，提交 task 方法给线程池中的线程执行。实际上执行的是 SearchTask 的 run()方法，其主要代码如下：

```
    public class SearchTask implements Runnable {
        …
        @Override
        public void run() {
            Bitmap bitmap = null;
            if (mImageCache != null && !stop && getAttachedImageView() != null &&
onScreen) {
                bitmap = mImageCache.getBitmapFromDiskCache(path, reqW, reqH);
            }
            if (bitmap == null && mImageCache != null && !stop
                    && getAttachedImageView() != null && onScreen) {
                try {
                    File file = downloadBitmap(path);
                    bitmap = mImageCache.getBitmapFromDiskCache(file, reqW, reqH);
                    if (bitmap == null) {
                        mIHandleCache.onError(getAttachedImageView());
                    }
                } catch (IOException e) {
                    mIHandleCache.onError(getAttachedImageView());
                    e.printStackTrace();
                }
            }
            if (bitmap != null && mImageCache != null && !stop && onScreen) {
                ImageView imageView = getAttachedImageView();
                mImageCache.addBitmapToCache(path, bitmap);
                if (imageView != null && !stop) {
                    mIHandleCache.onSetImage(imageView, bitmap);
                } else {
                    bitmap.recycle();
                    bitmap = null;
                }
            }
        }
    }
```

⑦ 在 SearchTask 的 run()方法中，调用 downloadBitmap(String urlString)方法从网络下载图片存到 SD 卡，再从 SD 卡缓存获取图片，接着把图片加入内存进行缓存。如果 ImageView 弱引用对象没有被回收，将回调监听器 OnHandleCacheListener 的 onSetImage(imageView, bitmap)方法，从而在手机屏幕上显示出图片。

downloadBitmap(String urlString)的主要代码如下：

```
    public enum ImageWorker {
       private File downloadBitmap(String urlString) throws IOException {
          DiskCache cache = DiskCache.openCache();
          final File cacheFile = new File(cache.createFilePath(urlString));
          if (cache.containsKey(urlString)) {
              return cacheFile;
          }
          HttpURLConnection urlConnection = null;
          BufferedOutputStream out = null;
          try {
              final URL url = new URL(urlString);
```

```
            urlConnection = (HttpURLConnection) url.openConnection();
            urlConnection.setReadTimeout(6 * 1000);
            urlConnection.setConnectTimeout(6 * 1000);
            InputStream    in    =    new    BufferedInputStream(urlConnection.
getInputStream(),
                IO_BUFFER_SIZE);
            out  =  new  BufferedOutputStream(new  FileOutputStream(cacheFile),
IO_BUFFER_SIZE);
            int b;
            while ((b = in.read()) != -1) {
                out.write(b);
            }
            cacheFile.setLastModified(System.currentTimeMillis());
        } catch (IOException e) {
            throw e;
        } finally {
            if (urlConnection != null) {
                urlConnection.disconnect();
            }
            if (out != null) {
                out.flush();
                out.close();
            }
        }
        return cacheFile;
    }
}
```

注意：如果生成 SearchTask 对象，会含有弱引用 mImageViewReference，指向 ImageView 对象，以及 boolean 变量 stop 被声明为 volatile。

8．技术要点小结

为了在手机上滑动屏幕显示网络中下载的大量图片，既快速流畅，又不至于最终导致 OOM，可以采取以下的处理方法。

（1）因为是从网络中下载多张图片，就需要多个线程，使用线程池就会大大提高执行效率，减少下载图片的时间，减小系统开销。

（2）使用 LruCache 类来创建图片缓存，指向缓存中图片对象的是强引用。因为图片是从网络上下载的，如果没有缓存，每一次都要重新加载，滑动手机屏幕时根本做不到在手机屏幕上快速流畅地显示多图片。当然即使采用图片缓存，第一次加载图片显示速度也要慢一些。

（3）ImageView 控件是显示图片的相框。在从网络上下载大量图片时，在本地会生成数量众多的 ImageView 控件及相应的图片下载任务，如果不及时回收，将会导致内存泄漏。因而为了内存的快速回收，将生成 ImageView 弱引用对象和图片下载任务的弱引用对象。

7.6 相关阅读：Android 的 Looper 与 ThreadLocal

7.6.1 Android 的 Looper

Looper 是用来运行消息循环的。默认创建的线程是没有 Looper 的，必须在线程中调用 Looper.prepare()方法和 Looper.loop()方法来创建 Looper。在这个消息循环中，有一个内置的

MessageQueue 消息队列，一个 Looper 对象对应一个内置的消息队列（一对一关系）。如果消息队列中没有消息，则该线程休眠；一旦有消息时，Looper 则开始工作。

下面是一个实现 Looper 线程的典型例子，在 Looper 线程中使用 prepare()方法和 loop()方法来初始化一个 Handler，用来与 Looper 进行通信。

```
class LooperThread extends Thread {
    public Handler mHandler;
    public void run() {
        Looper.prepare();
        mHandler = new Handler() {
            public void handleMessage(Message msg) {
                //process incoming messages here
            }
        };
        Looper.loop();
    }
}
```

小贴士：在处理 Android 多线程消息机制时，会因为 Handler 而了解到 Looper。有众多高手解释道: Looper 是“消息泵”。它内部维护了一个消息队列，即 MessageQueue。Looper 的职责就是负责“抽取” MessageQueue 中的消息让它去找宿主。对于一些普通的程序开发工作而言，只需要了解 Handler 的用法即可，几乎接触不到 Looper，更别说是 MessageQueue。但是如果想以后能写出优秀的程序或深入了解别人写的框架，了解 Looper 的原理是必备的。

以下是类 Looper 的源码摘要，须重点关注代码中的中文注释，对很多关键点都进行了说明。

```
public final class Looper {                    //这是个最终类，不能被继承
    /**
     *这里 ThreadLocal 用的是 final 的静态变量（常量），所有的 Looper 共用一个 ThreadLocal
就够了
     * sThreadLocal.get()方法会返回 null，除非调用了 prepare()方法。
     */
    static final ThreadLocal<Looper> sThreadLocal = new ThreadLocal<Looper>();
    private static Looper sMainLooper;    //类变量，Looper 的引用，指向一个 Looper 对象
    final MessageQueue mQueue;            //这就是传说中的消息队列，由 Looper 自己管理
    final Thread mThread;

    public static void prepare() {        //静态方法 prepare，初始化当前线程的 Looper
        prepare(true);                    //在这里又调用了重载的 prepare
    }
    private static void prepare(boolean quitAllowed) {    //重载的 prepare
        if (sThreadLocal.get() != null) { //如果 ThreadLocal 里面已经持有一个 Looper
对象
            throw new RuntimeException("Only one Looper may be created per thread");
                //抛出运行时异常（每一个线程只能创建一个 Looper 实例对象），
                //这样 ThreadLocal 只能持有一个与线程绑定的对象
        }
        sThreadLocal.set(new Looper(quitAllowed));
        //反之，如果 sThreadLocal.get()为 null 的话，设置 Looper 实例对象到 ThreadLocal 中
        //调用了 Looper 一个参数的构造方法，它会将 Looper 对象与当前的线程“绑定”在一起
        //因为 ThreadLocal 中里面有个哈希表 Map，就一个元素 Entry，其中 key 为当前线程对象，
        //value 就是 Looper 对象
    }

    /**
```

```
        * 初始化当前的线程作为一个 Looper，标记其为应用程序的 main Looper
        * 应用程序的主 Looper 是由 Android 环境创建的，所以没有必要自己主动地调用此方法
        * 请查看 prepare()方法
        */
        public static void prepareMainLooper() {
           prepare(false);
           synchronized (Looper.class) {
               if (sMainLooper != null) {
                  throw new IllegalStateException("The main Looper has already been
prepared.");
               }
               sMainLooper = myLooper();
               //调用 myLooper 静态方法，返回 Looper 对象
               //myLooper 静态方法中，从 ThreadLocal 获取 Looper 实例对象
           }
        }
        //获取 application UIThread 的 Looper，加了线程类锁
        public static Looper getMainLooper() {
           synchronized (Looper.class) {
               return sMainLooper;
           }
        }

        //在当前线程中运行消息队列循环，直到调用 quit()方法结束循环
       public static void loop() {
           final Looper me = myLooper();      //先从 ThreadLocal 中拿到 Looper 实例对象
           if (me == null) {                  //如果没有拿到 Looper 实例对象，则抛出异常
               //没有 Looper，说明在这个线程中还没有调用过 Looper.prepare()静态方法
               throw new RuntimeException("No Looper; Looper.prepare() wasn't called
on this thread.");
           }
           final MessageQueue queue = me.mQueue; //Looper 构造方法初始化了 MessageQueue 对象
           //这里获取该实例变量，赋值给局部变量 queue
           Binder.clearCallingIdentity();     //确保这个线程的身份是本地进程
           final long ident = Binder.clearCallingIdentity();
           for (; ; ) {                        //死循环，消息调度
               Message msg = queue.next();    //从 MessageQueue 中获取 Message
               if (msg == null) {             //这里是循环结束条件，Message 为 null，
                                              //证明 MessageQueue 里面没有 Message
                  return;                    //直接 return，注意这里，将整个 Loop 方法结束掉，
                                             //也就是说，该线程不再占用 CPU 时间片
               }
               Printer logging = me.mLogging; //获得 Looper 的实例变量 mLogging（Printer
实例）
               if (logging != null) {  //如果拿到 Printe 对象 r，就调用 Printer 的 println()
方法
                  logging.println(">>>>> Dispatching to " + msg.target + " " +
                          msg.callback + ": " + msg.what);
               }
               msg.target.dispatchMessage(msg);
               //这里调用 Message 里面的 target 实例变量是个 Handler 对象，然后调用
               //Handler 对象的一个实例方法 dispatchMessage，传入的还是当前的 Message 对象
               //handler 的 dispatchMessage(msg) 这个方法是在 Looper 线程中执行的
               if (logging != null) {  //如果有 Printer 对象，就调用它的实例方法 println()
```

```
            logging.println("<<<<< Finished to " + msg.target + " " +
msg.callback);
        }
        …
    }
}
//返回当前线程的 Looper 对象，如果当前线程还没有关联上 Looper，可能会返回 null
//例如，当前线程还没有执行到 Looper.prepare()方法时，就调用了 myLooper()方法
public static @Nullable Looper myLooper() {
    return sThreadLocal.get();
}
//返回与当前线程关联的 MessageQueue 对象
//调用此方法前必须确保当前线程已经运行了一个 Looper，否则会 NullPointerException
public static @NonNull MessageQueue myQueue() {
    return myLooper().mQueue;        //从 Looper 实例对象中得到 MessageQueue 实例对象
}
//在构造 Looper 的时候，内部构造了一个 MessageQueue 对象
private Looper(boolean quitAllowed) {
    mQueue = new MessageQueue(quitAllowed);
    mThread = Thread.currentThread();
}
//判断当前的线程，是否与初始化 Looper 时的线程是同一个实例对象（同一个线程）
//如果当前线程是 Looper 线程，则返回 true，否则返回 false
public boolean isCurrentThread() {
    return Thread.currentThread() == mThread;
}
public void setMessageLogging(@Nullable Printer printer) {
    mLogging = printer;              //在这里设置 Printer 实例对象
    //看来已给 Looper 传递进来一个 Printer 对象，即可打印日志了
}
public void quit() {                 //退出 Looper 循环
    mQueue.quit(false);              //MessageQueue 退出
}
public void quitSafely() {           //安全退出消息循环
    mQueue.quit(true);               //安全退出 MessageQueue
}
public @NonNull Thread getThread() {
    return mThread;                  //返回线程实例对象，所谓 Looper“绑定”的那个线程
}
public @NonNull MessageQueue getQueue() { //返回当前线程的 Looper 的 MessageQueue
    return mQueue;                        //返回消息队列对象
}
@Override
public String toString() { //重写了 toSting()方法，返回当前绑定的线程名、线程 ID，
                           //以及当前 Looper 实例对象的哈希值
    return "Looper (" + mThread.getName() + ", tid " + mThread.getId()
        + ") {" + Integer.toHexString(System.identityHashCode(this)) + "}";
}
…
}
```

Handler 的 dispatchMessage(Message msg)方法，其主要代码如下：

```
public void dispatchMessage(Message msg) {
    if (msg.callback != null) {
        handleCallback(msg);
```

```
    } else {
        if (mCallback != null) {
            if (mCallback.handleMessage(msg)) {
                return;
            }
        }
        handleMessage(msg);
    }
}
```

7.6.2 Handler 机制引出 ThreadLocal

关于 Handler 消息传递机制，由前面相关内容可知以下的结论是成立的。

- Handler 的处理过程运行在创建 Handler 的线程中。
- 一个 Looper 对应一个 MessageQueue，一个线程对应一个 Looper，一个 Looper 可以对应多个 Handler。
- 线程是默认没有 Looper 的，线程需要通过调用 Looper.prepare()，绑定 Handler 到 Looper 对象，并调用 Looper.loop()来建立消息循环。
- 主线程（UI 线程），也就是 ActivityThread，在被创建的时候就会初始化 Looper，所以主线程中可以默认使用 Handler。
- 通过 Looper 的 quitSafely()或者 quit()方法可以终结消息循环。quitSafely()方法相比于 quit()方法，其安全之处在于清空消息之前，它会派发所有的非延迟消息。

如何保证一个线程对应一个 Looper，同时各个线程之间的 Looper 又互不干扰，就引出了接下来要讨论的 ThreadLocal。

```
public final class Looper {
    private static final String TAG = "Looper";
    //sThreadLocal.get() will return null unless you've called prepare().
    static final ThreadLocal<Looper> sThreadLocal = new ThreadLocal<Looper>();
    ...//省略
}
```

小贴士：为了避免 ANR，通常会把耗时操作存放在子线程中去执行，因为子线程不能更新 UI，所以当子线程需要更新 UI 时就需要借助 Android 的消息机制，也就是 Handler 机制。

ThreadLocal 的作用是提供线程内的局部变量，这种变量在线程的生命周期内起作用，可减少同一个线程内多个函数或者组件之间一些公共变量传递的复杂度。

如果拿同步机制（如 synchronized）和 ThreadLocal 做对比，可以这么理解：

对于多线程资源共享的问题，前者仅提供一份变量，让不同的线程排队访问；后者为每一个线程都提供了一份变量，因此可以同时访问而互不影响。

但是 ThreadLocal 并不是为了解决并发或者多线程资源共享而设计的，也不是为了解决线程同步问题。ThreadLocal 的设计初衷就是为了提供线程内部的局部变量，方便在本线程内随时随地的读取，并且能与其他线程隔离。

小贴士：如何确保在每一个线程中只有一个 Looper 的实例对象呢？所谓 Looper 与 Thread 绑定，就是 Looper“拿到”Thread 的对象引用（这就用到了 ThreadLocal）。对于 ThreadLocal 可以这样理解：它自身就是一个 Key，然后可以用这个 Key 在不同的线程中存储/获取一个值，值的类型就是声明时尖括号中的类型。

以下是关于 Android 中的类 ThreadLocal 源码部分摘要和相应理解。

（1）构造函数。

```
public ThreadLocal() {} //创建一个线程的本地变量
```

（2）initialValue 函数。

```
protected T initialValue() {return null;}
```

该函数在调用 get 函数时会进行第一次调用，但是如果一开始就调用了 set 函数，则该函数不会被调用。通常该函数只会被调用一次，除非手动调用了 remove 函数之后又调用 get 函数，这种情况下，在 get 函数中还是会调用 initialValue 函数。该函数是 protected 类型，建议在子类重载该函数，所以通常该函数都会以匿名内部类的形式被重载，以指定其初始值。

```
//创建一个 Integer 型的线程本地变量
    public static final ThreadLocal<Integer> local = new ThreadLocal<Integer>()
{
    @Override
    protected Integer initialValue() {
        return 0;
    }
};
```

（3）get 函数。

该函数用来获取与当前线程关联的 ThreadLocal 值，如果当前线程没有该 ThreadLocal 值，则调用 initialValue 函数获取初始值返回。

```
public T get() {
    Thread t = Thread.currentThread();    //①首先获取当前线程
    ThreadLocalMap map = getMap(t);       //②根据当前线程获取一个 map
    //③如果获取的 map 不为空，则在 map 中以 ThreadLocal 的引用作为 key 来在 Map 中获取
    //对应的 Entry e，否则转到⑤
    if (map != null) {
        ThreadLocalMap.Entry e = map.getEntry(this);
        //④如果 e 不为 null，则返回 e.value，否则转到⑤
        if (e != null)
            return (T)e.value;
    }
    //⑤map 为空或者 e 为空，则通过 initialValue 函数获取初始值 value，然后用
    //ThreadLocal 的引用和 value 作为 firstKey 和 firstValue 创建一个新的 map
    return setInitialValue();
}

ThreadLocalMap getMap(Thread t) {
    return t.threadLocals;
}

private T setInitialValue() {
    T value = initialValue();
    Thread t = Thread.currentThread();
    ThreadLocalMap map = getMap(t);
    if (map != null)
        map.set(this, value);
    else
        createMap(t, value);
    return value;
}
void createMap(Thread t, T firstValue) {
    t.threadLocals = new ThreadLocalMap(this, firstValue);
}
```

值得注意的是，上面 getMap 方法中获取的 threadLocals 便是 Thread 中的一个成员变量。

（4）set 函数。

set 函数用来设置当前线程的 ThreadLocal 值，将其值设为 value。

```
public void set(T value) {
    Thread t = Thread.currentThread();      //①首先获取当前线程
    ThreadLocalMap map = getMap(t);         //②根据当前线程获取一个 map
    if (map != null)
        map.set(this, value);               //③如果 map 不为空，则把键值对保存到 map 中
    else
        createMap(t, value);
        //④如果 map 为空（第一次调用时 map 值为 null），则去创建一个
        //ThreadLocalMap 对象并赋值给 map，并把键值对保存到 map 中
}
```

（5）remove 函数。

remove 函数用来将当前线程 ThreadLocal 绑定的值删除，在某些情况下需要手动调用该函数，以防止内存泄漏。

```
public void remove() {
    ThreadLocalMap m = getMap(Thread.currentThread());
    if (m != null)
        m.remove(this);
}
```

（6）ThreadLocalMap。

可以看成一个 HashMap，但是它本身具体的实现却与 java.util.Map 没有关系。只是内部的实现同 HashMap 类似（通过哈希表的方式存储）。

```
static class ThreadLocalMap {
    static class Entry extend WeakReference<ThreadLocal> {
        /** The value associated with this ThreadLocal. */
        Object value;
        Entry(ThreadLocal k, Object v) {
            super(k);
            value = v;
        }
    }
    ...//省略
}
```

ThreadLocalMap 中定义了 Entry 数组实例 table，用于存储 Entry。相当于使用一个数组维护一张哈希表，负载因子是最大容量的 2/3。

```
private Entry[] table;
```

小贴士：ThreadLocal 中使用了一个存在弱引用的 map，当释放掉指向 ThreadLocal 对象的强引用以后，map 中的 value 对象却没有被回收，而这块 value 对象就永远不会被访问到了，所以存在着内存泄漏。最好的做法是采取调用 ThreadLocal 的 remove 方法，如图 7-5 所示。

每个 Thread（线程）对象中都存在一个 map，map 的类型是 ThreadLocal.ThreadLocalMap。map 中的 key 指向一个 ThreadLocal 实例。这个 map 的确使用了弱引用，不过弱引用只是针对 key。每个 key 的弱引用都指向 ThreadLocal 对象。当把 ThreadLocal Ref 设置为 null 以后，没有任何强引用指向 ThreadLocal 实例，所以 ThreadLocal 对象将会被 GC 回收。但是 value 对象却不能被回收，因为存在一条从 current thread 连接过来的强引用。只有当前 thread 结束以后，CurrentThread Ref 就不会存在栈中，强引用断开，Current Thread 对象（实例）、map、value 对象将全部被 GC 回收。

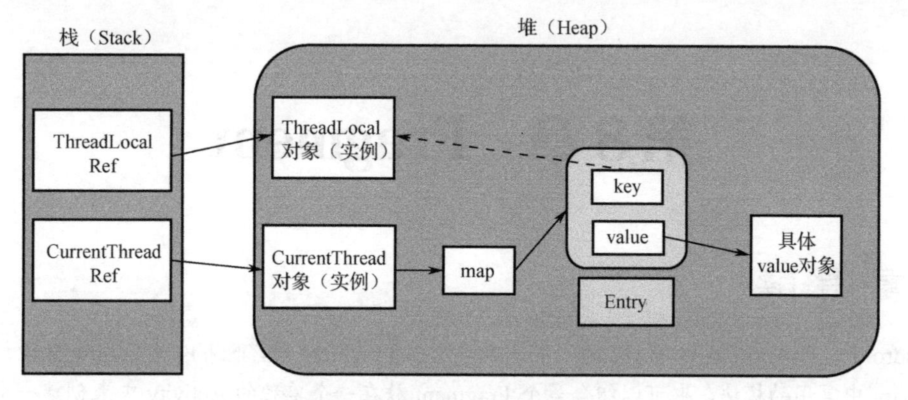

图 7-5　ThreadLocal 内存泄漏

可以得出这样一个结论：只要这个线程对象被 GC 回收，就不会出现内存泄漏问题。但在 ThreadLocal 设为 null 和线程结束这段时间是不会被回收的，也就是发生了所谓的内存泄漏。最要命的是由于线程对象不被回收的情况，而发生了真正意义上的内存泄漏。如使用线程池，线程结束后不会被销毁，而会被再次使用。

7.7　本章小结

本章以 Android 平台上的多线程程序设计的线程池如何使用为主线，介绍了 Executor 与 ExecutorService 接口、ThreadPoolExecutor 类和 Future 并发模式，最后使用线程池实现了网络多张图片的下载，且能在滑动手机屏幕时，快捷流畅地在手机屏幕上显示多张图片。这也结合了 LruCache 类及弱引用的使用。

使用线程池可以减少创建和销毁线程的次数，使得每个工作线程都可以被重复利用，以执行多个任务。同时也需要根据系统的承受能力，调整线程池中工作线程的数目，以防止内存消耗过多，可避免发生 OOM。

习　题　7

1. 为什么要使用线程池？
2. Executor、ExecutorService 和 Executors 的区别是什么？
3. 为什么说 Executors 类为创建 ExecutorService 提供了便捷的工厂方法？
4. 如何创建 Java 中的常用线程池？试编程加以说明。
5. 如何使用 LruCache 缓存图片？试编程加以说明。
6. Java 中的强引用、软应用和弱引用是怎么回事？试编程加以说明。
7. 简要解释 Java 关键字 volatile。
8. 试编程加以说明：在同等数量级的操作下，使用线程池的效率要远远高于单线程。
9. LruCache 底层实现原理是什么？
10. 内存泄漏和内存溢出各指的是什么？一般怎么处理内存泄漏？

第 8 章　Fragment

Android 是在 Android 3.0（API level 11）开始引入 Fragment 的，既可以把 Fragment 设计成在多个 Activity 中复用的模块，也可以组合多个 Fragment 放在一个单独的 Activity 中来创建一个多界面区域的 UI，并可以在多个 Activity 中重用某一个 Fragment。Fragment 既有自己的布局，也有自己的生命周期，可单独处理自己的输入，并在 Activity 运行时可以加载或者移除 Fragment 模块。本章主要内容有：（1）Fragment 的简介和生命周期；（2）Fragment 加入 Activity；（3）Fragment 与 Activity 的交互；（4）Fragment 间的数据传递；（5）ViewPager 和 PageAdapter；（6）使用 FragmentPageAdapter。

8.1　Fragment 简介

当开发的应用程序同时适用于平板电脑和手机时，可以利用 Fragment 实现灵活的布局，改善用户体验。把 Fragment 想象成一个 Activity 的模块化区域，有自己的生命周期，能接收和输入事件，并且可以在 Activity 运行期间进行添加和删除。

当一个 Fragment 作为 Activity 布局的一部分添加进来时，它处在 Activity 的 View Hierarchy 的 ViewGroup 中，并且定义有自己的 View 布局。通过在 Activity 的布局文件中声明插入一个 Fragment 到 Activity 布局中，或者可以写代码将它添加到一个已存在的 ViewGroup 中。

使用 Fragment 时，需要继承 Fragment 或者 Fragment 的子类。

android.support.v4.app. Fragment 的直接子类有 DialogFragment 和 ListFragment。

为了能够在旧版本的 Android 上使用一些新版本的 APIs，可以采取使用 Support Library 的方式。Support Library 是一个提供了 API 库函数的 JAR 文件，如 android-support-v4.jar。它的完整路径是 <sdk>/extras/android/support/v4/android-support-v4.jar。

因为它提供了 Fragment 的 APIs，就使 Android 1.6（API level 4）以上的系统都可以使用 Fragment。为了确定没有在旧版本系统上使用新版本的 APIs，需要导入语句如下。

```
import android.support.v4.app.Fragment;
import android.support.v4.app.FragmentManager;
```

当创建包含 Fragment 的 Activity 时，如果用的是 Support Library，那么继承的就应该是 FragmentActivity 而不是 Activity。创建 Fragment 需要实现的函数如下。

- onCreate()：系统在创建 Fragment 时调用这个方法，这里应该初始化相关的组件，一些被暂停或者被停止但依然需要保留的东西。
- onCreateView()：当第一次绘制 Fragment 的 UI 时系统调用这个方法，必须返回一个 View，如果 Fragment 不提供 UI 也可以返回 null。注意，如果继承自 ListFragment，onCreateView()默认的实现会返回一个 ListView，所以不用自己实现。
- onPause()：当用户离开 Fragment 时第一个调用这个方法，这里需要提交一些变化，因为用户很可能不再返回来。

提供 Fragment 的 UI，必须实现 onCreateView()方法。假设 Fragment 的布局设置写在 example_fragment.xml 资源文件中，那么 onCreateView()方法的代码如下：

```
    public static class ExampleFragment extends Fragment{
        @Override
        public View onCreateView(LayoutInflater inflater, ViewGroup container, Bundle
savedInstanceState) {
            // Inflate the layout for this fragment
            return inflater.inflate(R.layout.example_fragment, container, false);
        }
    }
```

其中，container 参数代表该 Fragment 在 Activity 中的父控件，savedInstanceState 则表示封装了上一个实例的数据。

关于 inflate()方法的三个参数说明如下：

- 第一个参数是 resource ID，指明了当前 Fragment 对应的资源文件；
- 第二个参数是父容器控件；
- 第三个布尔值参数表明是否连接该布局和其父容器控件，在这里的情况设置为 false，因为系统已经插入了这个布局到父控件，如设置为 true 则会产生一个多余的 View Group。

8.2 Fragment 的生命周期

宿主 Activity 的生命周期直接影响到 Fragment 的生命周期。Fragment 的生命周期如图 8-1 所示；Fragment 与宿主 Activity 的生命周期方法调用顺序，如图 8-2 所示。

（1）当一个 Fragment 被创建的时候，它会经历以下状态：onAttach()→onCreate()→onCreateView()→onActivityCreated()。

- onAttach()：当该 Fragment 被添加到 Activity 时调用，只调用一次。
- onCreate()：创建 Fragment 时调用，只调用一次。
- onCreateView()：每次创建、绘制该 Fragment 的 View 组件时回调，会将显示的 View 返回。
- onActivityCreated()：当 Activity 的 onCreate()方法返回时调用。

（2）当这个 Fragment 对用户可见的时候，它会经历以下状态：onStart()→onResume()。

- onStart()：启动 Fragment 时被回调。
- onResume()：恢复 Fragment 时被回调。onStart()方法后一定要回调 onResume()。由于 onStart()方法使可见，onResume()后才能交互。onResume()方法调用后，Fragment 处于运行状态，界面可见，并可获取焦点。Fragment 并被添加到“返回栈”。

（3）当这个 Fragment 进入后台的时候，它会经历以下状态：onPause()→onStop()。

- onPause()：当 Fragment 转到后台，或者 Fragment 被删除/替换的时候，会被首先回调。此时很可能（往往）其他的 Activity 位于前台可见。而对于原 Fragment 来说，则处于“不可获取焦点”的状态。
- onStop()：停止 Fragment 时被回调。此时 Fragment 不可见并失去焦点，处于停止状态。

（4）当这个 Fragment 被销毁了（或者持有它的 Activity 被销毁了），它会经历以下状态：onPause()→onStop()→onDestroyView()→onDestroy()→onDetach()。

- onDestroyView()：销毁该 Fragment 所包含的 View 组件时被回调。
- onDestroy()：销毁该 Fragment 时被回调。
- onDetach()：将该 Fragment 从宿主 Activity 中被删除/替换完成后回调该方法。onDestroy()方

法后一定会回调 onDetach()方法，该方法只能调用一次。此时 Fragment 被完全删除，处于销毁状态。

（5）一旦 Activity 进入 Resumed 状态（Running 状态），用户就可以自由地添加和删除 Fragment 了。因此，只有当 Activity 在 Resumed 状态时，Fragment 的生命周期才能独立地运转，其他时候是依赖于 Activity 的生命周期变化的。

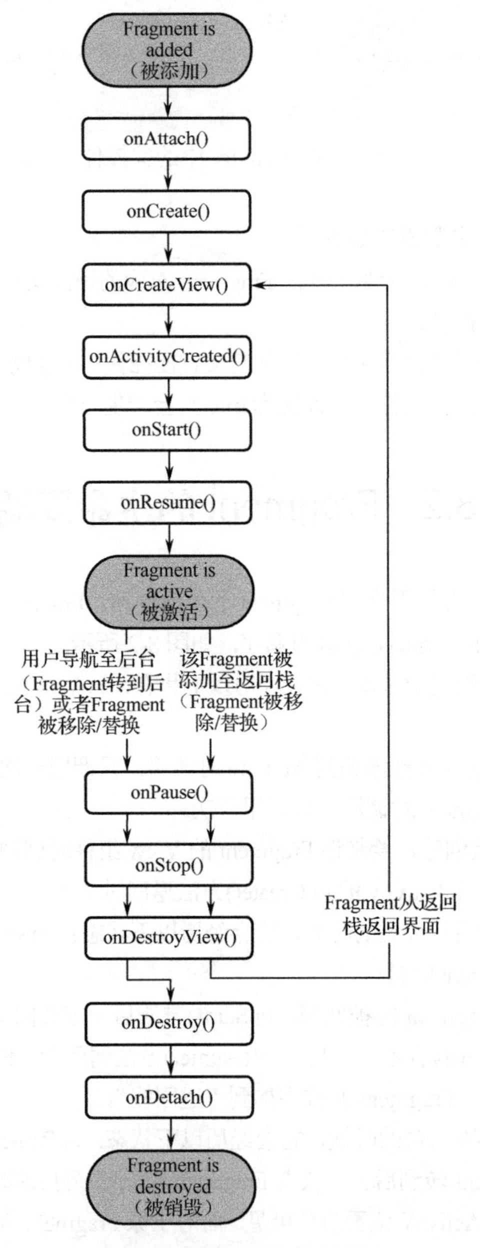

图 8-1　Fragment 的生命周期

小贴士：和 Activity 类似，可以用 Bundle 对象保存 Fragment 的状态，当 Activity 的进程被销毁后，需要重建 Activity 时，可以用于恢复 Fragment 的状态。存储时利用 onSaveInstanceState()回调函数，恢复时是在回调方法 onCreate()/onCreateView()/onActivityCreated()中进行的。

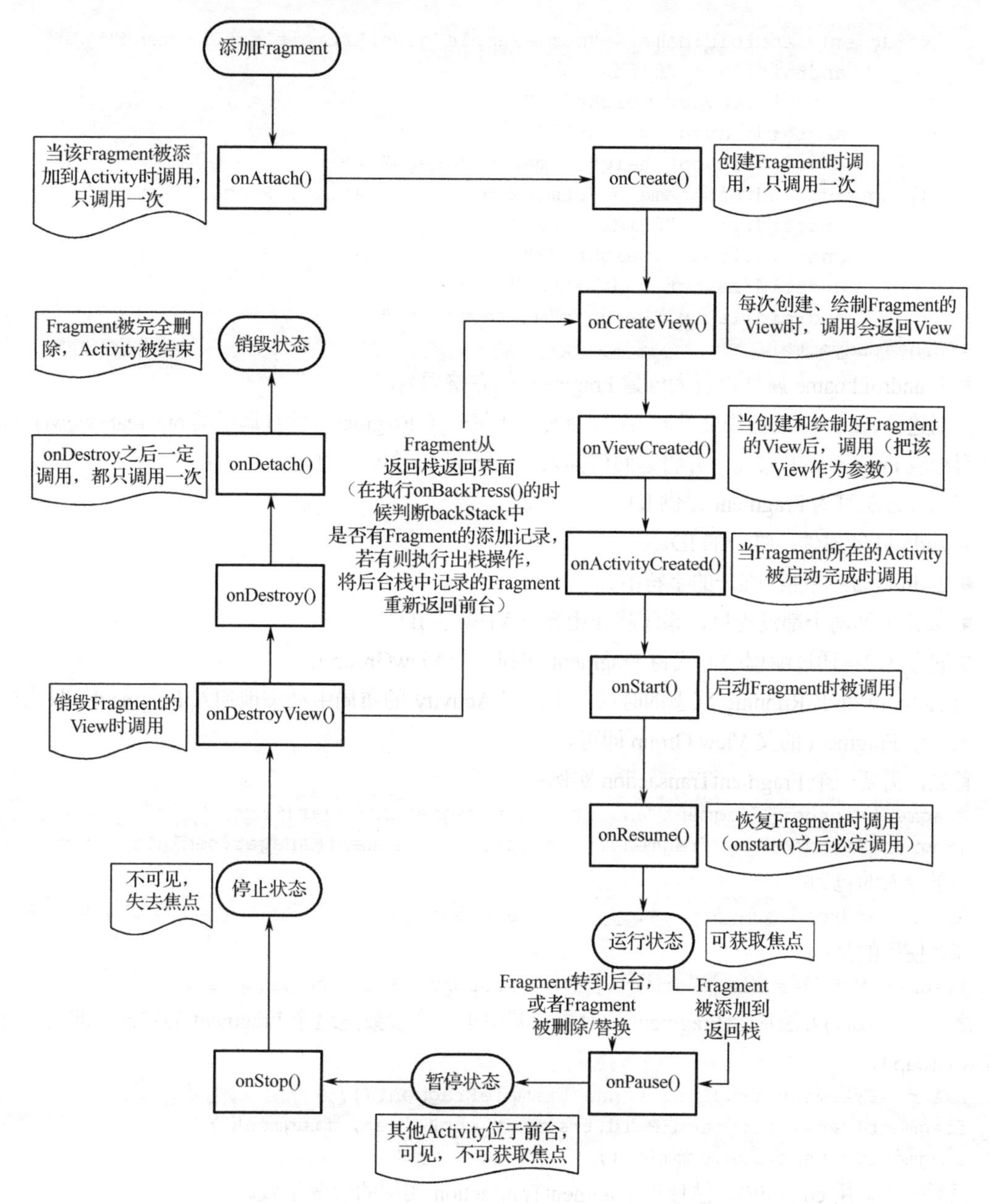

图 8-2　Fragment 与宿主 Activity 的生命周期方法调用顺序

8.3　Fragment 加入 Activity

当 Fragment 被加入 Activity 中时，它会处在对应的 View Group 中。

Fragment 有两种加载方式：①在 Activity 的 Layout 中使用标签<fragment>声明；②在代码中把它加到一个指定的 ViewGroup 中。

加载方式 1：通过 Activity 的布局文件将 Fragment 加入 Activity。

在 Activity 的布局文件中，将 Fragment 作为一个子标签加入即可，其代码如下：

```
<LinearLayout xmlns:android="http://schemas.android.com/apk/res/android"
    android:orientation="horizontal"
    android:layout_width="match_parent"
    android:layout_height="match_parent">
```

```
    <fragment  android:name = "com.example.news.ArticleListFragment"
            android:id = "@+id/list"
            android:layout_weight="1"
            android:layout_width="0dp"
            android:layout_height="match_parent" />
    <fragment  android:name = "com.example.news.ArticleReaderFragment"
            android:id = "@+id/viewer"
            android:layout_weight="2"
            android:layout_width="0dp"
            android:layout_height="match_parent" />
</LinearLayout>
```

其中 android:name 属性设置为创建 Fragment 的完整类名。

当创建该 Activity 的布局文件时，系统会实例化每一个 Fragment，并且调用其 onCreateView()方法，来获得相应 Fragment 的布局，并将返回值插入 Fragment 标签所在的地方。

有三种方法可为 Fragment 提供 ID。

- android:id 属性：唯一的 ID。
- android:tag 属性：唯一的字符串。
- 如果上面两个都没提供，系统将使用容器 View 的 ID。

加载方式 2：通过编程的方式将 Fragment 加到一个 ViewGroup 中。

当 Activity 处于 Running 状态的时候，可以在 Activity 的布局中动态地加入 Fragment，只需要指定加入这个 Fragment 的父 View Group 即可。

首先，需要一个 FragmentTransaction 实例：

```
FragmentManager fragmentManager = getFragmentManager();
FragmentTransaction fragmentTransaction = fragmentManager.beginTransaction();
```

如果导入的包为：

```
import android.support.v4.app.FragmentManager;
```

那么使用的是：

```
FragmentManager fragmentManager = getSupportFragmentManager();
```

之后，用 add()方法加上 Fragment 的对象，其中第一个参数是这个 Fragment 的容器，即父控件组（View Group）。

```
ExampleFragment fragment = new ExampleFragment();
fragmentTransaction.add(R.id.fragment_container, fragment);
fragmentTransaction.commit();
```

最后需要调用 commit()方法使得 FragmentTransaction 实例的改变生效。

小贴士：每个 FragmentTransaction 可以包含多个对 Fragment 的修改，如包含调用了多个 add()、remove()和 replace()操作，最后调用 commit()方法提交事务即可。在调用 commit()之前，也可调用 addToBackStack()将事务添加到返回栈中，该栈由 Activity 负责管理，这样允许用户按 Backspace 键返回到前一个 Fragment 状态。

【示例】 Fragment 加入 Activity 的两种方式：静态加载 Fragment 和动态加载 Fragment。

启动 Android Studio，在 Ch8 工程中创建 SimpleFragmentDemo 模块，在 SimpleFragmentDemo 模块中创建 MainActivity.java、SimpleFragment.java，以及相应的布局文件 activity_main.xml、simplefragment.xml 等。

小贴士：写一个类（SimpleFragment）继承自 Fragment 类，并且写好其布局文件（本例中是两个 TextView），在 SimpleFragment 类的 onCreateView()方法中加入该布局，之后用两种方法在 Activity 中加入这个 Fragment，第一种方法是在 Activity 的布局文件中加入<fragment>标签；第二种方法是在 Activity 的代码中使用 FragmentTransaction 的 add()方法加入 fragment。

SimpleFragmentDemo 模块的 SimpleFragment.java 主要代码如下：

```
public class SimpleFragment extends Fragment {
    String tag = "SimpleFragment";
    @Override
    public void onCreate(Bundle savedInstanceState) {
        super.onCreate(savedInstanceState);
        Log.d(tag, "SimpleFragment--onCreate");
    }
    @Override
    public View onCreateView(LayoutInflater inflater, ViewGroup container,
                             Bundle savedInstanceState) {
        Log.d(tag, "SimpleFragment--onCreateView");
        return inflater.inflate(R.layout.simplefragment, container, false);
    }
    @Override
    public void onPause() {
        super.onPause();
        Log.d(tag, "SimpleFragment--onPause");
    }
}
```

SimpleFragment 对应的布局文件 simplefragment.xml 主要代码如下：

```
<?xml version="1.0" encoding="utf-8"?>
<LinearLayout xmlns:android="http://schemas.android.com/apk/res/android"
    android:layout_width="match_parent"
    android:layout_height="match_parent"
    android:orientation="vertical" >
    <TextView
        android:layout_width="match_parent"
        android:layout_height="wrap_content"
        android:text="@string/num1"
        android:textColor="@android:color/holo_red_dark"
        android:textSize="30sp"
        android:textStyle="bold" />
    <TextView
        android:layout_width="match_parent"
        android:layout_height="wrap_content"
        android:text="@string/num2"
        android:textColor="@android:color/holo_red_dark"
        android:textSize="30sp"
        android:textStyle="bold" />
</LinearLayout>
```

SimpleFragmentDemo 模块的 MainActivity.java 主要代码如下：

```
public class MainActivity extends AppCompatActivity {
    @Override
    protected void onCreate(Bundle savedInstanceState) {
        super.onCreate(savedInstanceState);
        setContentView(R.layout.activity_main);
        FragmentManager fragmentManager = getSupportFragmentManager();
        FragmentTransaction fragmentTransaction = fragmentManager.beginTransaction();
        SimpleFragment fragment = new SimpleFragment();
        fragmentTransaction.add(R.id.linear,  fragment);
        fragmentTransaction.commit();
    }
}
```

MainActivity 对应的布局文件 activity_main.xml 主要代码如下：

```
<LinearLayout xmlns:android="http://schemas.android.com/apk/res/android"
    xmlns:tools="http://schemas.android.com/tools"
    android:layout_width="match_parent"
    android:layout_height="match_parent"
    android:orientation="vertical" >
    <Button
        android:id="@+id/btn1"
        android:layout_width="match_parent"
        android:layout_height="wrap_content"
        android:text="@string/btn1"
        android:textSize="18sp"
        android:textStyle="bold" />
    <fragment
        android:id="@+id/fragment1"
        android:name="com.mialab.simplefragmentdemo.SimpleFragment"
        android:layout_width="match_parent"
        android:layout_height="wrap_content" />
    <Button
        android:id="@+id/btn2"
        android:layout_width="match_parent"
        android:layout_height="wrap_content"
        android:text="@string/btn2"
        android:textSize="18sp"
        android:textStyle="bold" />
    <LinearLayout
        android:id="@+id/linear"
        android:layout_width="match_parent"
        android:layout_height="wrap_content"
        android:orientation="vertical" >
        <Button
            android:id="@+id/btn3"
            android:layout_width="match_parent"
            android:layout_height="wrap_content"
            android:text="@string/btn3"
            android:textColor="@android:color/holo_blue_dark"
            android:textSize="18sp"
            android:textStyle="bold" />
    </LinearLayout>
</LinearLayout>
```

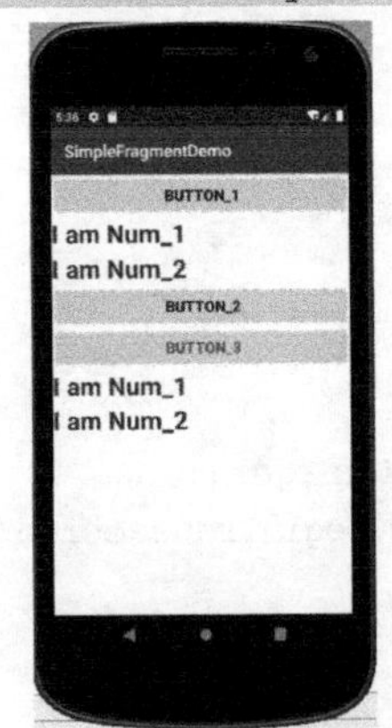

图 8-3　静态加载 Fragment 和动态加载 Fragment

运行 SimpleFragmentDemo 模块，其界面如图 8-3 所示。

可以看到第二种方法加入 Fragment 时，指定了父容器（线性布局容器）的 ID，父容器（线性布局容器）中已经有一个 Button 3，所以动态加载的 Fragment 加在了其后。

小贴士：通过 Activity 中 Fragment 的管理器，将 Fragment 添加到 Activity 指定的布局中，每一次添加 Fragment 都会在原界面的基础上添加 Fragment，可以理解为在原 Fragment 的基础上又覆盖了一层，如果 Fragment 之间背景透明，可以看到各个添加的 Fragment 内容重叠，所以，一般在第一个将 Fragment 显示到界面上时使用 addFragment 方式，要改变 Fragment 的内容，则通过 replaceFragment 来进行。若使用替换的方式，将不会出现内容重叠的情况。

Fragment 依附于 Activity 来做内容的显示，当在 Activity 中点击返回按钮时，这个 Activity 将会销毁。想通过点击实现 Fragment 的返回，而不是直接退出 Activity，则需要使用到返回栈。

表示入栈的操作是 addToBackStack(@Nullable String name)，其示例代码如下：

```
FragmentManager  mSupportFragmentManager =  getSupportFragmentManager();
mSupportFragmentManager.beginTransaction()
        .replace(R.id.frameLayout_content, fragments[0], "first")
        .addToBackStack("first")
        .commit();
```

出栈的操作有以下方法。

- popBackStack()：将返回栈的栈顶元素 pop 出栈。
- popBackStack(String name, int flags)：将标记为 name 之上的元素全部移除，flags 表示标记为 name 的这个元素是否需要出栈。0 表示不需要，POP_BACK_STACK_INCLUSIVE 表示该元素需要出栈。
- popBackStack(int id,int flags)：将栈中位置在 ID 上面的元素都移除，根据 flags 来判断是否移除该位置的元素。
- popBackStackImmediate()：和上面相比，只是多了个 Immediate，表示立即执行，其余的参数则和上面一致。

关于 Fragment 返回栈的示例可参见教学资源包中的 Ch8\BackStackDemo。

8.4　Fragment 与 Activity 的交互

将 Fragment 添加到 Activity 以后，Fragment 必须与 Activity 交互信息，这就需要 Fragment 能获取它所在的 Activity，Activity 也能获取它所包含的任意的 Fragment。通常 Fragment 与 Activity 之间还需要相互传递数据。

1．Fragment 获得 Activity 中的组件

Fragment 通过调用 getActivity()可获得它所在的 Activity。显然，Fragment 通过调用 getActivity().findViewById(R.id.<控件 id>)，便可获得所在 Activity 中的组件。

但是注意调用 getActivity()时，Fragment 必须和 Activity 关联（attached to an activity），否则将会返回一个 null。另外，当 Fragment 中需要使用 Context 对象时，也可以使用 getActivity()方法，这时获取到的 Activity 本身就是一个 Context 对象。

2．Activity 获得 Fragment 中的组件

Activity 通过调用所关联 FragmentManager 的 findFragmentById(int id)或者 findFragmentByTag (String tag)方法即可获取指定的 Fragment。

3．Activity 传递数据给 Fragment

在 Activity 中创建 Bundle 数据包，调用 Fragment 实例的 setArguments(Bundle bundle) 方法，将 Bundle 数据包传给 Fragment。然后在 Fragment 中调用 getArguments()方法获得 Bundle 对象，对 Bundle 对象进行解析即可得到传递过来的数据。

4．Fragment 传递数据给 Activity

在 Fragment 中定义一个内部回调接口，再让包含该 Fragment 的 Activity 实现该回调接口，这样 Fragment 即可使用该回调方法将数据传给 Activity。接口回调是 Java 不同对象之间数据交互的通用方法。

Activity 实现了 Fragment 中定义的回调接口后，又该怎么传给 Fragment 呢？

当 Fragment 添到 Activity 中时，就会回调 Fragment 的 onAttach()方法，可在此方法中检查 Activity 是否实现了 Fragment 中定义的接口，检查方法就是对传入的 Activity 实例进行类型转换，然后赋值给在 Fragment 中定义的接口引用。

当 Fragment 从 Activity 中"剥离"的时候，就会调用 onDetach()方法，这个时候要把传递进来的 Activity 对象释放掉，不然会影响 Activity 的销毁。

Fragment 传递数据给 Activity 的核心代码如下：

```
    //MainActivity 实现 ExampleFragment 中定义的接口
    public class MainActivity extends FragmentActivity implements ExampleFragment.
FragmentListener{
        @override
        public void processData(String data){…}
        …其他处理代码省略
    }

    public class ExampleFragment extends Fragment{
        public FragmentListener mListener;
        public static interface FragmentListener{
            public void processData(String data);
        }
        //接口回调
        public void getData(FragmentListener mListener){
            //获取文本框的信息,当然也可以传递其他类型的参数,主要看需求
            String msg = inputText.getText().toString();
            mListener.processData(msg);
        }

        @Override
        public void onAttach(Context context) {
            super.onAttach(context);
            if(context  instanceof  FragmentListener){ //对传递进来的Activity实例进行类
型转换
                mListener = ((FragmentListener)context);
            }
        }
        …其他处理代码省略
    }
```

图 8-4　Fragment 传递数据给 Activity 示例

【示例】 Fragment 传递数据给 Activity。

启动 Android Studio，在 Ch8 工程中创建 FragmentdataToActivity 模块，在 FragmentdataToActivity 模块中创建 MainActivity.java、ExampleFragment.java，及相应的布局文件 activity_main.xml、fragment_example.xml，以及用作线性布局边框的 border.xml（实际上是 ExampleFragment 的 UI 界面边框，用以区分 MainActivity 和加载 ExampleFragment 的不同区域，可方便演示），其代码参见教学资源 Ch8\FragmentdataToActivity。

运行 FragmentdataToActivity 模块，其界面如图 8-4 所示。

小贴士：回调的思想是，类 A 的 a()方法调用类 B 的 b()方法，类 B 的 b()方法执行完毕后，主动调用类 A 的 callback()方法（类 A 的实现 CallBack 接口中定义的方法）。关于接口回调机制，一般来说，①Class A 实现接口 CallBack；②Class A 中包含一个 Class B 的引用 b；③Class B 中定义了接口 CallBack；

④Class B 中有一个参数为 Callback 类型的方法 f(CallBack callback)，往往称之为回调方法；⑤A 的对象a调用B的方法f(CallBack callback)(A类调用B类的某个方法C)，然后b就可以在f(CallBack callback)方法中调用 A 的方法（B 类调用 A 类的某个方法 D）。相关示例可参见教学资源包中的 Ch8\FragmentdataToActivity_2。

8.5 Fragment 间的数据传递

这里“Fragment 间的数据传递”指的是同一个 Activity 托管的两个 Fragment 之间的数据传递。假设这两个 Fragment 分别为：FirstFragment 和 SecondFragment。

小贴士：要附加 Argument Bundle 给 Fragment，须调用 Fragment 实例的 setArguments(Bundle bundle)方法。而且还必须在 Fragment 创建后、添加给 Activity 前完成。习惯做法是，添加名为 newInstance()的静态方法给 Fragment 类。使用该方法完成 Fragment 实例及 Bundle 对象的创建，然后将传递的数据放入 Bundle 对象中，再附加 Argument Bundle 给 Fragment，其示例代码如下：

```
public class DemoFragment extends Fragment {
    public static DemoFragment newInstance(DataType data) {
        Bundle bundle = new Bundle();
        bundle.putSerializable(KEY, data);
        DemoFragment  fragment = new  DemoFragment();
        fragment.setArguments(bundle);
        return fragment;
    }
}
```

当 Fragment 需要获取 Argument 时，可以先调用 Fragment 类的 getArguments() 方法获得 Bundle 对象，再调用 Bundle 的限定类型 getxxx()方法，如 getString(String key)、getSerializable 等，相关示例可参考 Android 官方文档。

1. FirstFragment 传递数据给 SecondFragment

创建和设置 Fragment Argument 通常是在 newInstance()方法中完成的。在 SecondFragment.java 中，须添加 newInstance()方法，其示例代码如下：

```
public class SecondFragment extends Fragment {
    public staticSecondFragment newInstance(DataType data) {
        Bundle bundle = new Bundle();
        bundle.putSerializable(KEY, data);
        SecondFragment  fragment = new  SecondFragment();
        fragment.setArguments(bundle);
        return fragment;
    }
}
```

而在 FirstFragment 的 UI 界面的事件行为中，可调用 SecondFragment.newInstance (DataType data)方法，并把数据通过参数封装到 Bundle 对象中传递给 SecondFragment，SecondFragment 获取 Bundle 对象并解析，且得到 FirstFragment 传递过来的数据。

2. SecondFragment 返回数据给 FirstFragment

（1）设置目标 Fragment。

类似 Activity 间的关联，可将 FirstFragment 设置成 SecondFragment 的目标 Fragment。调用 Fragment 的 setTargetFragment ()方法可建立这种关联。

```
public void setTargetFragment (Fragment fragment, int requestCode)
```

该方法有两个参数：目标 fragment 和请求码 requestCode。

（2）传递数据给目标 Fragment。

Fragment 的 onActivityResult ()方法声明如下。

```
    public void onActivityResult (int requestCode,
int resultCode, Intent data)
```

其中，onActivityResult (int requestCode, int resultCode, Intent data)的三个参数解释如下。

- int requestCode：设置目标 Fragment 传入的请求码，告诉目标 Fragment 返回结果来自哪里。
- int resultCode：返回数据时传入的结果码，以判断处理结果是否成功，来决定下一步行动。
- Intent data：表示携带返回数据的 Intent。

【示例】 Activity 传递数据给 Fragment，以及同一宿主 Activity 的 Fragment 间的数据传递。

启动 Android Studio，在 Ch8 工程中创建 CourseDemo 模块，CourseDemo 模块（源码）目录结构如图 8-5 所示。

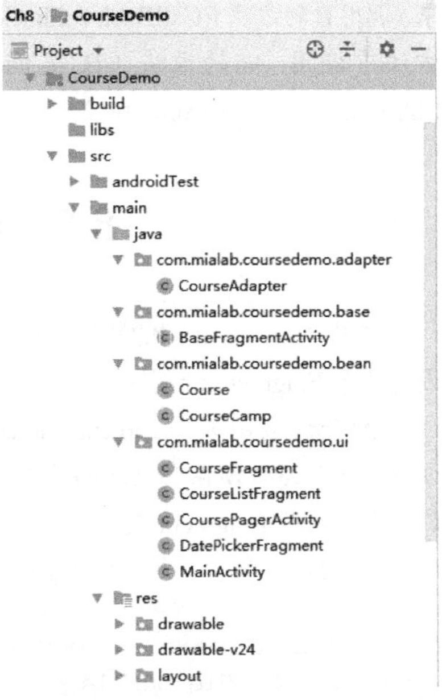

图 8-5 Ch8\CourseDemo 目录结构

运行 CourseDemo 模块并测试，其界面如图 8-6 所示。

CourseDemo 模块的 MainActivity 动态加载了 CourseListFragment，构成了主界面，如图 8-6（a）所示。MainActivity.java 的主要代码如下：

```
public class MainActivity extends BaseFragmentActivity {
    @Override
    protected Fragment createFragment() {
        return new CourseListFragment();
    }
}
```

其中，com.mialab.coursedemo.base 包中的 BaseFragmentActivity 是用来托管 Fragment 的抽象 Activity 类，MainActivity 扩展自 BaseFragmentActivity。同样如果有其他的 Activity 需要动态加载 Fragment 的话，也可以扩展 BaseFragmentActivity。

BaseFragmentActivity 类作为抽象基类，起到了减少代码冗余和重复的作用，提高了编码的效率。BaseFragmentActivity.java 的主要代码如下：

```
public abstract class BaseFragmentActivity extends FragmentActivity {
    protected abstract Fragment createFragment();
    @Override
    protected void onCreate(Bundle savedInstanceState) {
        super.onCreate(savedInstanceState);
        setContentView(R.layout.activity_base_fragment);
        FragmentManager fm = getSupportFragmentManager();
        Fragment fragment = fm.findFragmentById(R.id.fragment_container);
        if (fragment == null) {
            fragment = createFragment();
            fm.beginTransaction().add(R.id.fragment_container,fragment).commit();
        }
    }
}
```

其中，activity_base_fragment.xml 只提供了一个用来放置 Fragment 的 FrameLayout 容器视图，Fragment 可在 Activity 中使用代码获取。activity_base_fragment.xml 是一个通用的 Fragment 托管布局，

唯一所提供的 FrameLayout 容器 ID 为 fragment_container。

(a) (b) (c)

(d) (e) (f) (g)

图 8-6 运行 CourseDemo 模块的界面

com.mialab.coursedemo.ui 包中的 CourseListFragment 使用了强大的 RecyclerView 控件。

CourseListFragment.java 主要代码如下：

```
public class CourseListFragment extends Fragment {
    private RecyclerView recyclerView;
    private CourseAdapter courseAdapter;
    @Override
    public View onCreateView(LayoutInflater inflater, ViewGroup container, Bundle
savedInstanceState) {
        View view = inflater.inflate(R.layout.fragment_course_list, container,
false);
        recyclerView = (RecyclerView) view.findViewById(R.id.course_recycler_
view);
        recyclerView.setLayoutManager(new LinearLayoutManager(getActivity()));
        updateUI();
        return view;
    }
    @Override
```

```
    public void onResume() {
        super.onResume();
        updateUI();
    }
    private void updateUI() {
        CourseCamp courseCamp = CourseCamp.get(getActivity());
        List<Course> courseList = courseCamp.getCourseList();
        if (courseAdapter == null) {
            courseAdapter = new CourseAdapter(courseList, getActivity());
            recyclerView.setAdapter(courseAdapter);
        } else {
            courseAdapter.notifyDataSetChanged();
        }
    }
}
```

com.mialab.coursedemo.adapter 包中的 CourseAdapter.java 主要代码如下：

```
public class CourseAdapter extends RecyclerView.Adapter<CourseAdapter. Course
Holder> {
    private List<Course> mCourseList;
    private static Context mContext;
    public CourseAdapter(List<Course> courseList, Context context) {
        mCourseList = courseList;
        mContext = context;
    }
    @Override
    public CourseHolder onCreateViewHolder(ViewGroup parent, int viewType) {
        LayoutInflater layoutInflater = LayoutInflater.from(mContext);
        View view = layoutInflater.inflate(R.layout.list_item_course, parent,
false);
        CourseHolder courseHolder = new CourseHolder(view);
        return courseHolder;
    }
    @Override
    public void onBindViewHolder(CourseHolder holder, int position) {
        Course course = mCourseList.get(position);
        holder.bindCourse(course);
    }
    @Override
    public int getItemCount() {
        return mCourseList.size();
    }
    static class CourseHolder extends RecyclerView.ViewHolder implements View.
OnClickListener {
        private TextView cTitle;
        private TextView eDate;
        private CheckBox mCheck;
        private Course mCourse;
        public CourseHolder(View itemView) {
            super(itemView);
            itemView.setOnClickListener(this);
            cTitle = (TextView) itemView.findViewById(R.id.list_item_course_title_
text_view);
            eDate = (TextView) itemView.findViewById(R.id.list_item_course_date_
text_view);
```

```
            mCheck = (CheckBox) itemView.findViewById(R.id.list_item_course_completed_
check_box);
        }
        public void bindCourse(Course course) {
            mCourse = course;
            cTitle.setText(mCourse.getCourseTitle());
            eDate.setText("课程立项时间：" + mCourse.getEstablishedDate());
            mCheck.setChecked(mCourse.ismCompleted());
        }
        @Override
        public void onClick(View v) {
            Intent  intent  =  CoursePagerActivity.newIntent(mContext,  mCourse.
getCourseId());
            mContext.startActivity(intent);
        }
    }
}
```

CourseAdapter 是 CourseListFragment 中 RecyclerView 控件的适配器。

com.mialab.coursedemo.bean 包中的 Course.java 主要代码如下：

```
public class Course {
    private String courseId;
    private String courseTitle;
    private String establishedDate;
    private boolean mCompleted;
        …
}
```

com.mialab.coursedemo.bean 包中的 CourseCamp.java 主要代码如下：

```
public class CourseCamp {
    private static CourseCamp courseCamp;
    private ArrayList<Course> courseList;
    public static CourseCamp get(Context context) {
        if (courseCamp == null) {
            courseCamp = new CourseCamp(context);
        }
        return courseCamp;
    }
    private CourseCamp(Context context) {
        courseList = new ArrayList<>();
        for (int i = 1; i < 52; i++) {
            Course course = new Course();
            course.setCourseId("20161226-C" + i);
            course.setCourseTitle("立项课程编号：20161226-C" + i);
            course.setEstablishedDate("2016-12-26");
            course.setmCompleted(i%2 == 0);
            courseList.add(course);
        }
    }
    public List<Course> getCourseList() {
        return courseList;
    }
    public Course getCourse(String  newId) {
        for (Course course : courseList) {
            if (course.getCourseId().equals(newId)) {
```

```
                return course;
            }
        }
        return null;
    }
}
```

com.mialab.coursedemo.ui 包中的 CoursePagerActivity 和 CourseFragment，构成了“立项课程详情”界面，如图 8-6（b）所示。CoursePagerActivity.java 主要代码如下：

```
public class CoursePagerActivity extends FragmentActivity {
    private static final String EXTRA_COURSE_ID ="com.mialab.coursedemo.intent.
course_id";
    private ViewPager mViewPager;
    private List<Course> courseList;
    public static Intent newIntent(Context packageContext, String courseId) {
        Intent intent = new Intent(packageContext, CoursePagerActivity.class);
        intent.putExtra(EXTRA_COURSE_ID, courseId);
        return intent;
    }
    @Override
    protected void onCreate(Bundle savedInstanceState) {
        super.onCreate(savedInstanceState);
        setContentView(R.layout.activity_course_pager);
        String courseId = (String) getIntent().getSerializableExtra(EXTRA_
COURSE_ID);
        mViewPager = (ViewPager) findViewById(R.id.activity_course_pager_view_
pager);
        courseList = CourseCamp.get(this).getCourseList();
        FragmentManager fragmentManager = getSupportFragmentManager();
        mViewPager.setAdapter(new FragmentStatePagerAdapter(fragmentManager) {
            @Override
            public Fragment getItem(int position) {
                Course course = courseList.get(position);
                return CourseFragment.newInstance(course.getCourseId());
            }
            @Override
            public int getCount() {
                return courseList.size();
            }
        });
        for (int i = 0; i < courseList.size(); i++) {
            if (courseList.get(i).getCourseId().equals(courseId)) {
                mViewPager.setCurrentItem(i);
                break;
            }
        }
    }
}
```

com.mialab.coursedemo.ui 包中的 CourseFragment.java 主要代码如下：

```
public class CourseFragment extends Fragment {
    private static final String COURSE_ID = "course_id";
    private static final String DIALOG_DATE = "DialogDate";
    public static final String EXTRA_DATE = "com.mialab.coursedemo.intent.date";
    private static final int REQUEST_DATE = 0;
```

```
        private Course course;
        private EditText cTitle;
        private Button dateButton;
        private CheckBox checkBox;
        public static CourseFragment newInstance(String courseId) {
            Bundle args = new Bundle();
            args.putSerializable(COURSE_ID, courseId);
            CourseFragment fragment = new CourseFragment();
            fragment.setArguments(args);
            return fragment;
        }
        @Override
        public void onCreate(Bundle savedInstanceState) {
            super.onCreate(savedInstanceState);
            String courseId = (String) getArguments().getSerializable(COURSE_ID );
            course = CourseCamp.get(getActivity()).getCourse(courseId);
        }
        @Override
        public View onCreateView(LayoutInflater inflater, ViewGroup container, Bundle
savedInstanceState) {
            View v = inflater.inflate(R.layout.fragment_course, container, false);
            cTitle = (EditText) v.findViewById(R.id.course_title);
            cTitle.setText(course.getCourseTitle());
            cTitle.addTextChangedListener(new TextWatcher() {
                @Override
                public void beforeTextChanged(CharSequence s, int start, int count, int
after) {
                }
                @Override
                public void onTextChanged(CharSequence s, int start, int before, int
count) {
                    course.setCourseTitle(s.toString());
                }
                @Override
                public void afterTextChanged(Editable s) {
                }
            });
            dateButton = (Button) v.findViewById(R.id.crime_date);
            updateDate();
            dateButton.setOnClickListener(new View.OnClickListener() {
                @Override
                public void onClick(View v) {
                    FragmentManager manager = getFragmentManager();
                    SimpleDateFormat sdf = new SimpleDateFormat("yyyy-MM-dd");
                    String establishedDateStr = course.getEstablishedDate();
                    Date mDate = null;
                    try {
                        mDate = sdf.parse(establishedDateStr);
                    } catch (ParseException e) {
                        e.printStackTrace();
                    }
                    DatePickerFragment dialog = DatePickerFragment.newInstance(mDate);
                    dialog.setTargetFragment(CourseFragment.this, REQUEST_DATE);
                    dialog.show(manager, DIALOG_DATE);
```

```
            }
        });
        checkBox = (CheckBox) v.findViewById(R.id.course_completed);
        checkBox.setChecked(course.ismCompleted());
        checkBox.setOnCheckedChangeListener(new CompoundButton.OnCheckedChange
Listener() {
            @Override
            public void onCheckedChanged(CompoundButton buttonView, boolean isChecked) {
                course.setmCompleted(isChecked);
            }
        });
        return v;
    }
    @Override
    public void onActivityResult(int requestCode, int resultCode, Intent intent) {
        if (resultCode != Activity.RESULT_OK) {
            return;
        }
        if (requestCode == REQUEST_DATE) {
            Date eDate = (Date) intent.getSerializableExtra(DatePickerFragment.
EXTRA_DATE);
            SimpleDateFormat sdf = new SimpleDateFormat("yyyy-MM-dd");
            Bundle bundle = intent.getExtras();
            Date newDate = (Date) bundle.getSerializable(EXTRA_DATE);
            String eDateStr = sdf.format(newDate);
            course.setEstablishedDate(eDateStr);
            updateDate();
        }
    }
    private void updateDate() {
        dateButton.setText(course.getEstablishedDate());
    }
}
```

com.mialab.coursedemo.ui 包中的 DatePickerFragment 动态加载后的界面，如图 8-6（d）所示。DatePickerFragment.java 主要代码如下：

```
public class DatePickerFragment extends DialogFragment {
    public static final String EXTRA_DATE = "com.mialab.coursedemo.intent.date";
    private static final String COURSE_DATE = "course_date";
    private DatePicker datePicker;
    public static DatePickerFragment newInstance(Date date) {
        Bundle args = new Bundle();
        args.putSerializable(COURSE_DATE, date);
        DatePickerFragment fragment = new DatePickerFragment();
        fragment.setArguments(args);
        return fragment;
    }
    @Override
    public Dialog onCreateDialog(Bundle savedInstanceState) {
        Date date = (Date) getArguments().getSerializable(COURSE_DATE);
        Calendar calendar = Calendar.getInstance();
        calendar.setTime(date);
        int year = calendar.get(Calendar.YEAR);
        int month = calendar.get(Calendar.MONTH);
```

```
        int day = calendar.get(Calendar.DAY_OF_MONTH);
        View v = LayoutInflater.from(getActivity()).inflate(R.layout.dialog_date,
null);
        datePicker = (DatePicker) v.findViewById(R.id.dialog_date_date_picker);
        datePicker.init(year, month, day, null);
        return new AlertDialog.Builder(getActivity())
                .setView(v)
                .setTitle(R.string.date_picker_title)
                .setPositiveButton(android.R.string.ok,    new    DialogInterface.
OnClickListener() {
                        @Override
                        public void onClick(DialogInterface dialog, int which)
{
                            int year = datePicker.getYear();
                            int month = datePicker.getMonth();
                            int day = datePicker.getDayOfMonth();
                            Date date = new GregorianCalendar(year, month, day).
getTime();
                            Log.d("hand","" + year + "-" + (month+1) + "-" + day);
                            Log.d("hand","" + date);
                            sendResult(Activity.RESULT_OK, date);
                        }
                    })
                .create();
    }
    private void sendResult(int resultCode, Date date) {
        if (getTargetFragment() == null) {
            return;
        }
        Intent intent = new Intent();
        intent.putExtra(EXTRA_DATE, date);
        Log.d("hand","" + date);
        getTargetFragment().onActivityResult(getTargetRequestCode(), resultCode,
intent);
    }
}
```

其余文件代码参见教学资源包中的Ch8\CourseDemo模块，限于篇幅，这里不再赘述。

8.6 ViewPager和PageAdapter

ViewPager 用于实现多页面的切换效果，该类存在于 Google 的兼容包里面。ViewPager 是一个ViewGroup，包含多个View页，在手指横向滑动屏幕时，负责对View进行切换。

PageAdapter 是 ViewPager 的支持者，ViewPager 将调用它来取得所需显示的页。当 PageAdapter 也会在数据变化时，会通知ViewPager。这个类也是FragmentPagerAdapter及FragmentStatePagerAdapter的基类。

PageAdapter 提供一个适配器用于填充 ViewPager 页面。PageAdapter 的子类有 Fragment PagerAdapter和FragmentStatePagerAdapter，可以提供更加具体的实现。

当实现一个PagerAdapter时，至少需要覆盖以下几个方法：

- instantiateItem(ViewGroup, int);

- destroyItem(ViewGroup, int, Object)；
- getCount()；
- isViewFromObject(View, Object)。

使用 PagerAdapter 比 AdapterView 更加普通。ViewPager 使用回调函数来表示一个更新的步骤，而不是使用一个视图回收机制。在需要的时候 PagerAdapter 也可以实现视图的回收或者使用一种更为巧妙的方法来管理视图，如采用可以管理自身视图的 Fragment。

ViewPager 不直接处理每一个视图而是将各个视图与一个键联系起来。这个键用来跟踪且唯一代表一个页面，不仅如此，该键还独立于这个页面所在 Adapter 的位置。当 PagerAdapter 将要改变的时候，它就会调用 startUpdate 函数，接下来再调用一次或多次的 instantiateItem 或者 destroyItem，最后在更新后期会调用 finishUpdate。当 finishUpdate 返回时，instantiateItem 返回的对象应该添加到父 ViewGroup 中。

destroyItem 返回的对象会被 ViewGroup 删除。方法 isViewFromObject(View, Object)代表了当前的页面是否与给定的键相关联。

对于非常简单的 PagerAdapter 或许用户可以选择用 page 本身作为键，在创建并且添加到 ViewGroup 的 instantiateItem 方法里返回该 page 本身即可。

destroyItem 会将该 page 从 viewGroup 中移除。isViewFromObject 方法可直接可以返回 view == object。

PagerAdapter 支持数据集合的改变，数据集合的改变必须在主线程中执行，然后还要调用 notifyDataSetChanged 方法。

和 BaseAdapter 非常相似，数据集合的改变包括页面的添加、删除和修改位置。由于 ViewPager 要维持当前页面是活动的，所以必须提供 getItemPosition 方法。

【示例】 ViewPager 的简单使用。

启动 Android Studio，在 Ch8 工程中创建 PagerTitleDemo 模块，在 PagerTitleDemo 模块中创建 MainActivity.java、activity_main.xml、view1.xml、view2.xml、view3.xml 等。

在 activity_main.xml 的线性布局容器中放置了多页显示控件 ViewPager，ViewPager 控件又包含了 PagerTitleStrip，用于显示当前页面的标题。对应于 ViewGroup 中的 View、view1.xml、view2.xml、view3.xml 各放置了一幅图片。

activity_main.xml 的主要内容如下。

```
<LinearLayout xmlns:android="http://schemas.android.com/apk/res/android"
    android:layout_width="match_parent"
    android:layout_height="match_parent"
    android:orientation="vertical" >
    <android.support.v4.view.ViewPager
        android:id="@+id/viewpager"
        android:layout_width="wrap_content"
        android:layout_height="wrap_content"
        android:layout_gravity="center" >
        <android.support.v4.view.PagerTitleStrip
            android:id="@+id/pagertitle"
            android:layout_width="wrap_content"
            android:layout_height="wrap_content"
            android:layout_gravity="top"/>
    </android.support.v4.view.ViewPager>
</LinearLayout>
```

PagerTitleStrip 是 ViewPager 的一个关于当前页面、上一个页面和下一个页面的非交互的指示器。

它经常作为 ViewPager 控件的一个子控件被添加在 XML 布局文件中。

通常需要将 PagerTitleStrip 的 android:layout_ gravity 属性设置为 TOP 或 BOTTOM，以将其显示在 ViewPager 的顶部或底部。每个页面的标题是通过适配器的 getPageTitle(int)函数提供给 ViewPager 的。

PagerTitleDemo 模块的 MainActivity.java 主要代码如下：

```
public class MainActivity extends AppCompatActivity {
    private ViewPager mViewPager;
    private PagerTitleStrip mPagerTitleStrip;
    @Override
    public void onCreate(Bundle savedInstanceState) {
        super.onCreate(savedInstanceState);
        setContentView(R.layout.main);
        mViewPager = (ViewPager)findViewById(R.id.viewpager);
        mPagerTitleStrip = (PagerTitleStrip)findViewById(R.id.pagertitle);
        //将要分页显示的 View 装入数组中
        LayoutInflater mLi = LayoutInflater.from(this);
        View view1 = mLi.inflate(R.layout.view1, null);
        View view2 = mLi.inflate(R.layout.view2, null);
        View view3 = mLi.inflate(R.layout.view3, null);
        //每个页面的 Title 数据
        final ArrayList<View> views = new ArrayList<View>();
        views.add(view1); views.add(view2); views.add(view3);
        final ArrayList<String> titles = new ArrayList<String>();
        titles.add("Tab1"); titles.add("Tab2"); titles.add("Tab3");
        // 填充 ViewPager 的数据适配器
        PagerAdapter mPagerAdapter = new PagerAdapter() {
            @Override
            public boolean isViewFromObject(View arg0, Object arg1) {
                return arg0 = = arg1;
            }
            @Override
            public int getCount() {
                return views.size();
            }
            @Override
            public void destroyItem(View container, int position, Object object) {
                ((ViewPager)container).removeView(views.get(position));
            }
            @Override
            public CharSequence getPageTitle(int position) {
                return titles.get(position);
            }
            @Override
            public Object instantiateItem(View container, int position) {
                ((ViewPager)container).addView(views.get(position));
                return views.get(position);
            }
        };
        mViewPager.setAdapter(mPagerAdapter);
    }
}
```

运行 PagerTitleDemo 模块，其界面如图 8-7 所示。

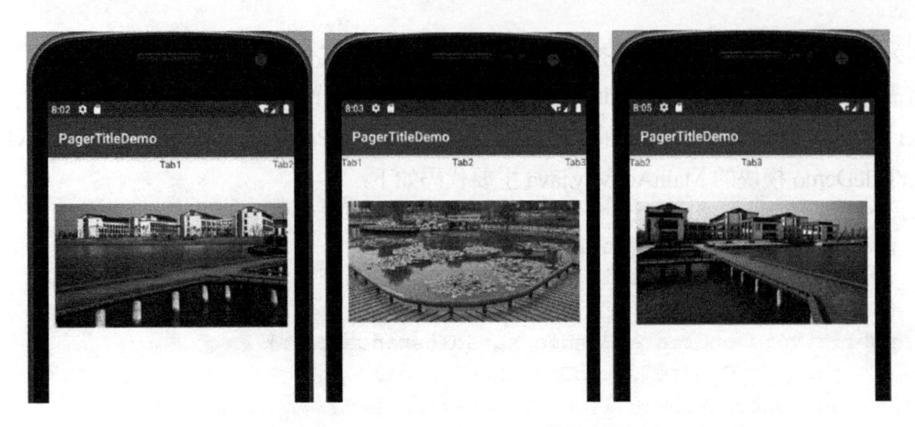

图 8-7　运行 PagerTitleDemo 模块的界面

8.7　使用 FragmentPageAdapter

FragmentPagerAdapter 继承自 PagerAdapter。相比通用的 PagerAdapter，它更专注于每一页均为 Fragment 的情况，其每一个生成的 Fragment 都将保存在内存之中，因此适用于那些相对静态、页面数量也比较少的情况。

如果需要处理很多页面，并且数据动态性较大、占用内存较多的情况，则应该使用 Fragment StatePagerAdapter。

FragmentPagerAdapter 重载实现了几个必须的函数，因此来自 PagerAdapter 的函数，只需实现 getCount()即可。由于 FragmentPagerAdapter.instantiateItem()的实现中，调用了一个新增的虚函数 getItem()，因此还至少需要实现一个 getItem()。总体上来说，相对于继承自 PagerAdapter，使用 FragmentPagerAdapter 则更方便一些。

FragmentStatePagerAdapter 和 FragmentPagerAdapter 一样都继承自 PagerAdapter。但和 FragmentPagerAdapter 不一样的是，该 PagerAdapter 的实现将只保留当前页面，当页面离开视线后，就会被消除，释放其资源。在页面需要显示时，将生成新的页面（就像 ListView 的实现一样）。这样实现的好处就是当拥有大量的页面时，不必占用大量的内存。

小贴士：在卸载不再需要的 Fragment 时，FragmentStatePagerAdapter 与 FragmentPagerAdapter 各自采用的处理方法有所不同。FragmentStatePagerAdapter 会销毁不需要的 Fragment。事务提交后，Activity 的 FragmentManager 中 Fragment 会被彻底移除。FragmentStatePagerAdapter 类名中的“State”表明，在销毁 Fragment 时，可在 onSaveInstanceState(Bundle bundle)方法中保存 Fragment 的 Bundle 信息。用户切换回来时，保存的实例状态可用来恢复生成新的 Fragment。但 FragmentPagerAdapter 有不同的做法。对于不再需要的 Fragment，FragmentPagerAdapter 会选择调用事务的 detach(Fragment fragment)来处理它，而非使用 remove(Fragment fragment)方法。FragmentPagerAdapter 只是销毁了 Fragment 的视图，Fragment 实例还保存在 FragmentManager 中。如果用户界面只需要少量固定的 Fragment，如每个 Activity 只需要加载三个左右的 Fragment，这时使用 FragmentPagerAdapter 是合适的选择。

【示例】 使用 FragmentPagerAdapter 实现仿 QQ 界面 。

启动 Android Studio，在 Ch8 工程中创建 FakeQQ 模块，在 FakeQQ 模块中创建 MainActivity.java、TestFragment.java、MyFragmentPagerAdapter.java、activity_main.xml、fragment_test.xml 等文件。

运行 FakeQQ 模块，其结果如图 8-8 所示，代码参见教学资源包中的 Ch8\FakeQQ 模块。

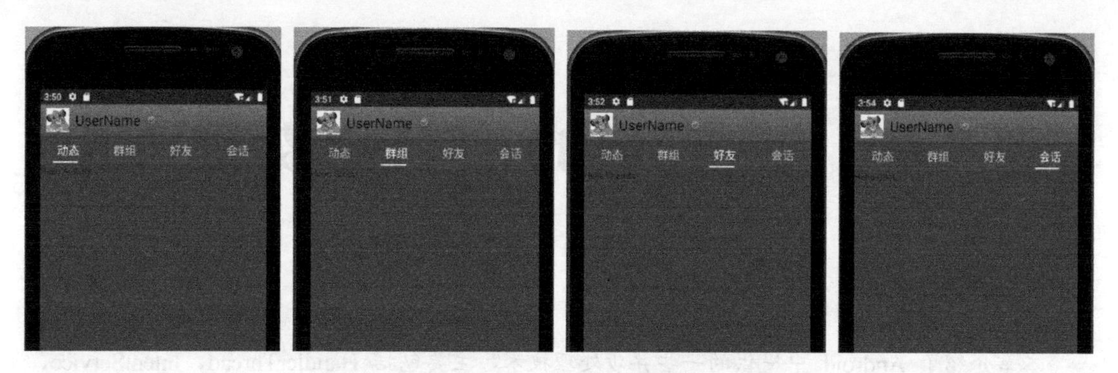

图 8-8　使用 FragmentPagerAdapter 实现仿 QQ 界面

8.8　本章小结

Fragment 通常作为 Activity 用户界面的一部分，并将其 Layout 提供给 Activity。为了给 Fragment 提供 Layout，必须实现 onCreateView()回调方法。此方法的实现代码必须返回一个能够表示 Fragment 的 Layout 的 View。

既可以把 Fragment 设计成在多个 Activity 中复用的模块，也可以组合多个 Fragment 放在一个单独的 Activity 中来创建一个多界面区域的 UI，并可以在多个 Activity 里重用某一个 Fragment。把 Fragment 想象成一个 Activity 的模块化区域，有自己的生命周期，接收输入事件，并且可以在 Activity 运行期间进行添加和删除。

因为 Fragment 必须嵌入在 Acitivity 中使用，所以 Fragment 的生命周期和它所在的 Activity 是密切相关的。如果 Activity 处于暂停状态，则其中所有的 Fragment 都是暂停状态；如果 Activity 处于 Stopped 状态，则这个 Activity 中所有的 Fragment 都不能被启动；如果 Activity 被销毁，那么其中的所有 Fragment 都会被销毁。但是，当 Activity 处于活动状态时，可以独立控制 Fragment 的状态，如进行添加或者移除 Fragment。

相关参考资源如下。

- Fragment 类文档：http://developer.android.com/reference/android/app/Fragment.html。
- Training：Building a Dynamic UI with Fragments。
- http://developer.android.com/training/basics/fragments/index.html。
- Fragments Develop Guide：http://developer.android.com/guide/components/fragments.html。

习　题　8

1．Fragment 的生命周期是什么？试编程加以说明。

2．如何把 Fragment 加入 Activity 中？动态加载 Fragment 和静态加载 Fragment 有何不同？试编程加以说明。

3．Activity 是如何传递数据给 Fragment 的？试编程举例说明。

4．Fragment 与 Fragment 之间是如何传递数据的？试编程举例说明。

5．如何使用 ViewPager 和 Fragment 实现 App 的导航框架？试编程举例加以说明。

6．FragmentStatePagerAdapter 与 FragmentPagerAdapter 的区别是什么？试编程加以说明。

第9章 Android的一些异步处理技术

本章介绍了 Android 中提供的一些异步处理技术，主要包括 HandlerThread、IntentService、AsyncTask。Android 应用的开发应正确处理主线程与子线程之间的关系，将耗时操作放到子线程中，以避免阻塞主线程，导致 ANR。异步处理技术是提高应用性能解决主线程和子线程之间通信问题的关键。本章主要内容有：（1）HandlerThread 的使用及源码分析；（2）IntentService 的使用及源码分析；（3）AsyncTask 的使用及工作原理。

9.1 HandlerThread

9.1.1 HandlerThread 的使用

HandlerThread 是 Android API 提供的一个便捷的类，使用它可以快速创建一个带有 Looper 的线程。Looper 可以用来创建 Handler 实例。注意：start()仍然必须被调用。

HanlderThread 类的声明：public class HandlerThread extends Thread {...}。

HanlderThread 继承自 Thread，使用 HandlerThread 能够创建拥有 Looper 的线程。在 HandlerThread 的 run()方法中，通过 Looper.prepare()创建了消息队列，并通过 Looper.loop()开启了消息循环。

使用时开启 HandlerThread，创建 Handler 与 HandlerThread 的 Looper 绑定，Handler 以消息的方式通知 HandlerThread 来执行一个具体的任务。

【示例】 使用 HandlerThread 获取随机数。

启动 Android Studio，在 Ch9 工程中创建 HandlerThreadDemo 模块，并在该模块中创建 MainActivity.java 和 activity_main.xml。

HandlerThreadDemo 模块的 MainActivity.java 主要代码如下：

```
public class MainActivity extends AppCompatActivity {
    private final static String TAG = "MainActivity";
    private Button btnGetNumber;
    private TextView mResult;
    protected final int MSG_GET = 1;
    protected final int MSG_RESULT = 2;
    private HandlerThread mHandlerThread;
    private Handler mSubThreadHandler;    //子线程中的 Handler 实例
    //与 UI 线程绑定的 Handler 实例
    private Handler mUiHandler = new Handler(){
        public void handleMessage(Message msg) {
            Log.i(TAG, "mUIHandler handleMessage thread:"+Thread.currentThread());
            switch (msg.what) {
                case MSG_RESULT:
                    mResult.setText((String)msg.obj);
                    mResult.setTextColor(Color.BLUE);
```

```
                mResult.setTextSize(20);
                break;
            default:
                break;
            }
        };
    };
    @Override
    protected void onCreate(Bundle savedInstanceState) {
        super.onCreate(savedInstanceState);
        setContentView(R.layout.activity_main);
        Log.i(TAG, "onCreate thread:"+Thread.currentThread());
        btnGetNumber = (Button) findViewById(R.id.button);
        mResult = (TextView) findViewById(R.id.textview);
        initHandlerThread();
        btnGetNumber.setOnClickListener(new View.OnClickListener() {
            @Override
            public void onClick(View v) {
                mSubThreadHandler.sendEmptyMessage(MSG_GET);
            }
        });
    }
    private void initHandlerThread() {
        //创建HandlerThread实例，参数字符串定义新线程的名称
        mHandlerThread = new HandlerThread("handler_thread");
        mHandlerThread.start(); //开始运行HandlerThread新线程
        Looper loop = mHandlerThread.getLooper(); //获取 HandlerThread 线程中的
Looper实例

            /**
         *创建与该线程绑定的Handler对象
         *将HandlerThread中的Looper作为参数，这样就完成了Handler对象与Handler
         Thread的Looper对象的绑定（这里的Handler对象可以看作是绑定在HandlerThread
         *子线程中，所以handlerMessage里的操作是在子线程中运行的）
         */
        mSubThreadHandler = new Handler(loop){
            public void handleMessage(Message msg) {
                Log.i(TAG, "mSubThreadHandler handleMessage thread: " + Thread.
currentThread());
                switch(msg.what){
                    case MSG_GET:
                        try { //模拟延时处理
                            Thread.sleep(1000);
                        } catch (InterruptedException e) {
                            e.printStackTrace();
                        }
                        double number = Math.random();
                        String result = "number: "+number;
                        //向UI线程发送消息，更新UI
                        Message message = new Message();
                        message.what = MSG_RESULT;
                        message.obj = result;
                        mUiHandler.sendMessage(message);
                        break;
```

```
                default:
                    break;
            }
        };
    };
}
@Override
protected void onDestroy() {
    super.onDestroy();
    Log.i(TAG, "onDestroy");
    mHandlerThread.quit();  //退出 HandlerThread 的 Looper 循环
}
}
```

在 MainActivity 的 onCreate()方法中调用 initHandlerThread()方法，在 initHandlerThread()方法中完成以下操作：创建 HandlerThread 线程并运行该线程，获取 HandlerThread 线程中的 Looper 实例，通过 Looper 实例创建 Handler 实例，从而使 mSubThreadHandler 与该线程绑定在一起。

运行 HandlerThreadDemo 模块，其界面如图 9-1 所示。

图 9-1　运行 HandlerThreadDemo 模块（获取随机数）

如图 9-1（a）所示，点击“获取随机数”按钮，向 mSubThreadHandler 发送消息，mSubThreadHandler 将接收到的消息进行处理，由 Logcat 可知 mSubThreadHandler 的 handleMessage()方法运行在子线程（HandlerThread 线程）中。在 mSubThreadHandler 的 handleMessage()方法中模拟耗时操作，生成随机数，然后向主线程（UI 线程）中的 mUiHandler 发送消息（Message）。mUiHandler 的 handleMessage()方法运行在主线程，可以用来更新 UI 界面。当 Activity 销毁时，回调 onDestroy()方法，于是调用 mHandlerThread.quit()，退出 HandlerThread 的 Looper 循环。

【示例】 使用 HandlerThread 异步加载网络图片。

启动 Android Studio，在 Ch9 工程中创建 HandlerThreadDemo2 模块，并在该模块中创建 MainActivity.java、activity_main.xml、Constants.java、ImageModel.java。相关示例代码参见教学资源包中的 Ch9\HandlerThreadDemo2。

运行 HandlerThreadDemo2 模块，其结果如图 9-2 所示。

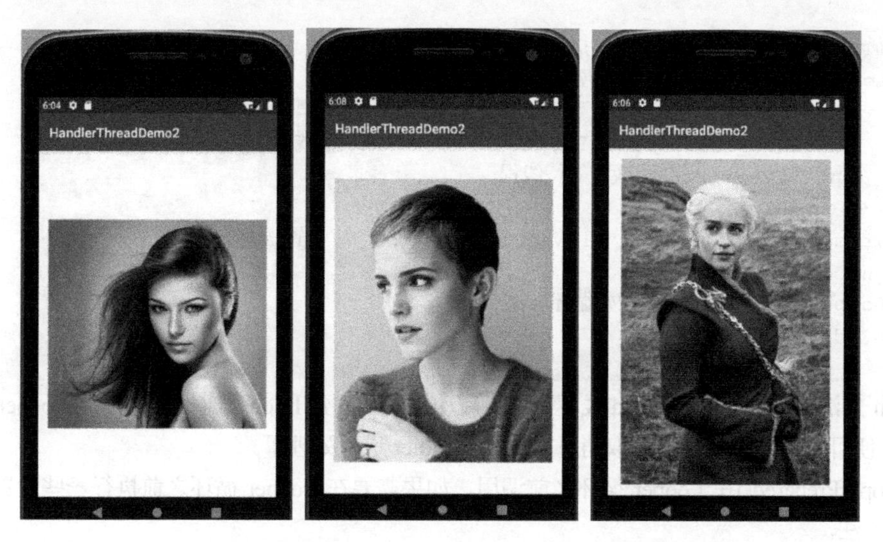

图 9-2　运行 HandlerThreadDemo2 模块（异步加载网络图片）

在 Ch9\HandlerThreadDemo2（示例）中，创建了两个 Handler，一个用于更新 UI 线程的 mUIHandler 和一个用于异步下载图片的 childHandler。childHandler 会每隔 3s，通过 sendEmptyMessageDelayed() 方法，通知 ChildCallback 的回调函数 handleMessage()去下载网络图片（ChildCallback 是实现 Handler.Callback 接口的），图片下载成功便会告知主线程的 mUIHandler 更新 UI 界面。

9.1.2　HandlerThread 的源码分析

先介绍 HandlerThread 的构造方法。

```
public class HandlerThread extends Thread {
    int mPriority;
    int mTid = -1;
    Looper mLooper;
    public HandlerThread(String name) {  //参数 name 表示线程名
        super(name);
        mPriority = Process.THREAD_PRIORITY_DEFAULT;
    }
    public HandlerThread(String name, int priority) {
        super(name);
        mPriority = priority;
    }
    ...
}
```

HandlerThread 是 Thread（线程）的子类。创建一个 HandlerThread 实例，也就是创建了一个特殊的线程实例。HandlerThread 提供了两个构造方法：

- HandlerThread(String name)：参数 name 表示线程名称，线程优先级设置为 Process.THREAD_PRIORITY_DEFAULT。
- HandlerThread(String name, int priority)：参数 name 为线程名称，priority 为设置的线程优先级。

其中，构造函数的传递参数有两个：name（线程名称）和 priority（线程优先级），根据需要调用即可。而成员变量 mLooper 就是 HandlerThread 自己持有的 Looper 对象。

通过 start()方法可以运行 HandlerThread（线程）的 run()方法。

```
protected void onLooperPrepared() { }
@Override
public void run() {
```

```
        mTid = Process.myTid();  //获取进程 id
        Looper.prepare();  //创建 Looper 实例
        synchronized (this) {
            mLooper = Looper.myLooper();   //获取当前线程的 Looper 实例
            notifyAll();  //唤醒等待线程
        }
        Process.setThreadPriority(mPriority);   //设置线程优先级
        onLooperPrepared();
        Looper.loop();   //开始循环
        mTid = -1;
    }
```

由 run 方法可知 HandlerThread 线程 start（启动）后，创建 Looper 实例，并开启 Looper 循环。这时 Looper 便可以从消息队列中获取消息，交给 Handler 进行处理了。

onLooperPrepared()在 Looper 循环之前调用，如果需要在 Looper 循环之前执行一些设置，可以显式覆盖此方法。

小贴士：onLooperPrepared()方法是一个空实现，是留给程序员去重写的，但是要注意重写的时机应在 Looper 循环启动前。

在 Ch9\HandlerThreadDemo（示例）的 MainActivity 的 initHandlerThread()方法中，开启 HandlerThread 线程后，便可以获取 Looper 实例，用来构建相应的 Handler 对象。

```
    Looper loop = mHandlerThread.getLooper(); //获取 HandlerThread 线程中的 Looper 实例
```

getLooper()方法的作用是：返回与此线程关联的 Looper。如果此线程未启动或由于任何原因致使 isAlive()返回值为 false，则此方法将返回 null。

HandlerThread 的 getLooper()方法的源码如下。

```
    public Looper getLooper() {
        if (!isAlive()) {
            return null;
        }
        // 如果这个线程已经启动，将会被阻塞，直到 mLooper 被初始化为止
        synchronized (this) {
            while (isAlive() && mLooper == null) {
                try {
                    wait();
                } catch (InterruptedException e) {
                }
            }
        }
        return mLooper;
    }
```

mHandlerThread.getLooper()获取与该线程绑定的 Looper 实例。mLooper 是在 HandlerThread 的 run()方法中赋值的（也就是在子线程中），getLooper()却是在主线程中调用的，该方法会阻塞直到 mLooper 赋值。

然后在 Ch9\HandlerThreadDemo（示例）中通过该 Looper 实例创建了 Handler 对象。

```
    mSubThreadHandler = new Handler(loop);    //创建 Handler 与该线程绑定
```

创建的 Handler 要与子线程绑定到一起，来处理子线程中的消息，所以应通过子线程中的 Looper 实例（有线程对应的消息队列）创建 Handler。这样通过 mSubThreadHandler 发送的消息就会添加到子线程中的消息队列中，然后 Looper“大管家”再对消息进行检索分发，交给 mSubThreadHandler 进行处理。

HandlerThread 提供的线程退出方法如下。

```
public boolean quit() {
    Looper looper = getLooper();
    if (looper != null) {
        looper.quit();
        return true;
    }
    return false;
}
public boolean quitSafely() {
    Looper looper = getLooper();
    if (looper != null) {
        looper.quitSafely();
        return true;
    }
    return false;
}
```

其中，quit()方法和 quitSafely()方法都是退出 HandlerThread 的消息循环。它们分别调用了 Looper 的 quit()方法和 quitSafely()方法。Looper 的 quit()方法会将消息队列中的所有消息移除（延迟消息和非延迟消息）。Looper 的 quitSafely()方法则会将消息队列所有的延迟消息移除，而将非延迟消息派发出去让 Handler 进行处理。

9.2 IntentService

IntentService 是一个基础类，用于处理 Intent 类型的异步任务请求。当客户端调用 android.content.Context#startService(Intent)发送请求时，Service 服务被启动，且在其内部构建一个工作线程来处理 Intent 请求。当工作线程执行结束时，Service 服务会自动停止。

IntentService 是一个抽象类，用户必须实现一个子类去继承它，且必须实现 IntentService 中的抽象方法 onHandleIntent()来处理异步任务请求。

9.2.1 IntentService 的使用

IntentService 是继承 Service 并处理异步请求的一个类，在 IntentService 内有一个工作线程来处理耗时操作，启动 IntentService 的方式和启动传统的 Service 一样。每一个耗时操作都会以队列的方式在 IntentService 的 onHandlerIntent()回调方法中执行，并且每一次只会执行一个工作线程，执行完第一个再执行第二个（注意此优先级要比普通 Service 的优先级高）。

小贴士： Service 是一个不可见的 Activity，它的几个方法（onCreate、onStartCommand、onBind 等）是运行在主线程中的，因此不要在 Service 中做一些重量级（耗时长）的操作，否则可能会导致 ANR。实际上，广播接收器 BroadcastReceiver 的 onReceive()方法也是运行在主线程中的，不能执行耗时长的操作。Service 与 Thread 是存在区别的，Service 不是一个单独的线程，也不是一个单独的进程。如果不是调用了 stopSelf()或者 stopService()，启动的 Service 可能会一直存在（占用系统内存和资源，且不会释放）。IntentService 执行完任务后会自动停止服务，而不需要用手动去控制。另外，可以启动 IntentService 多次。

1. IntentService 原理

（1）IntentService 本质是一个特殊的 Service，继承自 Service。它本身就是一个抽象类，封装了一个 HandlerThread 和一个 Handler，在内部通过 HandlerThread 和 Handler 实现异步操作。它可以用于在后台执行耗时的异步任务，当任务完成后会自动停止。

（2）IntentService 创建时启动一个 HandlerThread（线程），同时将 Handler 绑定 HandlerThread，所以通过 Handler 发送的消息都在 HandlerThread 中执行。

（3）IntentService 进入生命周期 onStartCommand()，再调用 onStart()，将传进的 Intent 对象以消息的形式使用 Handler 发送。

（4）Handler 收到消息后会调用 onHandleIntent()这样一个抽象方法，这个方法需要通过使用者实现来处理逻辑。最后所有任务都执行完成后，IntentService 会自动销毁。

IntentService 处理异步任务 Intent 请求的流程，如图 9-3 所示。

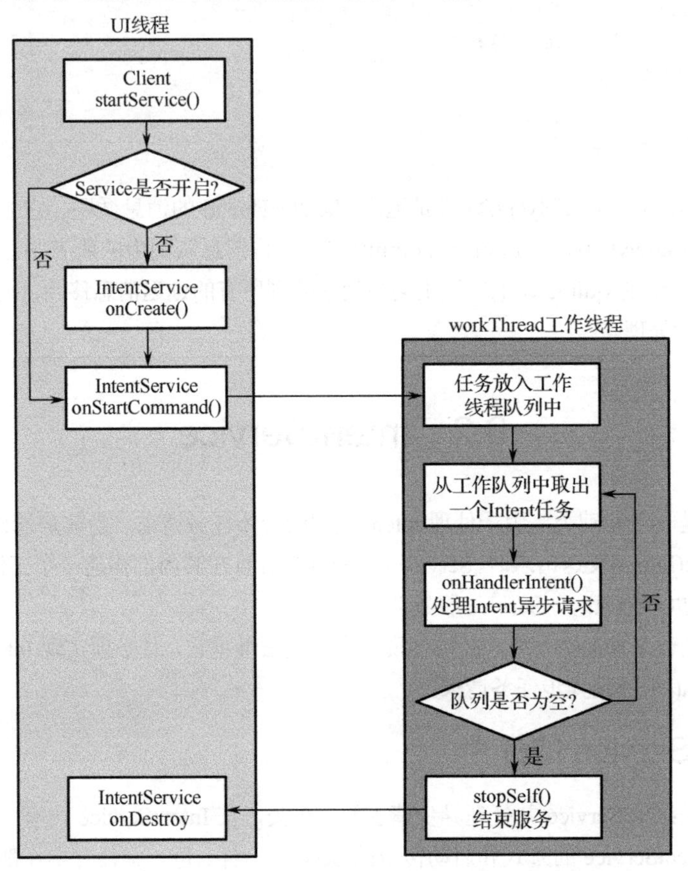

图 9-3　IntentService 处理异步任务 Intent 请求的流程

2. IntentService 特征

IntentService 作为对 Service 的补充(或者说 Service 的轻量级实现)，很好地弥补了 Service 的缺陷。IntentService 特征如下：

（1）创建并启动一个单独的线程（工作线程）来处理任务和请求（所有的 Intent 请求），所有的任务都在该工作线程中处理。因为是在单独的线程中处理任务和请求，其 onHandleIntent()方法运行在单独的线程中，而非主线程，因此可以执行异步操作。

（2）按照发送顺序处理任务和请求。所有请求处理完成后（当没有任务和请求时），IntentService 会自动停止并销毁（无须调用 stopSelf()方法停止 Service），因此它不会一直占用资源和内存。

（3）为 Service 的 onBind()方法提供默认实现，返回值为 null。因此不要尝试用 bindService 去调用 IntentService（IntentService 设计的目的是处理简单的异步任务）。

（4）为 Service 的 onStartCommand()方法提供默认实现，将请求 Intent 添加到队列中。

3. IntentService 使用

IntentService 使用步骤如下。

- 定义 IntentService 的子类，需复写 onHandleIntent()方法。
- 在 AndroidManifest.xml 文件中注册服务。
- 在 Activity 中开启 Service 服务，像使用 Service 一样使用 IntentService。

小贴士：扩展 IntentService 比较简单，提供一个构造方法和实现 onHandleIntent 方法就可以了，不用重写父类的其他方法。使用 Service 可以同时执行多个请求，但使用 IntentService 却只能同时执行一个请求。onHandleIntent()为异步方法，可以执行耗时操作，而普通的 Service 里是不能做耗时操作的。

【示例】 使用 IntentService 执行异步任务。

启动 Android Studio，在 Ch9 工程中创建 IntentServicDemo 模块，并在该模块中创建 DownLoadService.java、MainActivity.java 和 activity_main.xml。

DownLoadService.java 主要代码如下：

```
public class DownLoadService extends IntentService {
    public static final String TAG = "DownLoadService";
    public DownLoadService() {  //重写默认的构造方法
        super("DownLoadService");
    }
    @Override
    protected void onHandleIntent(Intent intent) {
        int key = intent.getIntExtra("key", 0);
        String value = intent.getStringExtra("value");
        switch (key) {
            case 168:
                try {  //模拟耗时任务1
                    Log.d(TAG,"---------------模拟耗时任务1----------------");
                    Thread.sleep(5 * 1000);
                } catch (InterruptedException e) {
                    e.printStackTrace();
                }
                break;
            case 218:
                try {   //模拟耗时任务2
                    Log.d(TAG,"---------------模拟耗时任务2----------------");
                    Thread.sleep(3 * 1000);
                } catch (InterruptedException e) {
                    e.printStackTrace();
                }
                break;
            default:
                break;
        }
        Log.e(TAG, "\nthe current time is: " + System.currentTimeMillis()/1000
                + "\nthe Thread id is " + Thread.currentThread().getId()
                + "\nthe current task is " + value);
    }
}
```

IntentServicDemo 模块的 MainActivity.jav 主要代码如下：

```
public class MainActivity extends AppCompatActivity {
    private Button btnGiveCommand;
    @Override
```

```
        protected void onCreate(Bundle savedInstanceState) {
            super.onCreate(savedInstanceState);
            setContentView(R.layout.activity_main);
            btnGiveCommand = (Button) findViewById(R.id.button);
            btnGiveCommand.setOnClickListener(new View.OnClickListener() {
                @Override
                public void onClick(View v) {
                    giveCommand();
                }
            });
        }
        public void giveCommand() {  //客户端同时发送两个任务到 IntentService 服务端执行
            Intent intent_1 = new Intent(this, DownLoadService.class);
            intent_1.putExtra("key", 168);
            intent_1.putExtra("value", "the first task_1");
            startService(intent_1);
            Intent intent_2 = new Intent(this, DownLoadService.class);
            intent_2.putExtra("key", 218);
            intent_2.putExtra("value", "the second task_2");
            startService(intent_2);
        }
    }
```

运行 IntentServicDemo 模块，Logcat 执行情况如图 9-4 所示。

图 9-4　运行 IntentServicDemo 模块

小贴士：IntentService 不建议使用 bindService()方法，因为 IntentService 的 onBind 会默认返回 null。如果用 Binder 或 Messenger 使其可以通信，则不会调用 onHandleIntent()方法，IntentService 就只是一个普通的 Service 了。

```
    public IBinder onBind(Intent intent) { return null;}
```

IntentService 一旦被停止，保存在消息队列中的任务都会被清除，也就不会执行了。

4．IntentService 与 Service 的区别

IntentService 与 Service 的区别如表 9-1 所示。

表 9-1　IntentService 与 Service 的区别

类　型	运行线程	结束服务（操作）	备　注
IntentService（继承自 Service 类）	创建 1 个工作线程处理多线程任务	不需要（在所有 Intent 被处理完后，系统会自动关闭服务）	IntentService 为 Service 的 onBind()方法提供了默认实现，返回 null
Service	主线程（不能处理耗时操作，否则会出现 ANR）	须主动调用 stopService()	IntentService 为 Service 的 onStartCommand()方法提供了默认实现，将请求的 Intent 添加到队列中

5. IntentService 与一般线程的区别

IntentService 与一般线程的区别如表 9-2 所示。

表 9-2　IntentService 与一般线程的区别

类　型	线程属性	作　用	线程优先级
IntentService	类似后台线程（采用 HandlerThread 实现）	后台服务（继承了 Service）	高（不容易被系统销毁）
一般线程	普通线程	普通多线程作用	低（若进程中无活动的四大组件，则该线程的优先级非常低，容易被系统销毁）

9.2.2 IntentService 的源码分析

IntentService.java 主要源码分析如下。

```
    public abstract class IntentService extends Service {
        // ①有一个 Looper 变量和一个 ServiceHandler 变量。ServiceHandler 继承 Handler 用于
处理消息
        private volatile Looper mServiceLooper;                    // 工作线程的 Looper
        private volatile ServiceHandler mServiceHandler;      // 结合 HandlerThread
来处理任务和请求
        private String mName;              // Service 名称
        private boolean mRedelivery;       // 进程销毁后，是否会重启
        private final class ServiceHandler extends Handler {
            public ServiceHandler(Looper looper) {
                super(looper);
            }
            @Override
            public void handleMessage(Message msg) {
                // 在工作线程中调用 onHandleIntent，子类根据 Intent 传递的数据执行具体的操作
                onHandleIntent((Intent)msg.obj);
                // 每次处理完任务后都会去尝试销毁。只有所有的任务都处理完时，才能被销毁
                stopSelf(msg.arg1);   // 任务执行完毕后，自动停止 Service
            }
        }
        public IntentService(String name) { //实现 IntentService 时构造方法要调用 Intent
Service 的构造方法
            super();
            mName = name;
        }
        /**
         * 如果 IntentService 存在待处理的任务或者正在处理任务时，IntentService 所处的进程
         * 被销毁，那么将根据 mRedelivery 的值来决定是否重启后分发 Intent
         *  mRedelivery 为 true 的情况：进程销毁会被重启，且分发未处理完毕的 Intent
         * 存在多个 Intent 的情况下，只会分发最近一次发送的 Intent
         *  mRedelivery 为 false 的情况：进程销毁后不会被重启，Intent 会随着进程消亡一般在构造
         *  方法中，会进行如下设置。
         */
        public void setIntentRedelivery(boolean enabled) {
            mRedelivery = enabled;
        }
        // ②在 OnCreate()方法中，创建了一个线程 HandlerThread 并启动。
        // 然后获取工作线程的 Looper，并用 Looper 初始化 Handler（Handler 的创建需要使用
Looper）
```

```
    @Override
    public void onCreate() {
        super.onCreate();
        // 创建、启动工作线程
        HandlerThread thread = new HandlerThread("IntentService[" + mName + "]");
        thread.start();
        // 关联处理任务的 Looper 和 Handler
        mServiceLooper = thread.getLooper();
        mServiceHandler = new ServiceHandler(mServiceLooper);
    } // 经过这一步，处理任务的工作线程和 Handler 就已经准备完毕，IntentService 处于就绪状态
      // 此时就可以处理发送过来的任务了
    // ③在 onStart()方法中发送消息给 Handler，并且把 Intent 传给了 Handler 处理
    @Override
    public void onStart(@Nullable Intent intent, int startId) {
        Message msg = mServiceHandler.obtainMessage();
        msg.arg1 = startId;
        msg.obj = intent;
        mServiceHandler.sendMessage(msg);
    }
    // ④onStartCommand() 直接调用的是 onStart()方法
    // 不要重写该方法，要去重写 onHnadleIntent()方法
    @Override
    public int onStartCommand(@Nullable Intent intent, int flags, int startId) {
        onStart(intent, startId);
        return mRedelivery ? START_REDELIVER_INTENT : START_NOT_STICKY;
    }
    @Override
    public void onDestroy() {
        mServiceLooper.quit();   // 退出消息循环
    }
    @Override
    @Nullable
    public IBinder onBind(Intent intent) {  // 一般情况下是不会重写该方法的
        return null;
    }
    // ⑤最后就是一个子类需要实现的抽象方法，这个方法在 handleMessage 中调用
    // 也就是在工作线程中执行
    @WorkerThread
    protected abstract void onHandleIntent(@Nullable Intent intent);
    // 该方法是需要重写的方法，它运行在工作线程中（启动的 HandlerThread 中）
    // 处理任务的地方（耗时操作等）
    // 该方法在 ServiceHandler 的 handleMessage()中被调用
}
```

在 onCreate()方法中，新建一个 HandlerThread 后，再创建一个 ServiceHandler，传入 HandlerThread（线程）中的 Looper。这样 ServiceHandler 的 handleMessage()方法将会运行在子线程中。ServiceHandler 的 handleMessage()中执行了 onHandleIntent()的方法，该方法是抽象方法，需要用户去实现，代码逻辑（需要实现的业务逻辑）写在此方法中即可。执行完毕后，将调用 stopSelf()方法来停止服务。

回调用 stopSelf(msg.arg1)，注意这个 msg.arg1 是 int 值，相当于一个请求的唯一标识。每发送一个请求，就会生成一个唯一的标识，然后将请求放入队列中，当全部执行完成（最后一个请求也就相当于 getLastStartId == startId），或者当前发送的标识是最近发出的那一个（getLastStartId == startId），则会销毁 Service。如果传入的是-1，则直接销毁。当任务完成销毁 Service 回调 onDestory()方法时，

就可以看到在onDestroy()中释放了Looper："mServiceLooper.quit()"。

9.3 AsyncTask

AsyncTask是一个抽象类，它是由Android封装的一个轻量级异步类（轻量体现在使用方便、代码简洁），它可以在线程池中执行后台任务，然后把执行的进度和最终结果传递给主线程，并在主线程中更新UI。

AsyncTask的内部封装了两个线程池（SerialExecutor和THREAD_POOL_EXECUTOR）和一个Handler（InternalHandler）。其中SerialExecutor线程池用于任务的排队，让需要执行的多个耗时任务按顺序排列，而THREAD_POOL_EXECUTOR线程池用于真正地执行任务。InternalHandler用于从工作线程切换到主线程。

9.3.1 AsyncTask的使用

1. AsyncTask的泛型参数

AsyncTask的类声明如下：

```
public abstract class AsyncTask<Params, Progress, Result>
```

AsyncTask是一个抽象泛型类，其三个泛型参数的含义如下。

（1）Params：开始执行异步任务时传入的参数类型。

（2）Progress：异步任务执行过程中返回下载进度值的类型。可将它视为在后台线程处理的过程中，能阶段性发布结果的数据类型。

（3）Result：执行异步任务完成后返回的结果类型。可将它视为任务全部完成后所返回的数据类型。

如果AsyncTask确定不需要传递具体参数，那么这三个泛型参数都可以用Void来代替。

有了这三个参数类型之后，也就控制了这个AsyncTask子类各个阶段的返回类型。如果有不同业务，需要再另写一个AsyncTask的子类进行处理。

2. AsyncTask的回调方法

通过调用AsyncTask的execute()方法传入参数并执行任务，然后AsyncTask就会依次调用以下四个方法。

（1）onPreExecute。

该方法的签名如下。

```
@MainThread
protected void onPreExecute() {}
```

它有MainThread注解，表示是运行在主线程中的。在AsyncTask执行了execute()方法后就会在UI线程上执行onPreExecute()方法。它在task真正执行前运行，通常可以进行一些界面上的初始化操作，如显示一个进度条或对话框等，从而告知用户后台任务即将开始。

（2）doInBackground。

该方法的签名如下。

```
@WorkerThread
protected abstract Result doInBackground(Params... params);
```

它有WorkerThread注解，表示运行在单独的工作线程中，而不是运行在主线程中。doInBackground会在onPreExecute()方法执行完成后立即执行。它用于在工作线程中执行耗时任务，并可以直接编写需要在后台线程中运行的逻辑代码。由于是运行在工作线程中，所以该方法不会阻塞UI线程。

它接收Params泛型参数，参数params是Params类型的不定长数组。它的返回值是Result泛型，由于doInBackgroud是抽象方法，在使用AsyncTask时必须重写。

在doInBackground中执行的任务可能要分解为很多步骤，每完成一个步骤就可以通过调用AsyncTask的publishProgress(Progress…)将阶段性的处理结果发布出去，阶段性处理结果是Progress泛型类型。

当调用了publishProgress方法后，处理结果会被传递到UI线程中，并在UI线程中回调onProgressUpdate方法。

根据具体需要，可以在doInBackground中不调用publishProgress方法，当然也可以多次调用。

doInBackgroud方法的返回值表示后台线程完成任务之后的结果。

（3）onProgressUpdate。

由上述介绍，可知当在doInBackground中调用publishProgress(Progress…)方法后，就会在UI线程上回调onProgressUpdate方法，该方法的签名如下。

```
@MainThread
protected void onProgressUpdate(Progress… values) {}
```

它也具有MainThread注解，表示在主线程上被调用，且传入的参数是Progress泛型定义的不定长数组。如果在doInBackground中多次调用了publishProgress方法，那么主线程就会多次回调onProgressUpdate方法。

（4）onPostExecute。

该方法的签名如下。

```
@MainThread
protected void onPostExecute(Result result) {}
```

它也具有MainThread注解，表示在主线程中被调用。当doInBackgroud方法执行完毕后，就表示任务完成了。doInBackgroud方法的返回值就会作为参数在主线程中传入到onPostExecute方法中，这样就可以在主线程中根据任务的执行结果更新UI。

上面几种方法的调用顺序为onPreExecute()→doInBackground()→publishProgress()→onProgressUpdate()→onPostExecute()。

如果不需要执行更新进度，则调用顺序为onPreExecute()→doInBackground()→onPostExecute()。

除了上面四种方法，AsyncTask还提供了onCancelled()方法，它同样在主线程中执行。当异步任务取消时，onCancelled()就会被调用。这时onPostExecute()则不会被调用。

注意：AsyncTask中的cancel()方法并不是真正取消任务，而是设置这个任务为取消状态，需终止任务还要在doInBackground()中判断。如要终止一个线程，调用interrupt()方法，只是标记为中断，还需要在线程内部进行标记判断，才能中断线程。

3. AsyncTask的使用

AsyncTask使用的示例代码如下：

```
class DownloadTask extends AsyncTask<Void, Integer, Boolean> {
    @Override
    protected void onPreExecute() {
        progressDialog.show();
    }
    @Override
    protected Boolean doInBackground(Void… params) {
        try {
            while (true) {
```

```
                int downloadPercent = doDownload();
                publishProgress(downloadPercent);
                if (downloadPercent >= 100) {
                    break;
                }
            }
        } catch (Exception e) {
            return false;
        }
        return true;
    }

    @Override
    protected void onProgressUpdate(Integer... values) {
        progressDialog.setMessage("当前下载进度：" + values[0] + "%");
    }
    @Override
    protected void onPostExecute(Boolean result) {
        progressDialog.dismiss();
        if (result) {
            Toast.makeText(context, "下载成功", Toast.LENGTH_SHORT).show();
        } else {
            Toast.makeText(context, "下载失败", Toast.LENGTH_SHORT).show();
        }
    }
}
```

以上代码模拟了一个下载任务，即在 doInBackground()方法中执行具体的下载逻辑，在 onProgressUpdate()方法中显示当前的下载进度，并在 onPostExecute()方法中提示任务的执行结果。如果想要启动这个任务，只需要简单地调用以下代码即可。

```
new DownloadTask().execute();
```

【示例】 使用 AsyncTask 下载图片并显示进度条功能。

启动 Android Studio，在 Ch9 工程中创建 AsyncTaskDemo2 模块，并在该模块中创建 MainActivity.java、activity_main.xml、Constants.java。

AsyncTaskDemo2 模块的 Constants.java 主要代码如下：

```
public class Constants {
    public static final String SiteURL = "http://192.168.0.104:8080/NetServer/";
    public static final String iconPath = SiteURL + "pic/";
}
```

AsyncTaskDemo2 模块的 MainActivity.java 主要代码如下：

```
public class MainActivity extends AppCompatActivity {
    private ImageView imageView;
    private ProgressDialog dialog;  //进度对话框
    private Button btnGetPic;
    private static String picUrl = Constants.iconPath + "suda6.jpg";
    @Override
    protected void onCreate(Bundle savedInstanceState) {
        super.onCreate(savedInstanceState);
        setContentView(R.layout.activity_main);
        btnGetPic = (Button) findViewById(R.id.button);
        imageView = (ImageView) findViewById(R.id.imageView);
        //第一步，新建一个对话框，不显示
        dialog = new ProgressDialog(this);
```

```
        dialog.setTitle("提示");
        dialog.setMessage("正在下载，请稍后...");
        dialog.setProgressStyle(ProgressDialog.STYLE_HORIZONTAL);//带有水平滚动条
        dialog.setCancelable(false);//不能取消
        btnGetPic.setOnClickListener(new View.OnClickListener() {
            @Override
            public void onClick(View v) {
                new DownloadTask().execute(picUrl);
            }
        });
    }

    //创建一个内部类，用来下载图片
    public class DownloadTask extends AsyncTask<String, Integer, Bitmap> {
        @Override
        protected void onPreExecute() {//第一步：准备阶段，让进度条显示
            super.onPreExecute();
            dialog.show();    //显示对话框
        }
        //第二步，一般进行复杂处理（后台），最重要的是接收 execute 传来的参数
        @Override
        protected Bitmap doInBackground(String... params) {
            String newurl = params[0];  //从可变参数的数组中拿到第 0 位的图片地址
            Bitmap bitmap = null;
            HttpURLConnection urlConnection;
            InputStream inputStream = null;  // 字节输入流
            ByteArrayOutputStream outputStream = new ByteArrayOutputStream();
            // 字节数组输出流
            try {  / /图片下载操作
                URL httpUrl = new URL(newurl);
                urlConnection = (HttpURLConnection) httpUrl.openConnection();
                long file_length = urlConnection.getContentLength();
                //根据响应获取文件大小
                inputStream = urlConnection.getInputStream();
                int len = 0;
                byte[] data = new byte[1024];//每次读取的字节
                int total_length = 0;
                while ((len = inputStream.read(data)) != -1) {
                // 以字节的方式读取图片数据
                    total_length += len;
                    int values = (int) ((total_length / (float) file_length) * 100);
                    // 计算进度
                    // 发布进度信息，AsyncTask 第二个参数类型，触发 onProgressUpdate 更新进度
                    publishProgress(values);
                    outputStream.write(data, 0, len);   // 写入输出流
                }
                byte[] result = outputStream.toByteArray();
                // 将字节数组输出流转换为字节数组
                bitmap = BitmapFactory.decodeByteArray(result, 0, result.length);
                // 生成二进制图片
                if (inputStream != null) {
                    inputStream.close();
                }
            } catch (IOException e) {
```

```
                e.printStackTrace();
            }
            return bitmap;    // 返回结果bitmap
        }
        //第三步，当有publishProgress(value)发来的数据时触发，更新UI
        @Override
        protected void onProgressUpdate(Integer... values) {
            super.onProgressUpdate(values);
            dialog.setProgress(values[0]); // 设置进度对话框的进度值，更新进度条，运行
                                           // 在UI中
        }
        @Override
        protected void onPostExecute(Bitmap result) {
        //第四步，下载结束后，隐藏对话框，更新UI
            super.onPostExecute(result);
            dialog.dismiss();
            imageView.setImageBitmap(result); //更新UI，显示图片，运行在UI中
        }
    }
}
```

运行 AsyncTaskDemo2 模块，其界面如图 9-5 所示。

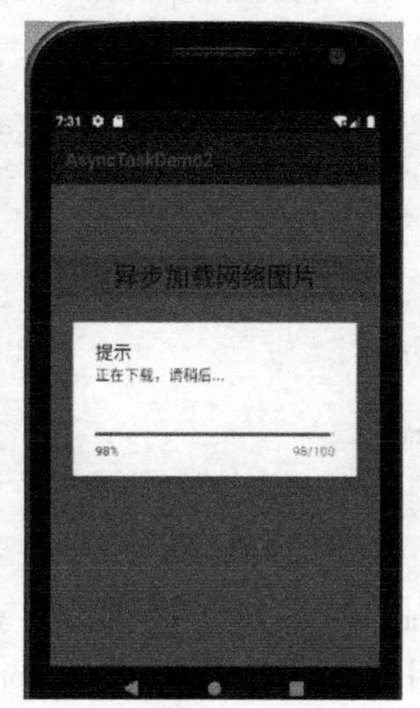

图 9-5　运行 AsyncTaskDemo2 模块（使用 AsyncTask 下载图片并显示进度）

4. 使用 AsyncTask 的注意事项

在使用 AsyncTask 的过程中，有以下几点需要注意。

（1）AsyncTask 对象必须在 UI 线程中创建。

（2）AsyncTask 的 execute(Params... params)方法必须在 UI 线程中调用。

（3）onPreExecute()、doInBackground(Params... params)、onProgressUpdate (Progress... values)和 onPostExecute(Result result)都是回调方法，Android 会自动调用，不用自己设置调用。

（4）不能在 doInBackground(Params... params)中更改 UI 组件的信息。

（5）对于一个 AsyncTack 的实例，只能执行一次 execute 方法，在该实例上第二次执行 execute 方法时就会抛出异常。

（6）AsyncTask 默认是串行执行，若要并发执行则要调用 executeOnExecutor(Executor exec, Params... params)方法，指定线程池执行也可以通过 setDefaultExecutor(Executor exec)设置默认线程池。

（7）AsyncTask 要与主线程有交互。

（8）AsyncTask 的四个关键方法，执行在两个不同的线程中。

① onPreExecute()、onProgressUpdate()，以及 onPostExecute() 都执行在 UI Thread 中。

② doInBackground() 则执行在一个独立的线程中。

9.3.2 AsyncTask 的工作原理

为了分析 AsyncTask 的工作原理，可以从 execute(Params... params)方法入手。

```
@MainThread
public final AsyncTask<Params, Progress, Result> execute(Params... params) {
    return executeOnExecutor(sDefaultExecutor, params);
}
```

execute(Params... params)方法又调用了 executeOnExecutor(Executor exec, Prams... params)方法。

```
@MainThread
public final AsyncTask<Params, Progress, Result> executeOnExecutor(Executor exec,
Params... params) {
    if (mStatus != Status.PENDING) {
        switch (mStatus) {
            case RUNNING:
                throw new IllegalStateException("Cannot execute task:" + " the task
is already running.");
            case FINISHED:
                throw new IllegalStateException("Cannot execute task:"
                        + " the task has already been executed "
                        + "(a task can be executed only once)");
        }
    }
    mStatus = Status.RUNNING;
    onPreExecute();
    mWorker.mParams = params;
    exec.execute(mFuture);
    return this;
}
```

从上面的源码可以看到，在 AsyncTask 执行的时候，就调用了 onPreExecute()方法，然后将 params 参数封装成 FutureTask 对象，调用了线程池 sDefaultExecutor 的 execute(mFuture)方法。线程池 sDefaultExecutor 实际上就是一个串行的线程池，以下是 sDefaultExecutor 线程池的实现。

```
private static volatile Executor sDefaultExecutor = SERIAL_EXECUTOR;
public static final Executor SERIAL_EXECUTOR = new SerialExecutor();
private static class SerialExecutor implements Executor {
    final ArrayDeque<Runnable> mTasks = new ArrayDeque<Runnable>();
    Runnable mActive;
    public synchronized void execute(final Runnable r) {
        mTasks.offer(new Runnable() {
            public void run() {
                try {
                    r.run();
                } finally {
                    scheduleNext();
                }
            }
        });
        if (mActive == null) {
```

```
            scheduleNext();
        }
    }

    protected synchronized void scheduleNext() {
        if ((mActive = mTasks.poll()) != null) {
            THREAD_POOL_EXECUTOR.execute(mActive);
        }
    }
}
```

在 SerialExecutor 中，FutureTask 作为 Runnable 对象传进来之后，会把 FutureTask 插入任务队列 mTasks 中。如果当前没有正在活动的 AsyncTask，就会调用 sheduleNext()方法；如果有正在活动的 AsyncTask，则在执行结束之后调用 sheduleNext()方法，直到所有的任务执行完成。因此，可以看出 SerialExecutor 是串行执行的，而真正执行的线程池为 THREAD_POOL_EXECUTOR，SerialExecutor 是对 THREAD_POOL_EXECUTOR 的封装。

在 THREAD_POOL_EXECUTOR 的 execute()方法中最后调用了 run 方法。

```
public void run() {
    …
    try {
        Callable<V> c = callable;
        if (c != null && state == NEW) {
            V result;
            boolean ran;
            try {
                result = c.call();
                ran = true;
    …
}
```

在 FutureTask 执行 run 方法时，会执行 Callable 的 call()方法，从 AsyncTask 的构造方法中可以发现，对应的是 mWorker 的 call()方法。

```
mWorker = new WorkerRunnable<Params, Result>() {
    public Result call() throws Exception {
        mTaskInvoked.set(true);
        Result result = null;
        try {
            Process.setThreadPriority(Process.THREAD_PRIORITY_BACKGROUND);
            result = doInBackground(mParams);
            Binder.flushPendingCommands();
        } catch (Throwable tr) {
            mCancelled.set(true);
            throw tr;
        } finally {
            postResult(result);
        }
        return result;
    }
};

mFuture = new FutureTask<Result>(mWorker) {…}
…
```

在 call()方法中，执行了 doInBackground(mParams)方法，然后将返回值传递给 postResult(result)方法。

```java
private Result postResult(Result result) {
    @SuppressWarnings("unchecked")
    Message message = getHandler().obtainMessage(MESSAGE_POST_RESULT,
            new AsyncTaskResult<Result>(this, result));
    message.sendToTarget();
    return result;
}
```

在 postResult(Result)方法中，将 result 包装成 Message，发送给 Handler，跟踪 getHandler()方法找到对应的 Handler 为 InternalHandler。

```java
private static class InternalHandler extends Handler {
    public InternalHandler(Looper looper) {
        super(looper);
    }
    @SuppressWarnings({"unchecked", "RawUseOfParameterizedType"})
    @Override
    public void handleMessage(Message msg) {
        AsyncTaskResult<?> result = (AsyncTaskResult<?>) msg.obj;
        switch (msg.what) {
            case MESSAGE_POST_RESULT:
                // There is only one result
                result.mTask.finish(result.mData[0]);
                break;
            case MESSAGE_POST_PROGRESS:
                result.mTask.onProgressUpdate(result.mData);
                break;
        }
    }
}
```

InternalHandler 收到 MESSAGE_POST_RESULT 消息之后，调用 AsyncTask 的 finish()方法。

```java
private void finish(Result result) {
    if (isCancelled()) {
        onCancelled(result);
    } else {
        onPostExecute(result);
    }
    mStatus = Status.FINISHED;
}
```

如果 AsyncTask 取消执行了，就调用 onCancelled()方法，否则调用 onPostExecute()方法。

从上述源码分析中得到，默认情况下 AsyncTask 的执行效果是串行的，因为有了 SerialExecutor 类来维持保证队列的串行。如果想使用并行执行任务，那么可以直接跳过 SerialExecutor 类，使用 executeOnExecutor(Executor exec, Params... params) 方法来执行任务。

小贴士：这里的“mFuture”是一个 FutureTask，也就是让一个 ThreadPoolExecutor 去 execute()的一个 FutureTask。

① 找到 AsyncTask 的构造方法，可以发现 mFuture 在这里构造时，所带的参数就是 mWorker。也就是说，当 exec 执行 mFuture 的时候，实际上就是新启动了一个线程，执行了 mWorker。

② 再看 mWorker，在它的 call()方法下面，果断调用了 doInBackground()方法。由此可知，doInBackground()方法实际执行在一个独立的非 UI 线程下面。

③ 在 doInBackground()方法调用之后，将结果返回。于是可以通过 mFuture 的 get()方法， 取到 doInBackground()方法的运行结果。get()方法到 result 后，通过 sHandler，将 result 发送给 AsyncTask.finish()

方法进行处理。

④ 终于，在 finish()方法中，看到了久违的 onPostExecute()方法。由于 AsyncTask 是由主线程创建的，从而可知执行在 AsyncTask.finish()方法中的 onPostExecute()方法，也属于主线程。

⑤ 随着对 mFuture 代码的读取，其他部分的逻辑也清晰了起来。关于 onProgressUpdate()方法又是何时被调用的？可查看源码，在 publishProgress()方法中，向 sHandler 发送了一条消息，从而调用了 onProgressUpdate()方法。显然，onProgressUpdate()方法是在主线程中被调用的。

9.4 本章小结

本章主要介绍了 Android 的一些异步处理技术，包括 HandlerThread 的使用及源码分析、IntentService 使用及源码分析、AsyncTask 的使用及工作原理。

HanlderThread 继承自 Thread，使用 HandlerThread 能够创建拥有 Looper 的线程。在 HandlerThread 的 run()方法中，通过 Looper.prepare()创建了消息队列，并通过 Looper.loop()开启了消息循环。

使用时开启 HandlerThread，创建 Handler 与 HandlerThread 的 Looper 绑定，Handler 以消息的方式通知 HandlerThread 来执行一个具体的任务。

1. IntentService 总结

（1）子类需继承 IntentService 并且实现 onHandlerIntent 抽象方法来处理 intent 类型的任务请求。

（2）子类需要重写默认的构造方法，且调用父类带参数的构造方法。

（3）IntentService 类内部利用 HandlerThread + Handler 构建了一个带有消息循环处理机制的后台工作线程，客户端只需调用 Context#startService(Intent)将 Intent 任务请求放入后台工作队列中，且客户端无须关注服务是否结束，非常适合可一次性处理的后台任务。如浏览器下载文件，当退出当前浏览器之后，下载任务依然存在后台，直到下载文件结束，服务自动销毁。

（4）只要当前 IntentService 服务没有被销毁，客户端就可以同时投放多个 Intent 异步任务请求，IntentService 服务端是顺序执行当前后台工作队列中 Intent 请求的，也就是每一时刻只能执行一个 Intent 请求，直到该 Intent 处理结束才开始下一个 Intent。因为 IntentService 类内部是利用 HandlerThread+ Handler 构建一个单线程来处理异步任务的。

2. AsyncTask 总结

（1）AsyncTask 定义了三种泛型类型，即 Params、Progress 和 Result。

- Params：指启动任务执行的输入参数，如 HTTP 请求的 URL。
- Progress：指后台任务执行的百分比。
- Result：指后台执行任务最终返回的结果，如 String。

（2）使用 AsyncTask 异步加载数据一般要重写以下两个方法。

- doInBackground(Params… params)：此方法在后台线程执行，完成任务的主要工作，比较耗时的操作都可以放在这里（注意这里不能直接操作 UI）。在执行过程中可以调用 publishProgress (Progress… values)方法来更新任务的进度。
- onPostExecute(Result result)：相当于 Handler 处理 UI 的方式，可以使用 doInBackground()得到的结果处理操作 UI。此方法在主线程（UI 线程）中执行。

（3）根据业务需要，还可以重写以下三种方法，但不是必须的。

- onProgressUpdate(Progress...values)：当 doInBackground(Params…params)中调用 publishProgress (Progress…values)方法后，就会在 UI 线程上回调 onProgressUpdate (Progress...values)方法，可

以使用进度条来增加用户的体验度。

- onPreExecute()：在任务执行之前调用此方法，可以显示出进度对话框。
- onCancelled()：用户调用取消时需要做的操作。

习 题 9

1．HandlerThread 的工作原理是什么？如何使用？试编程举例加以说明。

2．IntentService 的工作原理是什么？如何使用？试编程举例加以说明。

3．AsyncTask 的工作原理是什么？如何使用？试编程举例加以说明。

第 10 章　Android 网络应用

手机对网络的依赖性越来越大，几乎所有的 App 都需要使用网络，如何利用网络技术开发 Android 应用即为本章介绍的重点。所有的网络通信其核心任务是——Client（客户）端与 Server（服务器）端进行数据和交互操作。本章详细介绍了如何发送 HTTP 请求，并获取响应。客户端如何向远程服务器端发送请求，服务器端如何接收请求，并返回响应的数据。不同的通信方式，采用的技术也各不相同。本章主要内容有：（1）HTTP 协议；（2）使用 HttpURL Connection；（3）使用 OkHttp；（4）使用 Retrofit；（5）Socket 通信；（6）使用 WebView 显示网页等。

10.1　使用 HTTP 协议访问网络

10.1.1　HTTP 协议

1. HTTP 协议简介

协议是指计算机通信网络中两台计算机之间进行通信所必须共同遵守的规定或规则。超文本传输协议（HTTP）是一种通信协议，它允许将超文本标记语言（HTML）文档从 Web 服务器传送到客户端。

超文本传输协议（Hypertext Transfer Protocol，HTTP）是一种详细规定了客户端和万维网（WWW，World Wide Web）服务器之间互相通信的规则，通过因特网传送万维网文档的数据传送协议。

HTTP 协议属于 OSI 网络七层协议模型中的最上层“应用层”，由请求和响应构成，是一个标准的客户端服务器模型。在 Internet 中所有的传输都是通过 TCP/IP 进行的。HTTP 协议作为 TCP/IP 模型中应用层的协议也不例外。HTTP 协议通常承载于 TCP 协议之上，有时也承载于 TLS 或 SSL 协议层之上，就成了通常所说的 HTTPS。

小贴士：TCP/IP 协议是两个协议一起协同工作的简称。TCP（Transmission Control Protocol）指传输控制协议，IP（Internet Protocol）指网际协议。

IP 负责计算机之间的通信，IP 协议是计算机相互识别的一种机制，每台计算机在网络中都有一个 IP 地址用来标识。IP 负责计算机之间数据包的发送和接受，IP 协议将消息分装成比较小的包，并通过 Internet 在计算机之间传输。IP 负责计算机路由到目的地址，但是它只有发送和接受包的能力，对包的检测，只能对包的头部检测，而没有对整个包检测的能力。

TCP 能确保数据包正确到达，并且尝试检查包内容是否一致。TCP 在 IP 的应用端口上，将已经正确的数据传送到对应的程序。当一个程序想通过 TCP 与另外一个程序通信时，它会发送一个通信请求，这个请求有一个确定的地址，TCP 使两个程序建立全双工的通信，占用计算机的整个通信线路。TCP 在数据传输时负责将数据拆分成 IP 数据包，让它们到达指定位置后开始重组。

TCP/IP 协议是两个协议协同工作，具有一定的层次关系。TCP 负责应用程序的通信，IP 负责计算机之间的通信。TCP 负责将数据拆分成 IP 数据包和将 IP 数据包还原成原本的数据，IP 负责把 TCP 交给它的数据发送到指定的位置。

HTTP 和 FTP 等协议都处在于应用层，TCP 和 UDP 存在于传输层，IP 存在于网络层。

2. HTTP 工作原理

HTTP 是由请求和应答组成的，HTTP 的请求永远都是客户端发起请求，服务器端进行应答。HTTP 是一个无状态的协议，指 HTTP 连接不会持续太长时间，服务器端应答客户端后就会主动断开连接。HTTP 连接的工作过程是：客户端发送 HTTP 请求→服务器接收请求、回送响应→客户端解析、处理返回信息。在请求结束后，服务器处理完客户端的请求，就会主动释放连接。一次 HTTP 的请求简称为一次事务。对于移动开发来说，其具体过程如下。

（1）客户端执行网络请求，从 URL 中解析出服务器的主机名。

（2）将服务器的主机名转换成服务器的 IP 地址。

（3）将端口号从 URL 中解析出来。

（4）建立一条客户端与服务器之间的 TCP 连接。

（5）建立连接后，客户端通过输出流发送一条 HTTP 请求给服务器。通常一个 HTTP 请求报文由请求行、请求头、空行和可选的请求正文（请求数据）组成。

（6）服务器接到请求后，给予相应的响应信息，向客户端回送一条 HTTP 响应报文。一个完整的 HTTP 响应报文由响应行（状态行）、响应头（消息报头）、空行和响应正文组成。

（7）客户端从输入流获取报文。

（8）客户端解析报文，并关闭连接。

（9）客户端将结果显示在 UI 上。

3. HTTPS

HTTP 协议传输的数据都未加密（明文）如果 HTTP 请求被黑客拦截，并且里面含有银行卡密码等敏感数据的话，会非常危险。因此使用 HTTP 协议传输隐私信息非常不安全，为了保证这些隐私数据能加密传输，于是网景公司设计了 SSL（Secure Sockets Layer）协议用于对 HTTP 协议传输的数据进行加密，从而就诞生了 HTTPS。HTTPS 可以将数据加密传输，也就是传输的是密文，即便黑客在传输过程中拦截到数据也无法破译，这就保证了网络通信的安全。

HTTPS 协议是由 SSL（或 TLS）+ HTTP 协议构建的，可进行加密传输、身份认证的网络协议，要比 HTTP 协议安全。HTTPS 协议 =HTTP 协议 + SSL/TLS 协议。

在 HTTPS 数据传输的过程中，需要用 SSL/TLS 对数据进行加密和解密，用 HTTP 对加密后的数据进行传输，由此可以看出 HTTPS 是由 HTTP 和 SSL/TLS 一起合作完成的。

小贴士：SSL（Secure Sockets Layer，安全套接层协议）是为网络通信提供安全及数据完整性的一种安全协议。SSL 协议在 1994 年被 Netscape 发明，后来各个浏览器均支持 SSL，其最新的版本是 SSL3.0。TLS（Transport Layer Security，安全传输层协议）的最新版本 IETF（Internet Engineering Task Force，Internet 工程任务组）是制订的一种新协议，它建立在 SSL 3.0 协议的规范之上，是 SSL 3.0 的后续版本。在 TLS 与 SSL3.0 之间存在着显著的差别，主要是它们所支持的加密算法不同，所以 TLS 与 SSL3.0 不能相互操作。虽然 TLS 与 SSL3.0 在加密算法上不同，但是在理解 HTTPS 的过程中，可以把 SSL 和 TLS 视为同一个协议。

一个 HTTPS 请求实际上包含了两次 HTTP 传输，可以细分为以下 8 个步骤。

（1）客户端向服务器发起 HTTPS 请求，连接到服务器的 443 端口。

（2）服务器端有一个密钥对，即公钥和私钥，是用来进行非对称加密使用的。由服务器端保存着私钥，不能将其泄露，而公钥则可以发送给任何人。

（3）服务器将自己的公钥发送给客户端。

（4）客户端收到服务器端的公钥之后，会对公钥进行检查，验证其合法性，如果发现公钥有问题，那么 HTTPS 传输就无法继续。严格地说，在这里应该验证服务器发送数字证书的合法性。如果公钥合格，那么客户端会生成一个随机值，这个随机值就是用于进行对称加密的密钥，将该密钥称为 client key，即客户端密钥，这样在概念上可以和服务器端的密钥进行区分。然后用服务器公钥对客户端密钥进行非对称加密，这样客户端密钥就变成密文了，至此，HTTPS 中的第一次 HTTP 请求结束。

（5）客户端会发起 HTTPS 中的第二个 HTTP 请求，将加密之后的客户端密钥发送给服务器。

（6）服务器接收到客户端发来的密文之后，会用自己的私钥对其进行非对称解密，解密之后的明文就是客户端密钥，然后用客户端密钥对数据进行对称加密，这样数据就变成了密文。

（7）服务器将加密后的密文发送给客户端。

（8）客户端收到服务器发送来的密文，用客户端密钥对其进行对称解密，得到服务器发送的数据。这样 HTTPS 中的第二个 HTTP 请求结束，整个 HTTPS 传输完成。

HTTPS 和 HTTP 的主要区别如下：

（1）HTTPS 协议需要到 CA（Certificate Authority 负责发放和管理数字证书的权威机构）申请证书，一般免费证书较少，因而需要一定费用。

（2）HTTP 是超文本传输协议，信息为明文传输。HTTPS 则是具有安全性的 SSL 加密传输协议。

（3）HTTP 和 HTTPS 使用的是完全不同的连接方式，用的端口也不一样，前者是 80，后者是 443。

（4）HTTP 的连接很简单，是无状态的。HTTPS 协议是由 SSL+HTTP 协议构建的，可进行加密传输、身份认证的网络协议，比 HTTP 协议更安全。

小贴士：明文指的是未被加密过的原始数据。明文被某种加密算法加密之后，就会变成密文，从而确保原始数据的安全。密文也可以被解密，得到原始的明文。

密钥是一种参数，它是在明文转换为密文或将密文转换为明文的算法中输入的参数。密钥分为对称密钥与非对称密钥，分别应用在对称加密和非对称加密上。

对称加密：即私钥加密，指信息的发送方和接收方使用同一个密钥进行加密和解密数据。对称加密的特点是算法公开，且加密和解密速度快，适合于对大数据量进行加密，常见的对称加密算法有 DES、3DES、TDEA、Blowfish、RC5 和 IDEA。

其加密过程如下：明文 + 加密算法 + 私钥 => 密文 。

其解密过程如下：密文 + 解密算法 + 私钥 => 明文 。

对称加密中用到的密钥叫私钥，私钥表示个人私有的密钥，即该密钥不能被泄露。它在加密过程中使用的私钥与解密过程中用到的私钥是同一个密钥，这也是称加密为“对称”的原因。由于对称加密的算法是公开的，所以一旦私钥被泄露，那么密文就很容易被破解，所以对称加密的缺点是密钥安全管理困难。

非对称加密：即公钥加密。非对称加密与对称加密相比，其安全性更好。对称加密的通信双方使用相同的密钥，如果一方的密钥遭泄露，那么整个通信就会被破解。而非对称加密使用一对密钥，即公钥和私钥，且二者成对出现。私钥被自己保存，不能对外泄露。公钥指的是公共的密钥，任何人都可以获得该密钥。用公钥或私钥中的任何一个进行加密，并用另一个进行解密。

被公钥加密过的密文只能被私钥解密，其过程如下：

明文 + 加密算法 + 公钥 => 密文，密文 + 解密算法 + 私钥 => 明文。

被私钥加密过的密文只能被公钥解密，其过程如下：

明文 + 加密算法 + 私钥 => 密文，密文 + 解密算法 + 公钥 => 明文。

由于加密和解密使用了两个不同的密钥，这就是非对称加密的“非对称”原因。

非对称加密的缺点是加密和解密花费时间长、速度慢，只适合对少量数据进行加密。在非对称加密中使用的主要算法有：RSA、Elgamal、Rabin、D-H、ECC（椭圆曲线加密算法）等。

HTTPS 为了兼顾安全与效率，同时使用了对称加密和非对称加密。数据是被对称加密传输的，对称加密过程需要客户端的一个密钥，为了确保能把该密钥安全传输到服务器端，可采用非对称加密对该密钥进行加密传输。因此，对数据进行对称加密时，对称加密所要使用的密钥会通过非对称加密传输。

HTTPS 在传输的过程中涉及三个密钥：服务器端的公钥和私钥，可用来进行非对称加密；客户端生成的随机密钥，可用来进行对称加密。

4. HTTP 请求方式

HTTP 协议中共定义了 8 种方式或者叫“动作”来表明对 Request-URI 指定资源的不同操作方式，具体介绍如下。

（1）GET：向特定的资源发出请求。它的本质就是发送一个请求来取得服务器上的某个资源。资源通过一组 HTTP 头和呈现数据（如 HTML 文本、图片、视频等）返回给客户端。

（2）POST：向指定资源提交数据进行处理请求（如提交表单或者上传文件），数据被包含在请求体中。POST 请求可能会导致新资源的创建和/或已有资源的修改。

（3）PUT：向指定资源位置上传其最新内容。

（4）DELETE：请求服务器删除 Request-URI 所标识的资源。

（5）HEAD：向服务器索要与 GET 请求相一致的响应，只不过响应体将不会被返回。这种方法在不必传输整个响应内容的情况下，就可以获取包含在响应消息头中的元信息。

（6）TRACE：回显服务器收到的请求，主要用于测试或诊断。

（7）OPTIONS：返回服务器针对特定资源所支持的 HTTP 请求方法。也可以利用向 Web 服务器发送“*”请求来测试服务器的功能性。总之，通过请求 Web 服务器可告知其支持的各种功能。

（8）CONNECT：HTTP1.1 协议中预留给能够将连接改为管道方式的代理服务器。

虽然 HTTP 的请求有 8 种方式，但是在实际应用中常用的也就是 GET 和 POST，其他请求方式也可以通过这两种方式间接实现。

5. GET 与 POST 的区别

（1）GET 重点在从服务器上获取资源，POST 重点在向服务器发送数据。

（2）GET 传输数据是通过 URL 请求，以“field=value”的形式，置于 URL 后，并用“?”连接，若有多个请求数据可用“&”连接，如 http:// 192.168.0.104/Test/login.action?name=admin&password=admin，这个过程用户是可见的。POST 传输数据通过 HTTP 的 POST 机制，将字段与对应值封存在请求实体中发送给服务器，这个过程对用户是不可见的。

（3）GET 传输的数据量小，因为受 URL 长度限制，但效率较高。GET 方式提交的数据最多只能有 1024 字节，而 POST 则没有此限制。POST 可以传输大量数据，所以上传文件时只能用 POST 方式。

（4）GET 是不安全的，因为 URL 是可见的，可能会泄露私密信息，如密码等。POST 较 GET 安全性高。

（5）GET 只能支持 ASCII 字符，向服务器传送的中文字符可能会产生乱码。POST 支持标准字符集，可以正确传递中文字符。

小贴士： GET 是将传递的参数以某种数据串的方式显式追加到提交的 URL 后面，传递的数值封装在传值变量中，通过传值变量来获取传递的值。传值数据串与 URL 地址之间以“?”间隔，数据串以 name=value 的形式传递参数，name 为传值变量，value 为传递的值。一般采用这种方式发送的请求，传递参数数据量较小，安全性非常低。POST（POST 请求）可以向服务器传送数据，而且将数据放在 HTML HEADER 内一起传送到服务端 URL 地址，数据对用户不可见。POST 传输方式不在 URL 里传递，也正好解决了 GET 传输量小、容易篡改及不安全等一系列不足。

10.1.2 使用 HttpURLConnection

1. HttpURLConnection 简介

HttpURLConnection 继承于 URLConnection 类，二者都是抽象类。HttpURLConnection 类的声明如下：

```
public abstract class HttpURLConnection extends URLConnection {}
```

可通过 URL 的 openConnection()方法获得 HttpURLConnection 对象，其代码如下：

```
URL url = new URL(URL 地址);  //这里的“URL 地址”为字符串格式
HttpURLConnection urlConn = (HttpURLConnection)url.openConnection();
```

（1）HttpURLConnection 常用的 API 如下。

- void setDoInput()：设置 URLConnection 的 doInput 字段值，值为 true 或者 false。如果使用 URL 连接进行输入，则设为 true，反之，则设为 false。
- void setDoOutput()：设置 URLConnection 的 doOutput 字段值，值为 true 或者 false。如果使用 URL 连接进行输出，则设为 true，反之，则设为 false。
- InputStream getInputStream()：获取 URL 连接的输入流，从而获取响应的内容。
- OutputStream getOutputStream()：获取 URL 连接的输出流，从而传递参数给服务器。
- void setUseCaches()：设置 URL 连接的 useCaches 字段。
- void setInstanceFollowRedirects()：设置是否应该自动执行 HTTP 重定向。
- void setRequestProperty(String key,String value)：设置请求属性。
- int getResponseCode()：获取服务器的响应代码。
- String getResponseMessage()：获取服务器的响应信息。
- String getResquestMethod()：获取发送请求的方法。
- void setRequestMethod()：设置请求方式。
- setConnectTimeout()：设置连接主机超时（单位：ms）。
- setReadTimeout()：设置从主机读取数据超时（单位：ms）。

（2）HttpURLConnection 请求获取 URL 引用的资源步骤如下。

- 创建 HttpURLConnection 对象；
- 设置请求的属性和对象参数；
- 如果是 GET 方式请求，由于 HttpURLConnection 默认使用 GET 方式，因此直接调用 connect()方法即可建立连接。如果是以 POST 方式请求，则需要设置请求方式为 POST。

2. GET 请求的使用

HttpURLconnection 是同步的请求，所以必须放在子线程中，其使用示例如下。

```
new Thread(new Runnable() {
    @Override
    public void run() {
        try {
            String url = "https://www.baidu.com/";
            URL url = new URL(url);
            //得到 connection 对象
            HttpURLConnection connection = (HttpURLConnection) url.openConnection();
            connection.setRequestMethod("GET");  //设置请求方式
            connection.connect();  //连接
            int responseCode = connection.getResponseCode();  //得到响应码
            if(responseCode == HttpURLConnection.HTTP_OK){
                InputStream inputStream = connection.getInputStream(); //得到响应流
```

```
                String result = streamToStr(inputStream);  //将流转换为字符串
            }
        } catch (Exception e) {
            e.printStackTrace();
        }
    }
}).start();
```

GET 请求的使用方法同上。如果需要传递参数，则直接把参数拼接到 URL 后面。

```
String url = "https://www.baidu.com/?userName=admin&password=123456";
```

其中，URL 与参数之间用“?”隔开；键值对中的键与值之间用“=”连接；两个键值对之间用“&”连接。须注意的情况如下：

① 使用“connection.setRequestMethod("GET");”设置请求方式。

② 使用“connection.connect();”连接网络。请求行、请求头的设置必须放在网络连接前。

③ connection.getInputStream()只是得到一个流对象，并不是数据。不过可以从流中读出数据，从流中读取数据的操作必须放在子线程中。

④ connection.getInputStream()得到一个流对象，从这个流对象中只能读取一次数据，第二次读取时将会得到空数据。

3. 使用 POST 请求传递键值对参数

POST 传递参数的本质是：从连接中得到一个输出流，通过输出流把数据（参数）传给服务器。在 POST 中，数据（传递数据）的拼接采用键值对格式，键与值之间用“=”连接，每个键值对之间用“&”连接。使用 POST 传递键值对参数的使用示例如下。

```
new Thread(new Runnable() {
    @Override
    public void run() {
        try {
            URL url = new URL(getUrl);
            HttpURLConnection connection = (HttpURLConnection) url.openConnection();
            connection.setRequestMethod("POST");
            connection.setDoOutput(true);
            connection.setDoInput(true);
            connection.setUseCaches(false);
            connection.connect();
            String params = "userName=yunan&password=123456";
            BufferedWriter writer = new BufferedWriter(new OutputStreamWriter
(connection.getOutputStream(), "UTF-8"));
            writer.write(body);
            writer.close();
            int responseCode = connection.getResponseCode();
            if(responseCode == HttpURLConnection.HTTP_OK){
                InputStream inputStream = connection.getInputStream();
                String result = streamToStr(inputStream);   //将流转换为字符串
            }
        } catch (Exception e) {
            e.printStackTrace();
        }
    }
}).start();
```

4. 使用 POST 请求传递 JSON 格式参数

POST 请求也可以传递 JSON 格式的参数，其使用示例如下。

```
    new Thread(new Runnable() {
        @Override
        public void run() {
            try {
                URL url = new URL(getUrl);
                HttpURLConnection connection = (HttpURLConnection) url.openConnection();
                connection.setRequestMethod("POST");
                connection.setDoOutput(true);
                connection.setDoInput(true);
                connection.setUseCaches(false);
                    //设置参数类型是 JSON 格式
                connection.setRequestProperty("Content-Type",
"application/json;charset=utf-8");
                connection.connect();
                String params = "{userName:yunan, password:123456}";
                BufferedWriter writer = new BufferedWriter(new OutputStreamWriter
(connection.getOutputStream(), "UTF-8"));
                writer.write(body);
                writer.close();
                int responseCode = connection.getResponseCode();
                if(responseCode == HttpURLConnection.HTTP_OK){
                    InputStream inputStream = connection.getInputStream();
                    String result = streamToStr (inputStream);     // 将流转换为字符串
                }
            } catch (Exception e) {
                e.printStackTrace();
            }
        }
    }).start();
```

传递 JSON 格式的参数与传递键值对参数的区别在于：①传递 JSON 格式数据时，需要在请求头中设置参数类型为 JSON 格式；②params 是 JSON 格式的字符串。

5. 设置请求头

GET 请求与 POST 请求都可以设置请求头，设置请求头的方式也是相同的，可以设置多个请求头参数。请求头设置（包括对 HttpURLConnection 对象的一切配置），必须在 connect()函数执行之前完成。

```
    URLConnection connection = realUrl.openConnection();
    connection.setRequestMethod("POST");
    connection.setRequestProperty("version", "3.2.1");    //设置请求头
    connection.setRequestProperty("accept", "*/*");
    connection.setRequestProperty("connection", "Keep-Alive");
    connection.setRequestProperty("user-agent","Mozilla/4.0 (compatible; MSIE 6.0;
Windows NT 5.1; SV1)");
    connection.setRequestProperty("token", token);
    connection.connect();
```

6. 上传文件

在 POST 请求传递参数时，可以从连接中得到一个输出流，通过输出流可以向服务器写数据。同理，可以使用这个输出流将文件写到服务器，其示例代码如下：

```
    try {
        URL url = new URL(getUrl);
```

```
        HttpURLConnection connection = (HttpURLConnection) url.openConnection();
        connection.setRequestMethod("POST");
        connection.setDoOutput(true);
        connection.setDoInput(true);
        connection.setUseCaches(false);
        connection.setRequestProperty("Content-Type", "file/*");   //设置数据类型
        connection.connect();
        OutputStream outputStream = connection.getOutputStream();
        FileInputStream fileInputStream = new FileInputStream("file");   //把文件封
装成一个流
        int length = -1;
        byte[] bytes = new byte[1024];
        while ((length = fileInputStream.read(bytes)) != -1){
            outputStream.write(bytes,0,length);    //写的具体操作
        }
        fileInputStream.close();
        outputStream.close();
        int responseCode = connection.getResponseCode();
        if(responseCode == HttpURLConnection.HTTP_OK){
            InputStream inputStream = connection.getInputStream();
            String result = streamToStr(inputStream);//将流转换为字符串
        }
    } catch (Exception e) {
        e.printStackTrace();
    }
```

上传文件使用的是POST请求方式，其原理类似于POST请求中的上传参数。

7. 下载文件

从服务器下载文件是比较简单的操作，只要得到输入流，就可以从流中读出数据，其使用示例如下。

```
    try {
        String urlPath = "https://www.baidu.com/";
        URL url = new URL(urlPath);
        HttpURLConnection connection = (HttpURLConnection) url.openConnection();
        connection.setRequestMethod("GET");
        connection.connect();
        int responseCode = connection.getResponseCode();
        if(responseCode == HttpURLConnection.HTTP_OK){
            InputStream inputStream = connection.getInputStream();
            File dir = new File("fileDir");
            if (!dir.exists()){
                dir.mkdirs();
            }
            File file = new File(dir, "fileName");   //根据目录和文件名可得到file对象
            FileOutputStream fos = new FileOutputStream(file);
            byte[] buf = new byte[1024*8];
            int len = -1;
            while ((len = inputStream.read(buf)) != -1){
                fos.write(buf, 0, len);
            }
            fos.flush();
        }
    } catch (Exception e) {
        e.printStackTrace();
    }
```

8．使用 HttpURLConnection 的具体示例

【示例】 使用 HttpURLConnection 发送 GET 请求和 POST 请求。

启动 Android Studio，在 Ch10 工程中创建 HttpURLConnectionDemo 模块，并在该模块中创建 MainActivity.java、activity_main.xml、NetUtil.java、Constants.java、LoginUser.java，以及 res\drawable 文件夹下用线性布局（背景色）蓝色边框的 border.xml 等文件。

MainActivity 界面如图 10-1（a）所示，其主要功能有：

（1）可以发送 GET 请求获取网页内容，如图 10-1（b）所示。

（2）可以发送 GET 请求获取网络图片，如图 10-1（c）所示。

（3）可以发送 POST 请求进行登录账号的密码验证，它共有三种情况：①账号不得为空，如图 10-1（d）所示；②账号或密码错误，如图 10-1（e）所示；③账号和密码正确，如图 10-1（f）所示。

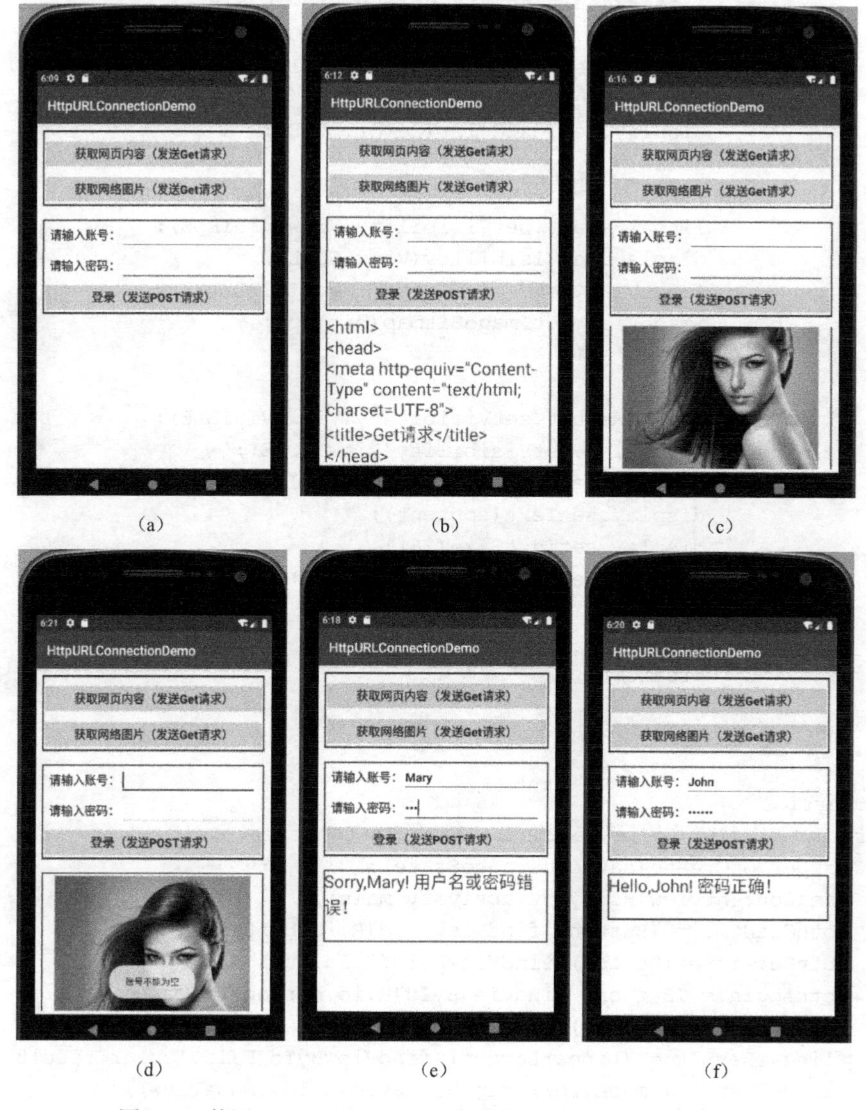

图 10-1 使用 HttpURLConnection 发送的 GET 请求和 POST 请求

HttpURLConnectionDemo 模块的 MainActivity.java 主要代码如下：

```
public class MainActivity extends AppCompatActivity {
    private Button btnGetHtml;
    private Button btnGetPic;
```

```
    private Button btnLogin;
    private EditText userName;
    private EditText passWord;
    private TextView display;
    private ImageView imageView;
    private String content;
    private LinearLayout linearResult;
    private Bitmap bitmap = null;
    // 创建Handler，处理Message，更新UI
    private Handler handler = new Handler() {
        public void handleMessage(android.os.Message msg) {
            switch (msg.what) {
                case 105:
                    linearResult.setVisibility(View.VISIBLE);
                    imageView.setVisibility(View.GONE);
                    display.setVisibility(View.VISIBLE);
                    display.setText(content);
                    display.setTextColor(Color.BLUE);
                    break;
                case 106:
                    linearResult.setVisibility(View.VISIBLE);
                    display.setVisibility(View.GONE);
                    imageView.setVisibility(View.VISIBLE);
                    imageView.setImageBitmap(bitmap);
                    break;
                case 108:
                    linearResult.setVisibility(View.VISIBLE);
                    imageView.setVisibility(View.GONE);
                    display.setVisibility(View.VISIBLE);
                    display.setText(content);
                    display.setTextSize(26);
                    display.setTextColor(Color.RED);
                    break;
                default:
                    break;
            }
        }
    };
    @Override
    protected void onCreate(Bundle savedInstanceState) {
        super.onCreate(savedInstanceState);
        setContentView(R.layout.activity_main);
        btnGetHtml = (Button) findViewById(R.id.button1);
        btnGetPic = (Button) findViewById(R.id.button2);
        btnLogin = (Button) findViewById(R.id.button3);
        display = (TextView) findViewById(R.id.txt_display);
        linearResult = (LinearLayout) findViewById(R.id.linear_result);
        imageView = (ImageView) findViewById(R.id.imageView);
        userName  = (EditText) findViewById(R.id.editText_1);
        passWord  = (EditText) findViewById(R.id.editText_2);
        btnGetHtml.setOnClickListener(new View.OnClickListener() {
            @Override
            public void onClick(View v) {
                new Thread(){
```

```
                @Override
                public void run() {
                    String url = Constants.SiteURL + "ch10/get.jsp";
                    content = NetUtil.getHtml(url);
                    handler.sendEmptyMessage(108);
                }
            }.start();
        }
    });
    btnGetPic.setOnClickListener(new View.OnClickListener() {
        @Override
        public void onClick(View view) {
            new Thread(){
                @Override
                public void run() {
                    String url = Constants.iconPath + "person1.jpg";
                    bitmap = NetUtil.downloadBitmap(url);
                    Log.d("hand","------------getImage-------------");
                    handler.sendEmptyMessage(106);
                }
            }.start();
        }
    });
    btnLogin.setOnClickListener(new View.OnClickListener() {
        @Override
        public void onClick(View view) {
            final LoginUser user = new LoginUser();
            String userNameStr = userName.getText().toString();
            if(userNameStr.equals("")){
                Toast.makeText(MainActivity.this,"账号不能为空",
                    Toast.LENGTH_LONG).show();
                return;
            }
            String userPwdStr = passWord.getText().toString();
            user.setUserName(userNameStr);
            user.setUserPwd(userPwdStr);
            new Thread(){
                @Override
                public void run() {
                    String url = Constants.SiteURL_2 + "Login";
                    content = NetUtil.sendPost(url,user);
                    handler.sendEmptyMessage(108);
                }
            }.start();
        }
    });
  }
}
```

HttpURLConnectionDemo 模块的 Constants.java 主要代码如下：

```
public class Constants {
    public static final String SiteURL = "http://192.168.0.104:8080/MyNetServer/";
    public static final String SiteURL_2 = "http://192.168.0.104:8080/NetServer2/";
    public static final String iconPath = SiteURL + "pic/";
}
```

HttpURLConnectionDemo 模块的 NetUtil.java 主要代码如下：

```
public class NetUtil {
    public static String getHtml(String url) {
        StringBuilder result = new StringBuilder();
        try {
            URL httpUrl = new URL(url);
            HttpURLConnection httpConn = (HttpURLConnection) httpUrl.open Connection();
            InputStreamReader inputStreamReader = new InputStreamReader(httpConn.
getInputStream());
            BufferedReader bufferedReader = new BufferedReader(inputStreamReader);
            String inputLine;
            while ((inputLine = bufferedReader.readLine()) != null)
                result.append(inputLine).append("\n");
            httpConn.disconnect();
            bufferedReader.close();
        } catch (Exception e) {
            e.printStackTrace();
        }
        return result.toString();
    }
    public static Bitmap downloadBitmap(String urlStr) {
        HttpURLConnection urlConnection = null;
        BufferedInputStream in = null;
        Bitmap bitmap = null;
        try {
            URL url = new URL(urlStr);
            urlConnection = (HttpURLConnection) url.openConnection();
            in = new BufferedInputStream(urlConnection.getInputStream(), 8 * 1024);
            bitmap = BitmapFactory.decodeStream(in);
        } catch (Exception e) {
            e.printStackTrace();
        } finally {
            if (urlConnection != null) {
                urlConnection.disconnect();
            }
            try {
                if (in != null) {
                    in.close();
                }
            } catch (final Exception e) {
                e.printStackTrace();
            }
        }
        return bitmap;
    }
    public static String sendPost(String url, LoginUser user) {
        StringBuilder result = new StringBuilder();
        HttpURLConnection httpConn = null;
        String userName = user.getUserName();
        String userPwd = user.getUserPwd();
        try {
            httpConn = (HttpURLConnection) new URL(url).openConnection();
            httpConn.setRequestMethod("POST");
            httpConn.setDoOutput(true);
```

```
            httpConn.setDoInput(true);
            httpConn.setUseCaches(false);  //Post 方式不能缓存,需手动设置为 false
            String params = "userName=" + userName + "&userPwd=" + userPwd;
            OutputStream out = httpConn.getOutputStream();  //获取输出流
            PrintWriter printWriter = new PrintWriter(out);
            printWriter.print(params);
            printWriter.flush();
            InputStream inputStream = httpConn.getInputStream();  // 获取响应的输
入流对象
            InputStreamReader inputStreamReader = new InputStreamReader(inputStream);
            BufferedReader bufferedReader = new BufferedReader(inputStreamReader);
            String inputLine = null;
            while ((inputLine = bufferedReader.readLine()) != null) {
               Log.d("hand", inputLine);
               result.append(inputLine).append("\n");
            }
            printWriter.close();
            bufferedReader.close();
            if (httpConn != null) {
               httpConn.disconnect();
            }
         } catch (Exception e) {
            e.printStackTrace();
         }
         return result.toString();
      }
   }
```

HttpURLConnectionDemo 模块的 LoginUser.java 主要代码如下：

```
public class LoginUser {
   private String userId;
   private String userName;
   private String userPwd;
   …
}
```

"获取网页内容"访问的服务端资源：

```
http://192.168.0.104:8080/MyNetServer/ch10/get.jsp
```

"获取网络图片"访问的服务端资源：

```
http://192.168.0.104:8080/MyNetServer/ pic/ person1.jpg
```

"登录（发送 POST 请求）"提交的服务端：

```
http://192.168.0.104:8080/NetServer2/Login
```

实际上，"登录（发送 POST 请求）"提交的请求是由服务端"192.168.0.104:8080 /NetServer2"中的一个 Servlet 来进行处理的，它的 URL-Mapping 就是"/Login"。

在 Eclipse 中创建"Dynamic Web Project"——NetServer2。

（这里使用的 Eclipse 版本是：eclipse-jee-oxygen-3a-win32-x86_64.zip，Eclipse Java EE IDE for Web Developers，Oxygen.3a Release (4.7.3a)。）

在 NetServer2 工程中创建 LoginServlet，LoginServlet.java 的主要代码如下：

```
@WebServlet("/Login")
public class LoginServlet extends HttpServlet {
      protected void doPost(HttpServletRequest request, HttpServletResponse
response) throws ServletException, IOException {
         response.setContentType("text/html;charset=utf-8");
```

```
            response.setCharacterEncoding("UTF-8");
            PrintWriter out = response.getWriter();
            String userName = request.getParameter("userName");
            String userPwd = request.getParameter("userPwd");
            if((userName!=null) && (userPwd!=null) && userName.equals("John") &&
userPwd.equals("168168")) {
                out.println("Hello," + userName + "! 密码正确! ");
            }else{
                out.println("Sorry," + userName + "! 用户名或密码错误! ");
            }
            out.close();
        }
    }
```

具体代码及文件可详见教学资源包的 Ch10\HttpURLConnectionDemo（客户端），以及服务端的 MyNetServer 和 NetServer2（均部署在 Tomcat 服务器下，其具体路径是：D:\tomcat8\webapps）。测试时，服务端的 IP 地址应更改为测试计算机的 IP 地址。

注意：本章所有内容都需要访问网络才能进行发送或者获取，因此需要在 Ch10\ HttpURLConnectionDemo 模块的配置文件 AndroidManifest.xml 中添加访问网络的权限。

```
<uses-permission android:name = "android.permission.INTERNET" />
```

9. Android 9.0/P http 网络请求的问题

Google 公司表示，为保证用户数据和设备的安全，针对下一代 Android 系统（Android 9.0/P）的应用程序，将要求默认使用加密连接，这意味着从 Android 9 开始，将禁止 App 使用所有未加密的连接，如果是 HTTPS 网络请求则不会受影响。

因此在 Android 9.0/P 使用 HttpUrlConnection 进行 HTTP 请求会出现以下异常。

```
W/System.err: java.io.IOException: Cleartext HTTP traffic to **** not permitted
```

如使用 OkHttp 请求则出现以下异常。

```
java.net.UnknownServiceException: CLEARTEXT communication ** not permitted by
network security policy
```

解决方案：在 res/ xml 目录下新增一个 network_security_config.xml 文件。

```
<?xml version="1.0" encoding="utf-8"?>
<network-security-config>
   <base-config cleartextTrafficPermitted="true" />
</network-security-config>
```

然后在该 Android 应用（如 Ch10\ HttpURLConnectionDemo）的 AndroidManifest.xml 文件的 application 标签下添加以下属性。

```
android:networkSecurityConfig="@xml/network_security_config"
```

其中，XML 文件的配置只是解决了项目中自己的接口请求使用 HTTP 的问题，如果有第三方的 HTTP 请求，则需要在 network_security_config.xml 文件中设置相关路径。

```
<?xml version="1.0" encoding="utf-8"?>
<network-security-config>
    <domain-config cleartextTrafficPermitted="true">
        <domain includeSubdomains="true">你的服务端接口地址</domain>
        <domain includeSubdomains="true">第三方接口路径</domain>
    </domain-config>
</network-security-config>
```

需要注意的是，接口地址不要加“http://”前缀，如果是 IP 地址，还不能加端口号。

10.1.3 使用 OkHttp

1. OkHttp 简介

Android SDK 中自带的 HttpURL Connection 虽然能基本满足网络访问需求，但在使用上有诸多不便。为此 Square 公司实现了一个 HTTP 客户端的类库 OkHttp，用于替代 HttpURL Connection 和 Apache HttpClient（Android API 23 中已移除 HttpClient）。

所以 OkHttp 是一个当前主流的网络请求开源框架，是安卓端最火热的轻量级框架。

OkHttp 是一个支持 HTTP 和 HTTP/2 的客户端，可以在 Android 和 Java 应用程序中使用，其具有以下特点。

（1）API 设计轻巧，通过几行代码的链式调用即可获取结果。

（2）既支持同步请求，也支持异步请求。同步请求会阻塞当前线程，异步请求不会阻塞当前线程，异步执行完成后可执行相应的回调方法。

（3）OkHttp 支持 HTTP/2 协议。通过 HTTP/2 协议可以使客户端中，同一服务器的所有请求共用同一个 Socket 连接。

（4）如果请求不支持 HTTP/2 协议，那么 OkHttp 会在内部维护一个连接池，通过该连接池，可以对 HTTP/1.x 的连接进行重用，从而减少了延迟。

（5）透明的 GZIP 处理降低了下载数据的大小。

（6）请求的数据会进行相应的缓存处理，下次再进行请求时，如果服务器告知 304（表明数据没有发生变化），那么就可直接从缓存中读取数据，从而降低了重复请求的数量。

OkHttp 的设计思路如图 10-2 所示。

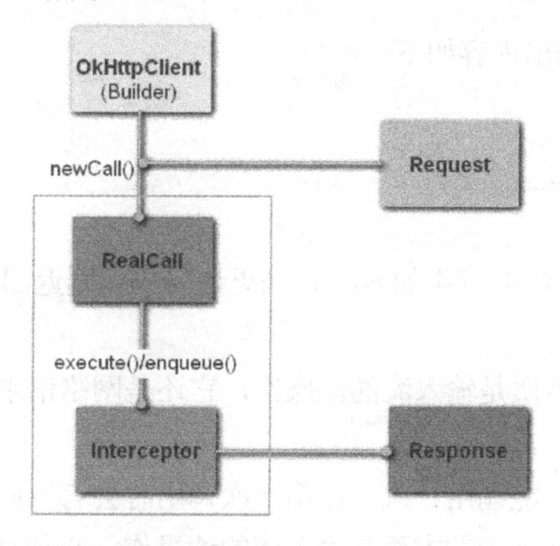

图 10-2　OkHttp 的设计思路

OkHttp 官网地址：https://square.github.io/okhttp/。

OkHttp GitHub 地址：https://github.com/square/okhttp。

在 AndroidStudio 中使用 OkHttp 不需要下载 jar 包，直接添加依赖即可，其示例如下。

```
implementation("com.squareup.okhttp3:okhttp:4.2.0")
```

OkHttp 开发主要步骤有：①创建 OkHttpClient 并添加配置；②创建 Request 对象；③创建 Call 对象；④执行网络请求（调用 Call 对象的 execute ()方法或者 enqueue ()方法）。

OkHttpClient 可通过 OkHttpClient.Builder 采用建造者模式构建，并通过 Builder 方便灵活地设置通用参数。

2. GET 请求

使用 OkHttp 进行网络请求支持两种方式：同步请求和异步请求，其中异步请求较为常用。

（1）GET 的同步请求。

在请求时开启子线程，请求成功后，需要跳转到 UI 线程修改 UI。

```
public void getDatasync(){
    new Thread(new Runnable() {
        @Override
        public void run() {
            try {
                OkHttpClient client = new OkHttpClient();//创建 OkHttpClient 对象
                Request request = new Request.Builder()
                        .url("http://www.baidu.com") //请求接口。如果需要上传参数则拼接
到接口后面
                        .build();   //创建 Request 对象
                Response response = null;
                response = client.newCall(request).execute();  //得到 Response 对象
                if (response.isSuccessful()) {
                    Log.d("OkHttp ","response.code()==" + response.code());
                    Log.d("OkHttp ","response.message()==" + response.message());
                    Log.d("OkHttp ","res==" + response.body().string());
                    // 此时的代码执行在子线程，修改 UI 的操作可使用 Handler 跳转到 UI 线程
                }
            } catch (Exception e) {
                e.printStackTrace();
            }
        }
    }).start();
}
```

Android Studio 的 Logcat 输出内容如下。

```
response.code()==200
response.message()==OK
res=={"code":200, "message":success}
```

有以下几点要加以说明。

① Response.code 是 HTTP 响应行中的 code，如果访问成功则返回 200。这个不是服务器设置的，而是 HTTP 协议中自带的。

② response.body().string()本质是输入流的读操作，它还是网络请求的一部分，所以这行代码必须放在子线程中。

③ response.body().string()只能调用一次，在第一次调用时会有返回值，第二次再调用时将会返回 null。原因是：response.body().string()的本质是输入流的读操作，必须有服务器输出流的写操作时，客户端的读操作才能得到数据。由于服务器的写操作只执行一次，所以客户端的读操作也只能执行一次，第二次将返回 null。

④ 响应体的 string()方法（response.body().string()），对于小文档来说十分方便、高效。但是如果响应体太大（超过 1MB），应避免使用 string()方法 ，因为它会把整个文档加载到内存中。对于超过 1MB 的响应 body，应使用流的方式来处理。

（2）GET 的异步请求。

GET 的异步请求方式不需要开启子线程，但回调方法是执行在子线程中的，所以在更新 UI 时还要跳转到 UI 线程中。

一般来说，异步 GET 请求有以下 4 个步骤。

- 创建 OkHttpClient 对象。
- 通过 Builder 模式创建 Request 对象，必须有个 URL 参数，可以通过 Request.Builder 设置更多

的参数，如 header、method 等。

- 通过 Request 对象构造得到一个 Call 对象，包括有 execute()和 cancel()等方法，这里假设用 call 表示 Call 对象。
- 以异步的方式去执行请求，调用的是 call.enqueue()方法，将 call（Call 对象）加入调度队列，任务执行完成后会在 Callback 中得到结果。

```
    private void getDataAsync() {
        OkHttpClient client = new OkHttpClient();
        Request request = new Request.Builder()
                .url("http://www.baidu.com")
                .build();
        client.newCall(request).enqueue(new Callback() {
            @Override
            public void onFailure(Call call, IOException e) {
            }
            @Override
            public void onResponse(Call call, Response response) throws IOException
            {
                if(response.isSuccessful()){   // 用回调的方法执行在子线程中
                    Log.d("OkHttp", "获取数据成功");
                    Log.d("OkHttp", "response.code()==" + response.code());
                    Log.d("OkHttp", "response.body().string()==" + response.body().
string());
                }
            }
        });
    }
```

异步请求的打印结果和注意事项与同步请求时相同。它们最大的不同点就是异步请求时不需要开启子线程，enqueue()方法会自动将网络请求部分放入子线程中执行，但仍有以下几点要进行说明。

- 回调接口的 onFailure 方法和 onResponse 执行在子线程中。
- 异步调用的回调函数是在子线程中的，因此不能在子线程中更新 UI，需要借助于 runOnUiThread() 方法或者 Handler 来处理。
- onResponse 回调有一个参数是 Response，如果想获得返回的是字符串，则可以通过 response.body().string()。如果获得返回的是二进制字节数组，则可以调用 response.body(). bytes()。如果想拿到返回的 inputStream，则可以调用 response.body().byteStream()，得到 inputStream 后就可以通过 IO 的方式写文件。
- response.body().string()方法也必须放在子线程中。当执行这行代码得到结果后，再跳转到 UI 线程修改 UI。

【示例】 使用 OkHttp 发送 GET 请求（同步方式和异步方式）。

启动 Android Studio，在 Ch10_OkHttp 工程中创建 OkHttpGetDemo 模块，并在该模块中创建 MainActivity.java、activity_main.xml 文件。

OkHttpGetDemo 模块中的 MainActivity.java 主要代码如下：

```
    public class MainActivity extends AppCompatActivity implements View.OnClick
Listener {
        TextView responseText;
        @Override
        protected void onCreate(Bundle savedInstanceState) {
            super.onCreate(savedInstanceState);
            setContentView(R.layout.activity_main);
```

```
        Button sendRequest = (Button) findViewById(R.id.send_request);
        responseText = (TextView) findViewById(R.id.response_text);
        sendRequest.setOnClickListener(this);
    }
    @Override
    public void onClick(View v) {
        if (v.getId() == R.id.send_request) {
            //getDataSync();
            getDataAsync();
        }
    }

    private void getDataSync() {
        new Thread(new Runnable() {
            @Override
            public void run() {
                try {
                    OkHttpClient client = new OkHttpClient();// 创建OkHttpClient对象
                    Request request = new Request.Builder()
                            .url("http://192.168.0.104:8080/MyNetServer/ch10/get.jsp")
                            .build();       // 创建Request对象
                    Response response = null;
                    response = client.newCall(request).execute();//得到Response对象
                    if (response.isSuccessful()) {
                        Log.d("suda_mialab", "response.code()==" + response.code());
                        Log.d("suda_mialab",  "response.message()=="  +  response.
message());
                        String responseData = response.body().string();
                        showResponse(responseData);
                    }
                } catch (Exception e) {
                    e.printStackTrace();
                }
            }
        }).start();
    }

    private void getDataAsync() {
        OkHttpClient client = new OkHttpClient.Builder()
                .connectTimeout(300, TimeUnit.SECONDS)      // 设置连接超时时间
                .readTimeout(300, TimeUnit.SECONDS)         // 设置读取超时时间
                .build();
        Request request = new Request.Builder().url("http://www.suda.edu.cn").
build();
        client.newCall(request).enqueue(new Callback() {
            @Override
            public void onFailure(Call call, IOException e) {
                Log.d("suda_mialab", "-----获取数据失败------");
            }
            @Override
            public void onResponse(Call call, Response response) throws IOException {
                if (response.isSuccessful()) {     //用回调的方法执行在子线程中
                    Log.d("suda_mialab", "获取数据成功了");
                    Log.d("suda_mialab", "response.code()==" + response.code());
```

```
                    String responseData = response.body().string();
                    Log.d("suda_mialab", "response.body().string()==" + responseData);
                    String responseData2 = null;
                    if ((responseData != null) && (responseData.length() > 601)) {
                        responseData2 = responseData.substring(0, 600);
                    }
                    showResponse(responseData2);
                }
            }
        });
    }
    private void showResponse(final String response) {
        runOnUiThread(new Runnable() {
            @Override
            public void run() {
                // 在这里进行 UI 操作，将结果显示界面上
                responseText.setText(response);
            }
        });
    }
}
```

在 MainActivity 的 onClick()方法中分别调用 getDataSync()方法（同步方式）和 getDataAsync()方法（异步方式）进行测试。

GET 请求（同步方式）访问的服务端是：

```
http://192.168.0.104:8080/MyNetServer/ch10/get.jsp
```

GET 请求（异步方式）访问的服务端是：

```
http://www.suda.edu.cn
```

运行 Ch10_OkHttp\OkHttpGetDemo，点击“Send Request”按钮，GET 请求（同步方式）的结果如图 10-3（a）所示，GET 请求（异步方式）的结果如图 10-3（b）所示。

（a）　　　　（b）

图 10-3　使用 OkHttp 发送 GET 请求（同步方式和异步方式）

【示例】 解析 JSON（使用 OkHttp 发送 GET 请求）。

启动 Android Studio，在 Ch10_OkHttp 工程中创建 OkHttpJsonDemo 模块，并在该模块中创建 MainActivity.java、activity_main.xml 文件。

OkHttpJsonDemo 模块的 MainActivity.java 主要代码如下：

```
    public class MainActivity extends AppCompatActivity implements View.OnClick
Listener {
        TextView responseText;
        @Override
        protected void onCreate(Bundle savedInstanceState) {
            super.onCreate(savedInstanceState);
            setContentView(R.layout.activity_main);
            Button sendRequest = (Button) findViewById(R.id.send_request);
            responseText = (TextView) findViewById(R.id.response_text);
            sendRequest.setOnClickListener(this);
        }
        @Override
        public void onClick(View v) {
            if (v.getId() == R.id.send_request) {
                sendRequestWithOkHttp();
            }
        }
        private void sendRequestWithOkHttp() {
            new Thread(new Runnable() {
                @Override
                public void run() {
                    try {
                        OkHttpClient client = new OkHttpClient();
                        Request request = new Request.Builder()
                                .url("http://192.168.0.104:8080/MyNetServer/json/book
.json")
                                .build();
                        Response response = client.newCall(request).execute();
                        String responseData = response.body().string();
                        parseJSONWithJSONObject(responseData);
                        showResponse(responseData);
                    } catch (Exception e) {
                        e.printStackTrace();
                    }
                }
            }).start();
        }
        private void parseJSONWithJSONObject(String jsonData) {
            try {
                JSONArray jsonArray = new JSONArray(jsonData);
                for (int i = 0; i < jsonArray.length(); i++) {
                    JSONObject jsonObject = jsonArray.getJSONObject(i);
                    String id = jsonObject.getString("bookId");
                    String name = jsonObject.getString("bookName");
                    String version = jsonObject.getString("bookPrice");
                    Log.d("suda_mialab", "bookId is:  " + id);
                    Log.d("suda_mialab", "bookName is:  " + name);
                    Log.d("suda_mialab", "bookPrice is:  " + version);
                }
            } catch (Exception e) {
                e.printStackTrace();
            }
        }
```

```
    private void showResponse(final String response) {
        runOnUiThread(new Runnable() {
            @Override
            public void run() {
                //在这里进行UI操作，将结果显示界面上
                responseText.setText(response);
            }
        });
    }
}
```

这里访问的服务端是：

```
http://192.168.0.104:8080/MyNetServer/json/book.json
```

book.json 的内容如下：

```
[{"bookId":"978-201","bookName":"Java Web技术及应用","bookPrice":"59.5"},
{"bookId":"978-202","bookName":"数据结构项目实训","bookPrice":"45"},
{"bookId":"978-206","bookName":"SCJP考试指南","bookPrice":"69"}]
```

运行 Ch10_OkHttp\OkHttpJsonDemo，点击 "Send Request"按钮，Android Studio 的 Logcat 将会有以下的输出。

```
…com.example.okhttpjsondemo D/suda_mialab: bookId is:  978-201
…com.example.okhttpjsondemo D/suda_mialab: bookName is:  Java Web技术及应用
…com.example.okhttpjsondemo D/suda_mialab: bookPrice is:  59.5
…com.example.okhttpjsondemo D/suda_mialab: bookId is:  978-202
…com.example.okhttpjsondemo D/suda_mialab: bookName is:  数据结构项目实训
…com.example.okhttpjsondemo D/suda_mialab: bookPrice is:  45
…com.example.okhttpjsondemo D/suda_mialab: bookId is:  978-206
…com.example.okhttpjsondemo D/suda_mialab: bookName is:  SCJP考试指南
…com.example.okhttpjsondemo D/suda_mialab: bookPrice is:  69
```

3. POST 方式提交键值对

POST 请求也分同步和异步两种方式，同步方式与异步方式的区别与 GET 请求类似，所以这里只讲解 POST 异步请求的使用方法。

为了在 POST 请求中传递参数，可以创建表单请求体对象，然后再把表单请求体对象作为 post(RequestBody body)方法的参数。POST 请求传递参数的方法还有其他方式，但都是通过 post(RequestBody body)方法传递的。

Request.Builder 类的 post(RequestBody body)方法定义如下。

```
public Builder post(RequestBody body) {
    return method("POST", body);
}
```

post(RequestBody body)方法接收的参数是 RequestBody 对象，所以只要是 RequestBody 类及子类对象都可以当作参数进行传递。FormBody 就是 RequestBody 的一个子类对象。

```
private void postDataWithParame() {
    OkHttpClient client = new OkHttpClient();    // 创建OkHttpClient对象
    FormBody.Builder formBody = new FormBody.Builder();    // 创建表单请求体
    formBody.add("username","Marry");      // 传递键值对参数
    Request request = new Request.Builder()     // 创建Request对象
            .url("http://192.168.0.104:8080/NetServer2/Login")
            .post(formBody.build())     // 传递请求体
            .build();
    client.newCall(request).enqueue(new Callback() {…});    // 此处省略回调方法
}
```

小贴士：由以上代码可以看到，POST 请求中并没有设置请求方式为 POST，回忆在 GET 请求中也没有设置请求方式为 GET，那么是怎么区分请求方式的呢？主要看 Request.Builder 类的 post()方法是否被调用。在 Request.Builder 对象创建最初默认是 GET 请求，所以在 GET 请求中不需要设置请求方式，当调用 post()方法时就会把请求方式修改为 POST。所以此时为 POST 请求。

【示例】 POST 请求使用 FormBody 提交键值对给服务器端的 Servlet 进行验证。

启动 Android Studio，在 Ch10_OkHttp 工程中创建 OkHttpPostDemo 模块，并在该模块中创建 MainActivity.java、activity_main.xml 文件。

OkHttpPostDemo 模块的 MainActivity.java 主要代码如下：

```
    public class MainActivity extends AppCompatActivity implements View.OnClick
Listener {
        TextView responseText;
        @Override
        protected void onCreate(Bundle savedInstanceState) {
            super.onCreate(savedInstanceState);
            setContentView(R.layout.activity_main);
            Button sendRequest = (Button) findViewById(R.id.send_request);
            responseText = (TextView) findViewById(R.id.response_text);
            sendRequest.setOnClickListener(this);
        }
        @Override
        public void onClick(View v) {
            if (v.getId() == R.id.send_request) {
                postDataWithParame();
            }
        }

        private void postDataWithParame() {
            OkHttpClient client = new OkHttpClient(); // 创建 OkHttpClient 对象
            FormBody.Builder formBody = new FormBody.Builder();   // 创建表单请求
            formBody.add("userName","John");              // 传递键值对参数
            formBody.add("userPwd","168168");             // 传递键值对参数
            Request request = new Request.Builder()    // 创建 Request 对象
                    .url("http://192.168.0.104:8080/NetServer2/Login")
                    .post(formBody.build())               // 传递请求体
                    .build();
            client.newCall(request).enqueue(new Callback() {
                @Override
                public void onFailure(Call call, IOException e) {
                }
                @Override
                public void onResponse(Call call, Response response) throws IOException {
                    if(response.isSuccessful()){          // 用回调方法执行到子线程中
                        Log.d("suda_mialab","获取数据成功了");
                        Log.d("suda_mialab","response.code()=="+response.code());
                        String responseData = response.body().string();
                        showResponse(responseData);
                    }
                }
            });
        }

        private void showResponse(final String response) {
```

```
            runOnUiThread(new Runnable() {
                @Override
                public void run() {
                    // 在这里进行 UI 操作，将结果显示到界面上
                    responseText.setText(response);
                }
            });
        }
    }
```

这里的服务端是：

```
http://192.168.0.104:8080/NetServer2/Login
```

运行 Ch10_OkHttp\OkHttpJsonDemo，点击“Send Request”按钮，其运行结果如图 10-4 所示。

图 10-4 使用 FormBody 提交键值对

4. 使用 RequestBody 传递 JSON 对象或 File 对象

RequestBody 是抽象类，故不能直接使用，但是 RequestBody 中有静态方法 create()，使用这个方法就可以得到 RequestBody 对象。在构造 RequestBody 时需要指定 MediaType，用于描述请求/响应 Body 的内容类型，关于 MediaType 的更多信息可以查看 RFC 2045。

通过 RequestBody 对象，可以携带要提交的数据。

上传 JSON 对象使用的示例如下。

```
    OkHttpClient client = new OkHttpClient();     // 创建 OkHttpClient 对象
    MediaType JSON = MediaType.parse("application/json; charset=utf-8");    // 数据
类型为 JSON 格式
    String jsonStr = "{\"username\":\"Mary\",\"nickname\":\"狗狗\"}";    // JSON 数据.
    RequestBody body = RequestBody.create(JSON, josnStr);
    Request request = new Request.Builder()
            .url("http://192.168.0.104:8080/NetServer2/Process")
            .post(body)
            .build();
    client.newCall(request).enqueue(new Callback() {…});       // 此处省略回调方法
```

OkHttp 框架中 RequestBody.java 的 create(MediaType contentType, String content)方法定义如下。

```
    public static RequestBody create(@Nullable MediaType contentType, String content) {
        Charset charset = UTF_8;
        if (contentType != null) {
            charset = contentType.charset();
            if (charset == null) {
                charset = UTF_8;
                contentType = MediaType.parse(contentType + "; charset=utf-8");
            }
        }
        byte[] bytes = content.getBytes(charset);
        return create(contentType, bytes);
    }
```

上传 File 对象使用的示例如下。

```
    MediaType mediaType = MediaType.parse("text/x-markdown; charset=utf-8");
    OkHttpClient okHttpClient = new OkHttpClient();
    File file = new File("test.md");
    Request request = new Request.Builder()
            .url("http://192.168.0.104:8080/NetServer2/Process")
            .post(RequestBody.create(mediaType, file))
```

```
            .build();
    okHttpClient.newCall(request).enqueue(new Callback() {
        @Override
        public void onFailure(Call call, IOException e) {
            Log.d(TAG, "onFailure: " + e.getMessage());
        }
        @Override
        public void onResponse(Call call, Response response) throws IOException {
            Log.d(TAG, response.protocol() + " " + response.code() + " " + response.
message());
            Headers headers = response.headers();
            for (int i = 0; i < headers.size(); i++) {
                Log.d(TAG, headers.name(i) + ":" + headers.value(i));
            }
            Log.d(TAG, "onResponse: " + response.body().string());
        }
    });
```

5. 使用 MultipartBody 同时传递键值对参数和 File 对象

FormBody 传递的是字符串型的键值对，RequestBody 传递的是多媒体，如果二者都要传递该怎么办呢？此时就需要使用 MultipartBody 类。MultipartBody 可以构建复杂的请求体，多块请求体中每块请求都是一个请求体，可以定义自己的请求头。

```
    OkHttpClient  okHttpClient = new OkHttpClient();    //①创建 OkHttpClient 对象
    File file = new File(Environment.getExternalStorageDirectory(), " yuxiaonan.png");
//上传的图片
    //②通过 new MultipartBody build() 创建 requestBody 对象，
    RequestBody  requestBody = new MultipartBody.Builder()
            .setType(MultipartBody.FORM)    // 设置类型是表单（MultipartBody.FORM）
            .addFormDataPart("username","yuxiaonan")
            .addFormDataPart("age","25")
            .addFormDataPart("image", "yuxiaonan.png", RequestBody.create(MediaType.
parse("image/png"), file))
            .build();
    //③创建 Request 对象，设置 URL 地址，将 RequestBody 作为 post()方法的参数传入
    Request request = new Request.Builder()
            .url("http://192.168.0.104:8080/NetServer2/Process")
            .post(requestBody)
            .build();
    //④创建一个 Call 对象，参数就是 Request 请求对象
    Call call = okHttpClient.newCall(request);
    //⑤请求加入调度，重写回调方法
    call.enqueue(new Callback() {
        @Override
        public void onFailure(Call call, IOException e) {
        }
        @Override
        public void onResponse(Call call, Response response) throws IOException {
        }
    });
```

6. 自定义 RequestBody 实现流的上传

由以上分析可知，只要是 RequestBody 类及子类都可以作为 post()方法的参数，下面就可以自定义一个类，继承 RequestBody 实现流的上传。

首先创建一个 RequestBody 类的子类对象。

```
RequestBody  myRequestBody = new RequestBody() {
    @Override
    public MediaType contentType() {
        return null;
    }
    @Override
    public void writeTo(BufferedSink sink) throws IOException {  // 重写writeTo()
方法
        FileInputStream fio = new FileInputStream(new File("fileName"));
        byte[] buffer = new byte[1024*8];
        if(fio.read(buffer) != -1){
            sink.write(buffer);
        }
    }
};
```

然后使用 myRequestBody 对象。

```
OkHttpClient client = new OkHttpClient();       // 创建OkHttpClient对象
Request request = new Request.Builder()
        .url("http://192.168.0.104:8080/NetServer2/Process")
        .post(body)
        .build();
client.newCall(request).enqueue(new Callback() {…});        // 此处省略回调方法
```

7. 设置请求头

OkHttp 中设置请求头特别简单，在创建 Request 对象时调用 header()方法即可。

```
Request request = new Request.Builder()
        .url("http://www.baidu.com")
        .header("User-Agent", "OkHttp Headers.java")
        .addHeader("token", "myToken")
        .build();
```

小贴士：在网络条件不好的情况下，用户通常会主动关闭页面，这时候就需要取消正在请求的 HTTP Request，OkHttp 为此提供了 cancel()方法。但是在使用过程中发现，如果调用 cancel()方法，就会回调到 CallBack 的 onFailure()方法中。通过测试发现不同的失败类型返回的 IOExceptione 不一样，所以可以通过 e.toString 中的关键字来区分不同的错误类型。

- 自己主动取消的错误提示信息是 java.net.SocketException: Socket closed。
- 超时的错误提示信息是 java.net.SocketTimeoutException。
- 网络出错的提示信息是 java.net.ConnectException: Failed to connect to xxxxx。

因此针对以上情况，可以进行如下处理。

```
call.enqueue(new Callback() {
    @Override
    public void onFailure(Call call, IOException e) {
        if(e.toString().contains("closed")) {
            … // 主动取消的情况
        }else{
            …  // 其他情况
        }
    }
    …
}
```

在 okhttp3.Callback 的回调方法中有个参数是 Call 类型。这个 call（表示 Call 实例）可以单独取消相应的请求，在 onFailure()方法或者 onResponse()方法内部执行 call.cancel()都可以。如果想取消所有的请求，则可以执行语句："okhttpclient.dispatcher().cancelAll();" 实现。

8．设置超时

没有响应时使用超时结束 call。没有响应的原因可能是客户点击链接、服务器可用性等问题。OkHttp 支持连接时，读取和写入超时。

```
Request request = new Request.Builder()
        .url("http://www.baidu.com")
        .connectTimeout(10, TimeUnit.SECONDS);    // 连接超时
        .readTimeout(10,TimeUnit.SECONDS);        // 读取超时
        .writeTimeout(10,TimeUnit.SECONDS);       // 写入超时
        .build();
```

9．Response Caching（响应缓存）

缓存主要是为了在没有网络的情况下，或者资源不需要后端进行实时更新时，可以直接从缓存中获取资源信息。一方面能够在断网的情况下仍然给用户良好的界面展示，另一个方面能够避免频繁请求网络给后端带来的压力。目前网络上介绍的在 OkHttp 中实现客户端的缓存机制，都是在自定义 Intercept 中设置 max-age 和 max-stale 的参数值来完成的。

为了缓存响应，需要一个可以读/写的缓存目录，此缓存目录当然会有缓存大小的限制。这个缓存目录应该是私有的，不信任的程序应不能读取缓存内容。

一个缓存目录同时拥有多个缓存访问是错误的。大多数程序只需要调用一次 new OkHttp()，在第一次调用时配置好缓存，然后其他地方只需要调用这个实例就可以了，否则两个缓存示例互相干扰，不仅会破坏响应缓存，而且还有可能导致程序崩溃。

响应缓存使用 HTTP 头作为配置。可以在请求头中添加 Cache-Control: max-stale=3600，OkHttp 缓存会支持。请求服务可通过响应头确定响应缓存多长时间，如使用 Cache-Control: max-age=9600。

```
    int cacheSize = 10 * 1024 * 1024; // 10 MiB
    Cache cache = new Cache(cacheDirectory, cacheSize);
    OkHttpClient.Builder builder = new OkHttpClient.Builder();
    builder.cache(cache);
    OkHttpClient client = builder.build();
    Request request = new Request.Builder().url("http://publicobject.com/helloworld.
txt").build();
    Call call = client.newCall(request);
    call.enqueue(new Callback() {…});
```

其中，max-age：告知缓存多长时间，在没有超过缓存时间的情况下，请求会返回缓存内的数据。在超出 max-age 的情况下向服务端发起新的请求，在请求失败时返回缓存数据，否则向服务端重新发起请求。

max-stale：指示客户机可以接收超出 max-age 时间的响应消息。max-stale 在请求设置中有效，在响应设置中无效。

因此，在请求中同时使用 max-age 和 max-stale，缓存的时间可以是 max-age 和 max-stale 的和。

10．下载文件和上传文件示例

在 OkHttp 中并没有提供下载文件的功能，但是在 Response 中可以获取流对象，有了流对象就可以实现文件的下载。

```
    try{
        InputStream is=response.body().byteStream();     // 从服务器得到输入流对象
```

```
        long sum=0;
        File dir=new File(mDestFileDir);
        if(!dir.exists()){
            dir.mkdirs();
        }
        File file=new File(dir,mdestFileName);     // 根据目录和文件名得到File对象
        FileOutputStream fos=new FileOutputStream(file);
        byte[]buf=new byte[1024*8];
        int len=0;
        while((len=is.read(buf))!=-1){
            fos.write(buf,0,len);
        }
        fos.flush();
        return file;
    }
```

以上这段代码是写在回调接口 CallBack 的 onResponse 方法中的。

【示例】 OkHttp 实现图片下载。

启动 Android Studio，在 Ch10_OkHttp 工程中创建 DownloadPic_2 模块，并在该模块中创建 MainActivity.java、activity_main.xml 文件。

DownloadPic_2 模块中 MainActivity.java 的主要代码如下：

```
    public class MainActivity extends AppCompatActivity implements View.OnClick
Listener {
        private Button btnDownload;
        private ImageView imgShow;
        private String Path = "http://192.168.0.104:8080/MyNetServer/pic/person8.jpg";
        //private String Path = "http://www.suda.edu.cn/images/index.jpg";
        private static final int SUCCESS = 1;
        private static final int FALL = 2;
        Handler handler = new Handler() {
            @Override
            public void handleMessage(Message msg) {
                switch (msg.what) {
                    // 加载网络成功进行UI的更新，处理得到的图片资源
                    case SUCCESS:
                        // 通过Imageview，设置图片
                        imgShow.setImageBitmap((Bitmap) msg.obj);
                        break;
                    // 当加载网络失败时执行的逻辑代码
                    case FALL:
                        Toast.makeText(MainActivity.this, "网络出现了问题",
                                Toast.LENGTH_SHORT).show();
                        break;
                }
            }
        };
        @Override
        protected void onCreate(Bundle savedInstanceState) {
            super.onCreate(savedInstanceState);
            setContentView(R.layout.activity_main);
            initView();
        }
        private void initView() {
            btnDownload = (Button) findViewById(R.id.download);
```

```
        imgShow = (ImageView) findViewById(R.id.imageView);
        btnDownload.setOnClickListener(this);
    }
    @Override
    // 根据点击事件获取网络上的图片资源，使用的是 OkHttp 框架
    public void onClick(View v) {
        switch (v.getId()) {
            case R.id.download:
                OkHttpClient okHttpClient = new OkHttpClient();
                Request request = new Request.Builder()
                        .get()
                        .url(Path)
                        .build();
                Call call = okHttpClient.newCall(request);
                call.enqueue(new Callback() {
                    @Override
                    public void onFailure(Call call, IOException e) {
                    }
                    @Override
                    public void onResponse(Call call, Response response) throws
IOException {
                        // 将响应数据转化为输入流数据
                        InputStream inputStream = response.body().byteStream();
                        // 将输入流数据转化为 Bitmap 位图数据
                        Bitmap bitmap = BitmapFactory.decodeStream(inputStream);
                        // 获取当前 App 的私有存储目录
                        String picPath = getExternalFilesDir
                                (Environment.DIRECTORY_DOWNLOADS).toString() + "/";
                        File file = new File(picPath + "beauty.jpg");
                        file.createNewFile();
                        // 创建文件输出流对象，用来向文件中写入数据
                        FileOutputStream out = new FileOutputStream(file);
                        // 将 bitmap 存储为 JPG 格式的图片
                        bitmap.compress(Bitmap.CompressFormat.JPEG, 100, out);
                        out.flush();  // 刷新文件流
                        out.close();
                        // 通过 Handler 更新 UI
                        Message message = handler.obtainMessage();
                        message.obj = bitmap;
                        message.what = SUCCESS;
                        handler.sendMessage(message);
                    }
                });
                break;
        }
    }
}
```

要访问的服务端是：

```
http://192.168.0.104:8080/MyNetServer/pic/person8.jpg
```

运行 Ch10_OkHttp\DownloadPic_2，点击“下载图片”按钮，其运行结果如图 10-5 所示。

将下载的图片重新命名为 beauty.jpg，它的保存路径如下。

/storage/emulated/0/Android/data/com.example.downloadpic_2/files/Download/beauty.jpg

通过点击 View→Tool Windows→Device File Explorer，便能找到这张 beauty.jpg 图片，可导出查看

进行验证，如图 10-6 所示，具体代码和文件参见教学资源包。

图 10-5 OkHttp 实现图片下载

Device File Explorer

Emulator Galaxy_Nexus_API_28 Android 9, API 28

Name	Permissions	Date	Size
▶ proc	dr-xr-xr-x	2019-09-01 10:59	0 B
▶ product	lrw-r--r--	2019-04-10 22:49	15 B
▶ sbin	drwxr-x---	2019-04-10 22:43	4 KB
▶ sdcard	lrw-r--r--	2019-04-10 22:49	21 B
▼ storage	drwxr-xr-x	2019-09-01 11:01	100 B
▶ 16E6-3507	drwxrwx--x	2019-09-01 11:01	2 KB
▼ emulated	drwx--x--x	2019-09-01 11:01	4 KB
▼ 0	drwxrwx--x	2019-09-01 11:02	4 KB
▶ Alarms	drwxrwx--x	2019-09-01 11:02	4 KB
▼ Android	drwxrwx--x	2019-09-01 11:01	4 KB
▼ data	drwxrwx--x	2019-09-01 11:21	4 KB
▼ com.example.downloadpic_2	drwxrwx--x	2019-09-01 11:21	4 KB
▼ files	drwxrwx--x	2019-09-01 11:21	4 KB
▼ Download	drwxrwx--x	2019-09-01 11:21	4 KB
beauty.jpg	-rw-rw----	2019-09-01 11:21	311.1 KB

图 10-6 beauty.jpg 的存储路径

【示例】 OkHttp 实现文件上传。

启动 Android Studio，在 Ch10_OkHttp 工程中创建 UploadDemo 模块，并在该模块中创建 MainActivity.java、activity_main.xml、OkHttpUtil.java、ProgressListener.java、ProgressRequestBody.java、ProgressResponseBody.java、ProgressModel.java 等文件。

UploadDemo 模块的 MainActivity.java 主要代码如下：

```
public class MainActivity extends AppCompatActivity implements View.OnClick
Listener {
    public static final String TAG = "MainActivity";
    private ProgressBar download_progress, post_progress;
    private TextView download_text, post_text;
    public String basePath;
    @Override
    protected void onCreate(Bundle savedInstanceState) {
        super.onCreate(savedInstanceState);
        setContentView(R.layout.activity_main);
        download_progress = (ProgressBar) findViewById(R.id.download_progress);
        download_text = (TextView) findViewById(R.id.download_text);
        post_progress = (ProgressBar) findViewById(R.id.post_progress);
        post_text = (TextView) findViewById(R.id.post_text);
        findViewById(R.id.ok_post_file).setOnClickListener(this);
        findViewById(R.id.ok_download).setOnClickListener(this);
        basePath = getExternalFilesDir(Environment.DIRECTORY_DOWNLOADS).toString()
+ "/";
    }
    @Override
    public void onClick(View view) {
        switch (view.getId()) {
            case R.id.ok_download:
                String url = "http://192.168.0.104:8080/MyNetServer/pic/suda6.jpg";
                final String fileName = url.split("/")[url.split("/").length - 1];
                OkHttpUtil.downloadFile(url, new ProgressListener() {
                    @Override
```

```
                    public void onProgress(long currentBytes, long contentLength,
boolean done) {
                        Log.i(TAG, "currentBytes==" + currentBytes + "==contentLength==" +
                                contentLength + "==done==" + done);
                        int progress = (int) (currentBytes * 100 / contentLength);
                        download_progress.setProgress(progress);
                        download_text.setText(progress + "%");
                    }
                }, new Callback() {
                    @Override
                    public void onFailure(Call call, IOException e) {
                    }
                    @Override
                    public void onResponse(Call call, Response response) throws
IOException {
                        if (response != null) {
                            InputStream is = response.body().byteStream();
                            FileOutputStream fos = new FileOutputStream(new File
(basePath + fileName));
                            int len = 0;
                            byte[] buffer = new byte[2048];
                            while (-1 != (len = is.read(buffer))) {
                                fos.write(buffer, 0, len);
                            }
                            fos.flush();
                            fos.close();
                            is.close();
                        }
                    }
                });
                break;
            case R.id.ok_post_file:
                File file = new File(basePath + "OkHttp3基本使用.pptx");
                String postUrl = "http://192.168.0.104:8080/NetServer2/Upload";
                OkHttpUtil.postFile(postUrl, new ProgressListener() {
                    @Override
                    public void onProgress(long currentBytes, long contentLength,
boolean done) {
                        Log.i(TAG, "currentBytes==" + currentBytes + "==contentLength==" +
                                contentLength + "==done==" + done);
                        int progress = (int) (currentBytes * 100 / contentLength);
                        post_progress.setProgress(progress);
                        post_text.setText(progress + "%");
                    }
                }, new Callback() {
                    @Override
                    public void onFailure(Call call, IOException e) {
                    }
                    @Override
                    public void onResponse(Call call, Response response) throws
IOException {
                        if (response != null) {
                            String result = response.body().string();
                        }
```

```
                }
            }, file);
            break;
        }
    }
}
```

build.gradle 的 dependencies 节点须添加以下代码。

```
dependencies {
    implementation("com.squareup.okhttp3:okhttp:3.14.1")
    implementation 'com.squareup.okio:okio:1.7.0'
}
```

UploadDemo 模块的 OkHttpUtil.java 主要代码如下：

```
public class OkHttpUtil {
    private static OkHttpClient okHttpClient = new OkHttpClient.Builder(). connect
Timeout(10000,
            TimeUnit.MILLISECONDS)
            .readTimeout(10000,TimeUnit.MILLISECONDS)
            .writeTimeout(10000, TimeUnit.MILLISECONDS).build();
    public static final MediaType JSON = MediaType.parse("application/json;
charset=utf-8");
    public static final MediaType MEDIA_TYPE_MARKDOWN = MediaType.parse("text/
x-markdown; charset=utf-8");
     // 文件上传
    public static void postFile(String url, final ProgressListener listener,
Callback callback, File…files){
        MultipartBody.Builder builder = new MultipartBody.Builder();
        builder.setType(MultipartBody.FORM);
        builder.addFormDataPart("filename",files[0].getName());
        builder.addFormDataPart("position","0");
        builder.addFormDataPart("file",files[0].getName(),
                RequestBody.create(MediaType.parse("application/octet-stream"),
files[0]));
        MultipartBody multipartBody = builder.build();
        Request request  = new Request.Builder().url(url).post(new
                ProgressRequestBody(multipartBody,listener)).build();
        okHttpClient.newCall(request).enqueue(callback);
    }
    // 文件下载
    public static void downloadFile(String url, final ProgressListener listener,
Callback callback){
        OkHttpClient client = okHttpClient.newBuilder().addNetworkInterceptor
(new Interceptor() {
            @Override
            public Response intercept(Interceptor.Chain chain) throws IOException {
                Response response = chain.proceed(chain.request());
                return response.newBuilder().body(new
                        ProgressResponseBody(response.body(),listener)).build();
            }
        }).build();
        Request request  = new Request.Builder().url(url).build();
        client.newCall(request).enqueue(callback);
    }
    …
```

```
}
```

ProgressListener 是监听进度的接口。ProgressListener.java 主要代码如下：

```
public interface ProgressListener {
    void onProgress(long currentBytes, long contentLength, boolean done);
}
```

ProgressRequestBody 是自定义 RequestBody 类，其成员包含有进度监听器。

UploadDemo 模块的 ProgressRequestBody.java 主要代码如下：

```
public class ProgressRequestBody extends RequestBody {
    public static final int UPDATE = 0x01;
    private RequestBody requestBody;
    private ProgressListener mListener;
    private BufferedSink bufferedSink;
    private MyHandler myHandler;
    public ProgressRequestBody(RequestBody body, ProgressListener listener) {
        requestBody = body;
        mListener = listener;
        if (myHandler == null) {
            myHandler = new MyHandler();
        }
    }
    class MyHandler extends Handler {
        public MyHandler() {
            super(Looper.getMainLooper());
        }
        @Override
        public void handleMessage(Message msg) {
            switch (msg.what) {
                case UPDATE:
                    ProgressModel progressModel = (ProgressModel) msg.obj;
                    if (mListener != null)
                        mListener.onProgress(progressModel.getCurrentBytes(),
                                progressModel.getContentLength(), progressModel. isDone());
                    break;
            }
        }
    }
    @Override
    public MediaType contentType() {
        return requestBody.contentType();
    }
    @Override
    public long contentLength() throws IOException {
        return requestBody.contentLength();
    }
    @Override
    public void writeTo(BufferedSink sink) throws IOException {
        if (bufferedSink == null) {
            bufferedSink = Okio.buffer(sink(sink));
        }
        requestBody.writeTo(bufferedSink);  // 写入
        bufferedSink.flush();   // 刷新
    }
    private Sink sink(BufferedSink sink) {
```

```
            return new ForwardingSink(sink) {
                long bytesWritten = 0L;
                long contentLength = 0L;
                @Override
                public void write(Buffer source, long byteCount) throws IOException {
                    super.write(source, byteCount);
                    if (contentLength == 0) {
                        contentLength = contentLength();
                    }
                    bytesWritten += byteCount;
                    //回调
                    Message msg = Message.obtain();
                    msg.what = UPDATE;
                    msg.obj = new ProgressModel(bytesWritten, contentLength, bytesWritten
== contentLength);
                    myHandler.sendMessage(msg);
                }
            };
        }
    }
```

UploadDemo 模块的 ProgressModel.java 主要代码如下：

```
    public class ProgressModel implements Parcelable {
        private long currentBytes;
        private long contentLength;
        private boolean done = false;
        ...
    }
```

如图 10-7 所示，上传（服务器）文件在手机客户端的存储路径如下。

```
    /storage/emulated/0/Android/data/com.example.uploaddemo/files/Download/OkHttp
3 基本使用.pptx
```

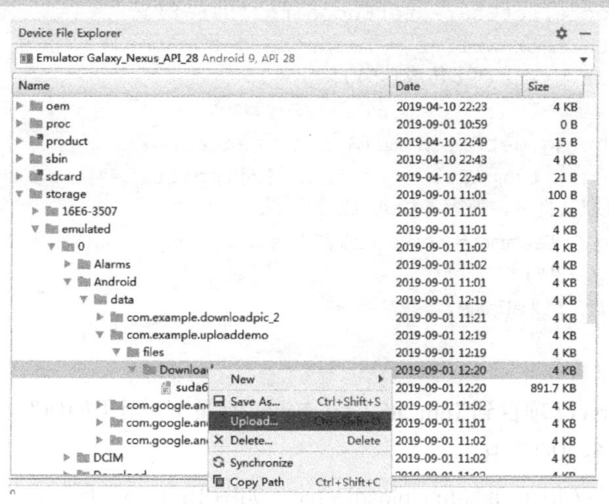

图 10-7　加载欲上传（服务器）文件至手机客户端

在“Device File Explorer”窗口中，可以右击“/storage/emulated/0/.../files/Download/”目录，从弹出的快捷菜单中选择“Upload”选项，就可以把计算机中的文件“OkHttp3 基本使用.pptx”加载到手机客户端。

以上是 Android 客户端的构建。

访问的服务器端是：

http://192.168.0.104:8080/NetServer2/Upload。

关于服务器端的构建如下。

打开 Eclipse，在 NetServer2 工程的 src 目录下 com.mialab.ch10.servlet 包中创建 UploadFileServlet.java。NetServer2 工程的 UploadFileServlet.java 主要代码如下：

```
    @WebServlet("/Upload")
    @MultipartConfig
    public class UploadFileServlet extends HttpServlet {
        protected void doGet(HttpServletRequest request, HttpServletResponse response)
                    throws ServletException, IOException {
                doPost(request, response);
            }
            protected void doPost(HttpServletRequest request, HttpServletResponse
response)
                    throws ServletException, IOException {
                System.out.println("doPost==");
                request.setCharacterEncoding("utf-8");
                // 获取file命名的part，注意要与Android客户端一样
                Part part = request.getPart("file");
                // 获取请求头，请求头的格式：form-data; name="file"; filename="OkHttp3
基本使用.pptx"
                String header = part.getHeader("content-disposition");
                System.out.println(header);
                String fileName = getFileName(header);
                // 存储路径
                String root = getServletContext().getRealPath("/");
                System.out.println("-----------" + root + "------------");
                String savePath = root + "upload";
                // 把文件写到指定路径
                part.write(savePath + File.separator + fileName);
                response.setCharacterEncoding("UTF-8");
                PrintWriter writer = response.getWriter();
                writer.print("上传成功");
            }
            public String getFileName(String header) {
                String[] tempArr2 = tempArr1[2].split("=");
                // 获取文件名，兼容各种浏览器的写法
                String fileName = tempArr2[1].substring(tempArr2[1].lastIndexOf("\\")
+ 1).replaceAll("\"", "");
                return fileName;
            }
    }
```

发布（部署）NetServer2 项目到 Tomcat8 的 webapps 目录下。至于如何发布（部署）“Dynamic Web Project”到 Tomcat，可参见附录 B。

运行 Android 客户端 Ch10_OkHttp\UploadDemo，如图 10-8（a）所示，点击“下载文件”按钮，便可下载一幅图片至 Android 客户端的/storage/emulated/0/Android/data/com.example.uploaddemo/files/Download/目录下。如前所述，右击“... /Download/”目录，把计算机中的文件“OkHttp3 基本使用.pptx”加载到手机客户端。这时再点击“上传文件”按钮，便可以把 Android 客户端的文件“OkHttp3 基本使用.pptx”上传到服务器了。在服务器的存储路径是：“D:\tomcat8\webapps\NetServer2\upload\ OkHttp3 基本使用.pptx”。

上传文件的进度显示如图 10-8（b）所示，其具体代码和文件可参见教学资源包。

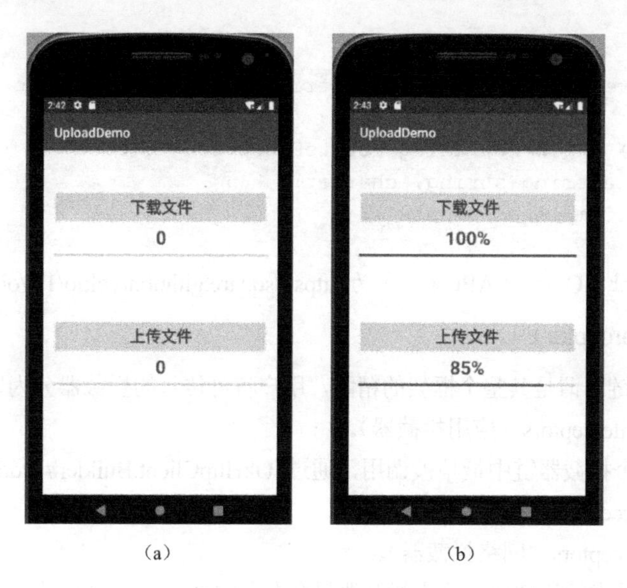

（a）　　（b）

图 10-8　上传文件和下载文件并显示进度

小贴士：自从 Google 官方将 OkHttp 作为底层的网络请求之后，OkHttp 底层 IO 操作的 Okio 也走进了开发者的视野，这个甚至可以取代 Java 的原生 IO 库到底有什么特殊的本领呢？Okio 的代码比较精巧，其核心代码大约有 5000 行。Okio 相当于一个工具类，可以返回各种需要的流工具。

Sink 接口负责写数据，Source 接口负责读数据。

```
Okio.sink(new File(arg0));                 // 填入文件返回 Sink 写接口
Okio.sink(new OutputStream()）;            // 填入字节流返回 Sink 写接口
Okio.sink(socket);                         // 填入套接字返回 Sink 写接口
```

最基本的写用法代码如下：

```
String filePath = "D:/abc.txt";
try{
    Sink sink=Okio.sink(new File(filePath));
    BufferedSink buffer=Okio.buffer(sink);
    buffer.writeString("hello,boy",Charset.forName("UTF-8"));
    buffer.flush();
}catch(Exception e){
    e.printStackTrace();
}
```

最基本的读用法代码如下：

```
String filePath = "D:/abc.txt";
try {
    Source source = Okio.source(new File(filePath));
    BufferedSource buffer = Okio.buffer(source);
    String myStr = buffer.readString(Charset.forName("UTF-8"));
    System.out.println(myStr);
} catch (Exception e) {
    e.printStackTrace();
}
```

Okio.sink()方法包装了 FileOutputStream()方法。

```
public static Sink sink(File file) throws FileNotFoundException {
    if (file == null) throw new IllegalArgumentException("file == null");
    return sink(new FileOutputStream(file));
}
```

装饰者模式，可以把 sink 用 RealBufferedSink 包装起来，那么最后执行写方法的就是它了。

```
    @Override
    public  BufferedSink  writeString(String  string,  Charset  charset)  throws
IOException {
        if (closed) throw new IllegalStateException("closed");
        buffer.writeString(string, charset);
        return emitCompleteSegments();
    }
```

Okio 帮助文档地址（Okio 的 API 地址）为 https://square.github.io/okio/1.x/okio/。

11. 拦截器（Interceptor）

OkHttp 的拦截器链可谓是其整个框架的精髓，用户将可传入的拦截器分为以下两类。

（1）Application Interceptors（应用拦截器）。

该类拦截器在整个拦截器链中最早被调用，通过 OkHttpClient.Builder# addInterceptor(Interceptor) 传入，比 NetworkInterceptor 先执行。

（2）Network Interceptors（网络拦截器）。

另一类是非网页请求的拦截器，这类拦截器只会在非网页请求中被调用，并且是在组装完请求之后，真正发起网络请求前被调用，通过 OkHttpClient.Builder# addNetworkInterceptor (Interceptor)传入。

小贴士：一个复杂点的 OkHttpClient 配置。

```
    File cacheDir = new File(getCacheDir(), "okhttp_cache");
    //File cacheDir = new File(getExternalCacheDir(), "okhttp");
    Cache cache = new Cache(cacheDir, 10 * 1024 * 1024);
    OkHttpClient  okHttpClient = new OkHttpClient.Builder()
           .connectTimeout(5*1000, TimeUnit.MILLISECONDS)     // 连接超时
           .readTimeout(10*1000,TimeUnit.MILLISECONDS)        // 读取超时
           .writeTimeout(10*1000,TimeUnit.MILLISECONDS)       // 写入超时
           .addInterceptor(new HttpHeadInterceptor())  // 应用拦截器：统一添加消息头
           .addNetworkInterceptor(new NetworkspaceInterceptor())   // 网络拦截器
           .addInterceptor(loggingInterceptor)                // 应用拦截器：打印日志
           .cache(cache)  //设置缓存
           .build();
```

【示例】 假如有这样一个需求，需要监控 App 通过 OkHttp 发出的所有原始请求，以及整个请求所耗费的时间。针对这样的需求就可以使用应用拦截器。

可以自定义 LoggingInterceptor 实现拦截器接口。

```
    public class LoggingInterceptor implements Interceptor {
        private static final String TAG = "LoggingInterceptor";
        @Override
        public Response intercept(Chain chain) throws IOException {
            Request request = chain.request();
            long startTime = System.nanoTime();
            Log.d(TAG, String.format("Sending request %s on %s%n%s",
                    request.url(), chain.connection(), request.headers()));
            Response response =  chain.proceed(request);
            long endTime = System.nanoTime();
            Log.d(TAG, String.format("Received response for %s in %.1fms%n%s",
                    response.request().url(), (endTime - startTime) / 1e6d, response.
headers()));
            return response;
        }
    }
```

发送网络请求的核心代码如下。（在 MainActivity 中写个方法 getDataAsync()，然后调用它。）

```
    private void getDataAsync() {
        OkHttpClient okHttpClient = new OkHttpClient.Builder()
                .addInterceptor(new LoggingInterceptor())
                .build();
        Request request = new Request.Builder()
                .url("http://www.publicobject.com/helloworld.txt")
                .header("User-Agent", "OkHttp Example")
                .build();
        okHttpClient.newCall(request).enqueue(new Callback() {
            @Override
            public void onFailure(Call call, IOException e) {
                Log.d("hand", "onFailure: " + e.getMessage());
            }
            @Override
            public void onResponse(Call call, Response response) throws IOException {
                ResponseBody body = response.body();
                if (body != null) {
                    Log.d("hand", "onResponse: " + response.body().string());
                    body.close();
                }
            }
        });
    }
```

具体代码可参见教学资源包中的 Ch10_OkHttp\LoggingInterceptorDemo 模块。

运行 Ch10_OkHttp\LoggingInterceptorDemo 模块，在 Android Studio 的 Logcat 将会有以下输出。

```
    D/LoggingInterceptor: Sending request http://www.publicobject.com/helloworld.
txt on null
    User-Agent: OkHttp Example
    D/LoggingInterceptor: Received response for https://publicobject.com/helloworld.
txt in 2660.0ms
    Server: nginx/1.10.0 (Ubuntu)
    Date: Sun, 01 Sep 2019 17:24:01 GMT
    Content-Type: text/plain
    Content-Length: 1759
    Last-Modified: Tue, 27 May 2014 02:35:47 GMT
    Connection: keep-alive
    ETag: "5383fa03-6df"
    Accept-Ranges: bytes
```

注意这个请求进行了重定向。更多的关于拦截器的使用方法，以及应用拦截器和网络拦截器各自的优/缺点，可参考 OkHttp 官方的说明文档。

12. OkHttp 使用上的一些缺点

（1）对于 GET 请求，如果请求参数较多，则拼接 URL 较为麻烦，其示例代码如下：

```
    HttpUrl  httpUrl = new HttpUrl.Builder()
            .scheme("http")
            .host("www.baidu.com")
            .addPathSegment("user")
            .addPathSegment("login")
            .addQueryParameter("username", "yuxiaonan")
            .addQueryParameter("password","123456")
            .build();
```

拼接结果如下。

```
http://www.baidu.com/user/login/username=yuxiaonan&password=123456
```

如果能做一些封装，直接使用 addParam(key,value)形式则会简单很多。

（2）Callback 在子线程中回调，很多时候都是需要更新 UI 的，还需要 POST 到主线程中进行处理。

（3）构建请求步骤比较多，因此 Square 提供了针对 OkHttp 的封装库 Retrofit。

10.1.4 使用 Retrofit

1. Retrofit 简介

在 Android 开发中，网络请求十分常见。在 Android 网络请求库中，Retrofit 是十分流行的一个网络请求库。Retrofit 是 Square 公司开发的一款针对 Android 网络请求的框架，底层基于 OkHttp 实现，而 OkHttp 已经得到了 Google 官方的认可。

在处理 HTTP 请求的时候，因为不同场景或者边界等情况比较难处理，需要考虑网络状态、在请求失败后重试、处理 HTTPS 等问题，而 Retrofit 则可以避免这些麻烦。

有了 Retrofit 之后，对于一些请求只需要一行代码或者一个注解就可完成。如果看源码会发现，其本质就是 OkHttp 的封装，使用面向接口的方式进行网络请求，利用动态生成的代理类封装了网络接口请求的底层，使其请求返回 JavaBean，并对网络认证 REST API 进行很友好的支持。使用 Retrofit 将会极大地提高人们的网络体验。

Retrofit 的优点：①请求的方法参数注解可以编制；②支持同步、异步和 RxJava；③超级解耦；④可以配置不同的反序列化工具来解析数据，如 JSON、XML 等。

2. Retrofit 原理

Retrofit 其实是一个网络请求框架的封装，因为网络请求的工作本质是由 OkHttp 完成的，而 Retrofit 仅负责网络请求接口的封装。Retrofit 原理如图 10-9 所示。

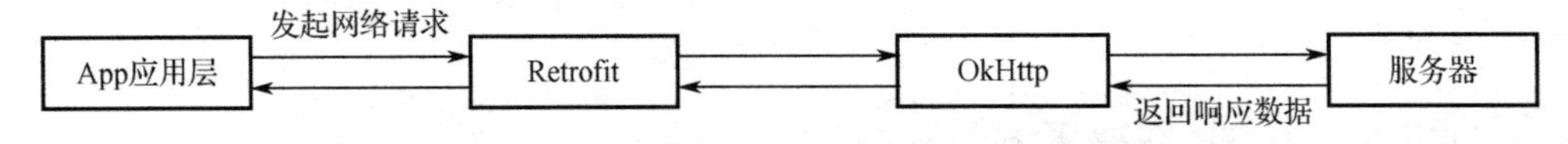

图 10-9　Retrofit 原理

App 应用程序通过 Retrofit 请求网络，使用 Retrofit 接口层封装请求参数、Header、URL 等信息，由 OkHttp 完成后续的请求操作。

在服务端返回数据之后，OkHttp 将原始结果交给 Retrofit，Retrofit 再根据用户的需求对结果进行解析。

3. Retrofit 使用

使用 Retrofit 的步骤如下。

（1）添加 Retrofit 库的依赖。

可在 build.gradle 文件的 dependencies 节点中添加对 Retrofit 库的依赖，如果对网络响应使用 Gson 自动转换，也要加入对 converter-gson 的依赖。

```
dependencies {
    implementation 'com.squareup.retrofit2:retrofit:2.1.0'
    implementation 'com.squareup.retrofit2:converter-gson:2.1.0'
}
```

（2）创建接收服务器返回数据的类。

根据返回数据的格式和数据解析方式定义，实际上可视为对 JavaBean 的定义。

（3）创建用于描述网络请求的接口。

Retrofit 将 HTTP 请求抽象成 Java 接口，采用注解描述网络请求参数和配置网络请求参数。接口中的每个方法的参数都需要使用注解标注，否则会报错。

```
public interface GetRequest_Interface {
        @GET("studentQuery")
    Call<List<Student>> getCall();
    // @GET 注解的作用：采用 GET 方法发送网络请求
    // getCall()：接收网络请求数据的方法
    // 其中返回类型为 Call<*>，*是接收数据的类（在以下的示例中是 List<Student>）
    // 如果想直接获得 Responsebody 的内容，可以定义网络请求返回值为 Call<ResponseBody>
}
```

（4）创建 Retrofit 实例。

创建 Retrofit 实例的示例代码如下：

```
Retrofit retrofit = new Retrofit.Builder()
        baseUrl("http://192.168.0.104:8080/MyNetServer/")       //设置网络请求 URL
        .addConverterFactory(GsonConverterFactory.create())    //设置使用 Gson 解析
        .build();
```

Retrofit 支持多种数据解析方式，使用时需要在 Gradle 中添加依赖，如表 10-1 所示。

表 10-1　数据解析器和 Gradle 依赖

数据解析器	Gradle 依赖
Gson	com.squareup.retrofit2:converter-gson:2.0.2
Jackson	com.squareup.retrofit2:converter-jackson:2.0.2
Simple XML	com.squareup.retrofit2:converter-simplexml:2.0.2

（5）创建网络请求接口实例。

创建网络请求接口实例（对象）的示例代码如下：

```
GetRequest_Interface request = retrofit.create(GetRequest_Interface.class);
Call< List<Student>> call = request.getCall();    //对发送请求进行封装
```

（6）发送网络请求（异步/同步）。

这个过程中封装了数据转换及线程切换。

```
call.enqueue(new Callback<List<Student>>() {
    //请求成功时的回调
    @Override
    public void onResponse(Call<List<Student>> call, Response<List<Student>>
response) {
        //请求处理，输出结果
        Log.d("hand","--------" + response.body() + "---------");
            //asyncText.setText(" 异 步 请 求 结 果 ： " + response.body().get(0).
getStuName());
    }
    //请求失败时的回调
    @Override
    public void onFailure(Call<List<Student>> call, Throwable throwable) {
        Log.d("hand","连接失败");
        Log.d("hand","------" + throwable.getMessage() + "------");
    }
});
// 发送网络请求（同步）。这里需要注意的是网络请求一定要在子线程中完成，不能直接在 UI 线程中
// 执行，否则会 Crash
```

```
    Response< List<Student>> response = call.execute();
    response.body().show();   // 对返回数据进行处理
```

（7）处理服务器返回的数据。

通过 response 类的 body()对返回的数据进行处理，其代码同上。

4. Retrofit 使用示例

【示例】 使用 Retrofit 分别发送 GET 请求和 POST 请求查询学生信息。

启动 Android Studio，在 Ch10 工程中创建 RetrofitDemo 模块，并在该模块中创建 MainActivity.java、activity_main.xml、GetStudentService.java、PostStudentService.java 和 Student.java 等文件。

RetrofitDemo 模块的 GetStudentService.java 的主要代码如下：

```
public interface GetStudentService {
    @GET("studentQuery")
    Call<List<Student>>  getCall();
}
```

RetrofitDemo 模块的 PostStudentService.java 的主要代码如下：

```
public interface PostStudentService {
    @POST("studentQuery")
    @FormUrlEncoded
    Call<List<Student>> getCall(@Field("stuId") String stuId);
}
```

RetrofitDemo 模块的 MainActivity.java 的主要代码如下：

```
public class MainActivity extends AppCompatActivity {
    @Override
    protected void onCreate(Bundle savedInstanceState) {
        super.onCreate(savedInstanceState);
        setContentView(R.layout.activity_main);
        getStudentRequest();
        //postStudentRequest();
    }
    private void postStudentRequest() {
        //（4）创建 Retrofit 对象
        Retrofit retrofit = new Retrofit.Builder()
            .baseUrl("http://192.168.0.104:8080/NetServer2/")  // 设置网络请求
baseUrl
            .addConverterFactory(GsonConverterFactory.create()) //设置使用 Gson 解
析，须加入依赖
            .build();
        //（5）创建网络请求接口实例（对象）
        PostStudentService request = retrofit.create(PostStudentService.class);
        //对发送请求进行封装
        Call<List<Student>> call = request.getCall("9801");
        //（6）发送网络请求（异步）
        call.enqueue(new Callback<List<Student>>() {
            //请求成功时回调
            @Override
            public void onResponse(Call<List<Student>> call, Response<List <Student>>
response) {
                // 请求处理,输出结果
                Log.d("hand","----" + response.body() + "------");
            }
            //请求失败时回调
            @Override
            public void onFailure(Call<List<Student>> call, Throwable throwable) {
```

```
                Log.d("hand","请求失败");
                Log.d("hand",throwable.getMessage());
            }
        });
    }
    private void getStudentRequest() {
        //（4）创建Retrofit对象
        Retrofit retrofit = new Retrofit.Builder()
            .baseUrl("http://192.168.0.104:8080/NetServer2/")  // 设置网络请求
baseUrl
            .addConverterFactory(GsonConverterFactory.create()) //设置使用Gson解
析，必须加入依赖
            .build();
        //（5）创建网络请求接口实例（对象）
        GetStudentService request = retrofit.create(GetStudentService.class);
        //对发送请求进行封装
        Call<List<Student>> call = request.getCall();
        //（6）发送网络请求（异步）
        call.enqueue(new Callback<List<Student>>() {
            //请求成功时回调
            @Override
            public void onResponse(Call<List<Student>> call, Response<List <Student>>
response) {
                //请求处理，输出结果
                Log.d("hand","--------" + response.body() + "---------");
                //Log.d("hand","--------" + response.body().get(0).getStuName() +
"---------");
            }
            //请求失败时回调
            @Override
            public void onFailure(Call<List<Student>> call, Throwable throwable) {
               Log.d("hand","连接失败");
               Log.d("hand","------" + throwable.getMessage() + "------");
            }
        });
    }
}
```

RetrofitDemo模块的Student.java的主要代码如下：

```
public class Student {
    private  String stuId;
    private String stuName;
    private float stuScore;
    …
    @Override
    public String toString() {
        return "Student{" +
                "stuId='" + stuId + '\'' +
                ", stuName='" + stuName + '\'' +
                ", stuScore=" + stuScore +
                '}';
    }
}
```

关于服务端的构建如下。

打开 Eclipse，在 NetServer2 工程 src 目录的 com.mialab.ch10.servlet 包中创建 StudentServlet.java，并在 com.mialab.ch10.bean 包中创建 Student.java。

NetServer2 工程的 StudentServlet.java 主要代码如下：

```
@WebServlet("/studentQuery")
public class StudentServlet extends HttpServlet {
    protected void doGet(HttpServletRequest request, HttpServletResponse response)
            throws ServletException, IOException {
        doPost(request, response);
    }
    protected void doPost(HttpServletRequest request, HttpServletResponse response)
            throws ServletException, IOException {
        response.setContentType("text/html;charset=utf-8");
        request.setCharacterEncoding("utf-8");
        response.setCharacterEncoding("utf-8");
        PrintWriter out = response.getWriter();
        String stuId = request.getParameter("stuId");
        String result = null;
        if (stuId == null) {
            List<Student> stuList = queryDataAll();
            result = JSON.toJSONString(stuList);  // 用到了 FastJson
            System.out.println("queryDataAll():" + result);
        } else if (stuId.equals("9801")) {
            List<Student> stuList = queryDataById(stuId);
            result = JSON.toJSONString(stuList);
            System.out.println("queryDataById():" + result);
        } else if (stuId.equals("9802")) {
            List<Student> stuList = queryDataById(stuId);
            result = JSON.toJSONString(stuList);
            System.out.println("queryDataById():" + result);
        }
        out.println(result);
        out.flush();
        out.close();
    }
    //此处只是模拟。如何从数据库中取出数据，并转换成 JSON，将在后续讲述
    List<Student> queryDataAll(){
        List<Student> stuList = new ArrayList<Student>();
        Student stu_1 = new Student("9801","余小男",408);
        Student stu_2 = new Student("9802","李白鹭",396);
        stuList.add(stu_1);
        stuList.add(stu_2);
        return stuList;
    }
    List<Student> queryDataById(String stuId){
        List<Student> stuList = new ArrayList<Student>();
        if(stuId.equals("9801")) {
            Student stu = new Student("9801","余小男",408);
            stuList.add(stu);
        } else if(stuId.equals("9802")) {
            Student stu = new Student("9802","李白鹭",396);
            stuList.add(stu);
        }
        return stuList;
    }
}
```

NetServer2 工程的 StudentServlet.java 代码与客户端同。

由于在 NetServer2 工程中用到了 FastJson，使 ArrayList 容器对象转换成 JSON 字符串更为方便。这里将 fastjson-1.2.9.jar 复制到 NetServer2 工程的 WEB-INF\lib\目录下。

发布(部署)NetServer2 项目到 Tomcat8 的 webapps 目录下，如何发布(部署)"Dynamic Web Project"到 Tomcat，可参考附录 B。

在浏览器中输入 http://192.168.0.104:8080/NetServer2/studentQuery 后，浏览器输出内容为:

```
    [{"stuId":"9801","stuName":"余小男","stuScore":408},{"stuId":"9802","stuName":
"李白鹭","stuScore":396}]
```

在浏览器中输入 http://192.168.0.104:8080/NetServer2/studentQuery?stuId=9801 后，浏览器输出内容为:

```
    [{"stuId":"9801","stuName":"余小男","stuScore":408}]
```

在浏览器中输入 http://192.168.0.104:8080/NetServer2/studentQuery?stuId=9802 后，浏览器输出内容为:

```
    [{"stuId":"9802","stuName":"李白鹭","stuScore":396}]
```

运行 Ch10_Retrofit\RetrofitDemo（Android 客户端），Android Studio 的 Logcat 输出为:

```
    2019-08-31 02:42:02.376 6190-6190/commialab.retrofitdemo D/hand:
    --------[Student{stuId='9801', stuName='余小男', stuScore=408.0}, Student
{stuId='9802', stuName='李白鹭', stuScore=396.0}]---------
```

把 RetrofitDemo 模块的 MainActivity.java 的 onCreate()方法的 getStudentRequest()方法注释掉，改为调用 postStudentRequest()方法。

```
    protected void onCreate(Bundle savedInstanceState) {
            super.onCreate(savedInstanceState);
            setContentView(R.layout.activity_main);
            //getStudentRequest();
            postStudentRequest();
        }
    }
```

运行 Ch10_Retrofit\RetrofitDemo（Android 客户端），Android Studio 的 Logcat 输出为:

```
    /commialab.retrofitdemo D/hand: ----[Student{stuId='9801', stuName='余小男',
stuScore=408.0}]------
```

10.2　Socket 通信

Android 与服务器的通信方式主要有两种: HTTP 通信和 Socket 通信。两者的最大差异在于，HTTP 通信连接使用的是"请求—响应方式"，即在请求时建立连接通道，当客户端向服务器发送请求后，服务器端才能向客户端返回数据。而 Socket 通信则是在双方建立起连接后就可以直接进行数据传输，在连接时可实现信息的主动推送，而不需要每次都由客户端向服务器发送请求。

1. Socket 简介

Socket 是一种抽象层，通信两端通过它来发送和接收数据，使用 Socket 可以将应用程序添加到网络中，与处于同一网络中的其他应用程序进行通信。简单来说，Socket 提供了应用程序与外界通信的端口并为通信双方提供了数据传输通道。Socket 的主要特点有数据丢失率低，使用简单且易于移植等。

Socket 有两种传输模式：面向连接和无连接。面向连接模式使用 TCP 协议，在通信两端建立通信链路，进行发送和接收数据。无连接模式使用 UDP 协议，将数据进行打包发送，与面向连接模式不同的是，接收端接收数据包的顺序与发送端发送的顺序是不一样的，而且数据安全性欠佳。如果希望确保数据能按预期有序地、正确地发送，那么建议采用面向连接模式，但相对来说，内存消耗要比无连接模式大；如果希望能快速、高效地传输数据，那么可以采用无连接模式。

2. 使用基于 TCP 协议的 Socket 进行通信

TCP（Transmission Control Protocol，传输控制协议）是一种面向连接的、可靠的、基于字节流的运输层通信协议。在简化的计算机网络 OSI 模型中，它可以完成传输层所指定的功能。应用层向传输层发送用于网络间传输的数据流，然后 TCP 把数据流分割成适当长度的报文段，之后 TCP 再把结果包传给 IP 层，由它来通过网络将包传送给接收端实体的 TCP 层。

当两个用户端进行连接时，TCP 协议会让它们建立一个用于收发数据的虚拟链路，由 TCP 协议负责收集信息，通过这个链路才可以进行数据传输。TCP 提供超时重发、丢弃重复数据、检验数据、流量控制等功能，确保数据能从一端传到另一端。

当 Java 建立了两个 Socket 对象时，分别代表链路两端实体的通信接口，通过 Socket 产生 I/O 流进行网络通信，此时这两个通信实体并没有服务器端和客户端的区分。在虚拟链路尚未建立起来之前，还需要有一个通信实体先做出“主动”姿态，主动接收来自其他通信实体的连接请求。Java 提供了 ServerSocket 类，用于监听来自客户端 Socket 的连接，如果没有连接请求，则一直处于等待状态。

基于 TCP 的 Socket 通信步骤如下。

（1）创建 TCP 服务器端。

利用 ServerSocket 创建 TCP 服务端，创建 ServerSocket 对象有以下几个构造器。

- ServerSocket(int port)：指定端口号（port）来创建 ServerSocket 对象。
- ServerSocket(intport,inbacklog)：在上面一个构造器的基础上，增加了改变连接队列的参数 backlog。
- ServerSocket(int port,int backlog,InetAddress localAddr)：在上面一个构造器的基础上增加了 IP 地址参数 localAddr，用来指定将 ServerSocket 绑定到指定的 IP 地址。

（2）等待客户端请求。

一个服务器可以接收多个客户端请求，通过循环调用 accept()方法来不断接收客户端请求。

```
ServerSocket server=new ServerSocket(50000);
while(true) {
   Socket socket=server.accept();
}
```

（3）服务器端接收请求，返回响应。

accept()方法在没有数据进行接收时处于堵塞状态。一旦接收到数据，就可以通过 InputStream 读取接收的数据，并通过 OutputStream 返回响应数据。

（4）客户端发送请求，获取返回信息。

服务器端创建 ServerSocket 对象等待客户端请求，而客户端创建一个指定服务器端 IP 地址和端口号的 Socket 对象，并通过 InputStream 读取数据、OutputStream 写入数据，即可实现使用 TCP 协议进行 Socket 数据传输。

相关示例可参见相关教学资源包，限于篇幅，不再赘述。

10.3 使用 WebView 显示网页

Android 提供了一个控件 WebView，是内置的浏览器控件，用来显示 Web 应用、浏览网页等。

WebView 控件本身就是一个浏览器实现，它的内核基于开源的 WebKit 引擎。但 WebView 仅加载 Web 页面，并不能够实现 Web 浏览器的所有功能，如地址栏、导航栏等就没有。WebView 继承于 View。

1. 加载网页

使用 WebView 加载网页相当于浏览器加载 Web 页面。要使用 WebView 控件访问 HTML 页面，需要在 AndroidManifest.xml 中添加访问网络的权限。

```
<uses-permission android:name="android.permission.INTERNET" />
```

调用 loadUrl(String url)方法加载 URL 对应的网页，其代码如下：

```
WebView webView = (WebView) findViewById(R.id.webView);
webView.loadUrl("http://www.baidu.com");
```

WebView 可以加载某个 URL 网络地址指向的网页，也可以加载本地文件，此时的 URL 是文件路径，本地文件存放在 assets 文件中，如“file:///android_asset/XX.html”。

与此同时，由于 WebView 仅加载网页的内容，并不是 Web 浏览器，为了使得 Android 界面能达到浏览页面的效果，WebView 控件提供了大量的方法，如 goback()后退、goForward()前进等，具体可以查阅 Android 官方提供的相关 API。

【示例】 使用 WebView 控件浏览网页，实现页面的加载及回退功能。

本例用户界面采用线性布局。一个编辑框用于接收要访问页面的网址。其右上方的两个按钮，分别用于“前进（加载页面）”和“后退”返回上一次加载的页面。一个 WebView 用来加载 Web 网页。

在编辑框输入网址后，如输入“http://192.168.0.104:8080/”，并点击“前进（加载页面）”按钮，则可以在 WebView 中加载网页，如图 10-10 所示，显示了 Tomcat 服务器的首页。点击“后退”按钮，则退回到上一次打开的页面。

启动 Android Studio，在 Ch10 工程中创建 WebBrowser 模块，并在该模块中创建 MainActivity.java 和 activity_main.xml 文件。

WebBrowser 模块的 MainActivity.java 主要代码如下：

```
public class MainActivity extends AppCompatActivity {
    @Override
    protected void onCreate(Bundle savedInstanceState) {
        super.onCreate(savedInstanceState);
        setContentView(R.layout.activity_main);
        Button btngo = (Button) findViewById(R.id.btngo);
        Button btnback = (Button) findViewById(R.id.btnback);
        final EditText txtUrl = (EditText) findViewById(R.id.txt_url);
        final WebView webView = (WebView) findViewById(R.id.webbrowser);
        btngo.setOnClickListener(new View.OnClickListener() {
            @Override
            public void onClick(View arg0) {
                webView.setWebViewClient(new WebViewClient() {
                    @Override
                    public boolean shouldOverrideUrlLoading(WebView view, String
url) {
                        view.loadUrl(url);
                        return true;
                    }
                });
                webView.loadUrl(txtUrl.getText().toString());
            }
        });
        btnback.setOnClickListener(new View.OnClickListener() {
            @Override
```

```
            public void onClick(View arg0) {
                webView.goBack();
            }
        });
    }
}
```

调用 WebView 的 loadUrl()方法，通常会启动 Android 系统自带的浏览器进行加载页面，如果希望页面直接显示在 WebView 中而不用另外使用其他浏览器，则需要设置 WebView 的 WebViewClient 对象，并重写 shouldOverrideUrlLoading()方法。

图 10-10　使用 WebView 控件浏览网页

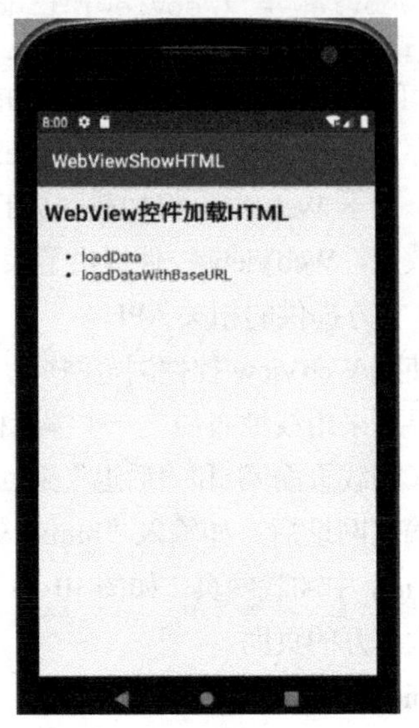

图 10-11　WebView 加载 HTML

2. WebView 加载 HTML 代码

当使用 TextView 来显示 HTML 文本时，TextView 控件并不会对 HTML 的标记进行解析而是直接以字符显示输出。WebView 提供一个 loadData()方法可以直接载入 HTML 字符串，并对该字符串进行 HTML 解析。

WebView 的 loadData()方法提供了 3 个参数，包括加载的 HTML 字符串、HTML 代码的 MIME 类型和 HTML 代码编码所用的字符集。随着 Android 版本的升级，在 Android 4.0 以上的版本中，WebView 若使用 loadData 就会出现中文乱码，为解决这个问题，引入了一个新的加载 HTML 代码的方法 loadDataWithBaseURL()，调用该方法需要 5 个参数，其签名如下。

```
    void  loadDataWithBaseURL(String baseUrl, String data, String mimeType, String
encoding,  String historyUrl)
```

loadDataWithBaseURL()比 loadData()多了两个参数，是 loadDate()的加强版。这两个参数可以指定该页面的 baseUrl，如果为 null，则默认值为“about blank”，也可以指定历史 URL。

【示例】 实现 WebView 加载 HTML 代码。

启动 Android Studio，在 Ch10 工程中创建 WebViewShowHTML 模块，并在该模块中创建 MainActivity.java 和 activity_main.xml 文件。

运行完成后的 Ch10\WebViewShowHTML，其界面如图 10-11 所示。

WebViewShowHTML 模块的 MainActivity.java 的主要代码如下：

```
public class MainActivity extends AppCompatActivity {
    @Override
    protected void onCreate(Bundle savedInstanceState) {
```

```
        super.onCreate(savedInstanceState);
        setContentView(R.layout.activity_main);
        WebView webView  = (WebView)findViewById(R.id.webview);
        StringBuilder data = new StringBuilder("<html><body bgcolor=\"#F2F6F8\
">");
        data.append("<h2>WebView控件加载HTML</h2>");
        data.append("<ul>");
        data.append("<li>loadData</li>");
        data.append("<li>loadDataWithBaseURL</li>");
        data.append("</ul>");
        data.append("</body></html>");
        webView.loadDataWithBaseURL("", data.toString(), "text/html", "UTF-8","");
    }
}
```

相关代码参见教学资源包中的Ch10\WebBrowser和Ch10\WebViewShowHTML。

10.4 项目实战：查询学生信息

10.4.1 任务说明

在手机界面中输入要查询的学生学号，点击“查询”按钮，向服务器发送网络请求，并将服务器返回的查询结果显示在手机界面中，如图10-12所示。

图10-12 查询学生信息

10.4.2 项目讲解

1．总体框架

（1）客户端构建。

启动Android Studio，打开Ch10_Retrofit工程创建StudentManager模块，并在该模块中创建相应的包及文件。

① src\main\java\目录下的com.example.studentmanager包：MainActivity.java。

② src\main\java\目录下的com.example.studentmanager.service包：StudentService.java。

③ src\main\java\目录下的 com.example.studentmanager.bean 包：Student.java。

④ src\main\res\layout 目录下：activity_main.xml 文件。

⑤ src\main\res\drawable 目录下：border.xml 文件。

（2）服务器端构建。

需要访问的服务器端 URL 如下。

```
http://192.168.0.104:8080/NetServer2/studentQuery
```

实际上处理 Android 客户端提交请求（查询学生信息的请求）的服务器端组件，就是在 NetServer2 工程中创建的 Servlet 组件：StudentServlet.java。

2. 程序主要流程

程序的入口是 MainActivity 的 onCreate 方法，MainActivity 的界面见图 10-12。输入学生学号以后，点击“查询”按钮，将触发按钮的点击监听。这里使用了 Retrofit 网络请求框架，返回响应的数据。StudentService 接口的 getStuAll()方法和 getStuById(@Field("stuId") String stuId)方法返回的数据类型是 Call<List<Student>>，通过 response.body()便可以得到 List<Student>容器对象（封装了学生查询信息），在 Android 客户端手机界面显示即可。

10.4.3 典型代码及技术要点

1. 布局界面

res\drawable 目录的 border.xml 内容如下。

```
<?xml version="1.0" encoding="utf-8"?>
<shape xmlns:android="http://schemas.android.com/apk/res/android" >
   <solid android:color="#FFFFFF" />
   <stroke
      android:width="2dp"
      android:color="#0000EE" />
   <padding
      android:bottom="1dp"
      android:left="0.5dp"
      android:right="0.5dp"
      android:top="0dp" />
</shape>
```

res\layout 目录的 activity_main.xml 内容如下。

```
<TableLayout xmlns:android="http://schemas.android.com/apk/res/android"
   android:layout_width="match_parent"
   android:layout_height="match_parent" >
   <TableRow
      android:layout_width="wrap_content"
      android:layout_height="wrap_content"
      android:layout_marginRight="25dp"
      android:layout_marginLeft="25dp"
      android:layout_marginTop="25dp">
      <TextView
         android:layout_width="wrap_content"
         android:layout_height="wrap_content"
         android:text="请输入学号："
         android:textSize="18sp"
         android:textStyle="bold" />
      <EditText
         android:id="@+id/edt_stuId"
```

```
            android:layout_width="wrap_content"
            android:layout_height="wrap_content"
            android:ems="10"
            android:textStyle="bold"></EditText>
    </TableRow>
    <Button
        android:id="@+id/btn_search"
        android:layout_width="wrap_content"
        android:layout_height="wrap_content"
        android:layout_marginBottom="20dp"
        android:layout_marginLeft="25dp"
        android:layout_marginRight="25dp"
        android:layout_marginTop="20dp"
        android:text="查询"
        android:textColor="@android:color/holo_blue_dark"
        android:textSize="24sp"
        android:textStyle="bold" />
    <ScrollView
        android:layout_width="match_parent"
        android:layout_height="wrap_content"
        android:padding="5dp">
        <LinearLayout
            android:id="@+id/linear_result"
            android:layout_width="match_parent"
            android:layout_height="wrap_content"
            android:layout_marginLeft="5dp"
            android:layout_marginTop="5dp"
            android:layout_marginRight="5dp"
            android:layout_marginBottom="5dp"
            android:background="@drawable/border"
            android:orientation="vertical"
            android:visibility="gone">
            <TextView
                android:id="@+id/display"
                android:layout_width="wrap_content"
                android:layout_height="wrap_content"
                android:layout_marginLeft="15dp"
                android:layout_marginTop="15dp"
                android:layout_marginRight="15dp"
                android:layout_marginBottom="15dp"
                android:text=""
                android:textColor="#1A2CC2"
                android:textSize="18sp"
                android:textStyle="bold" />
        </LinearLayout>
    </ScrollView>
</TableLayout>
```

2. 功能实现

com.example.studentmanager.MainActivity.java 的主要代码如下：

```
public class MainActivity extends AppCompatActivity {
    private Button btnSearch;
    private ListView listStu;
    private SimpleAdapter simpleAdapter;
```

```
    private TextView display;
    private EditText edtStuId;
    private LinearLayout linearResult;
    @Override
    protected void onCreate(Bundle savedInstanceState) {
        super.onCreate(savedInstanceState);
        setContentView(R.layout.activity_main);
        display = (TextView) findViewById(R.id.display);
        edtStuId = (EditText) findViewById(R.id.edt_stuId);
        btnSearch = (Button) findViewById(R.id.btn_search);
        linearResult = (LinearLayout) findViewById(R.id.linear_result);
        btnSearch.setOnClickListener(new View.OnClickListener() {
            @Override
            public void onClick(View view) {
                String newStuId = edtStuId.getText().toString();
                queryStudentRequest(newStuId);
            }
        });
    }
    private void queryStudentRequest(String stuId) {
        //创建Retrofit对象
        Retrofit retrofit = new Retrofit.Builder()
                .baseUrl("http://192.168.0.104:8080/NetServer2/") //设置网络请求URL
                .addConverterFactory(GsonConverterFactory.create()) // 设 置 使 用
Gson解析
                .build();
        //创建网络请求接口的实例（对象）
        StudentService request = retrofit.create(StudentService.class);
        //对发送请求进行封装
        Call<List<Student>> call = null;
        if(stuId.equals(""))
            call = request.getStuAll();
        else
            call= request.getStuById(stuId);
        //发送网络请求（异步）
        call.enqueue(new Callback<List<Student>>() {
            // 请求成功时回调
            @Override
            public void onResponse(Call<List<Student>> call, Response<List <Student>>
response) {
                // 请求处理,输出结果
                List<Student> stuList = response.body();
                if(stuList == null){
                    linearResult.setVisibility(View.VISIBLE);
                    display.setText("无此学生信息");
                }
                else{
                    StringBuilder helpStr = new StringBuilder();
                    for(int i=0;i<stuList.size();i++){
                        Student stu = stuList.get(i);
                        helpStr.append("学生学号: " + stu.getStuId() + "\n");
                        helpStr.append("学生姓名: " + stu.getStuName() + "\n");
                        helpStr.append("学生成绩: " + stu.getStuScore() + "\n");
                    }
```

```
                linearResult.setVisibility(View.VISIBLE);
                display.setText(helpStr.toString());
            }
            Log.d("hand","----" + response.body() + "------");
        }
        //请求失败时回调
        @Override
        public void onFailure(Call<List<Student>> call, Throwable throwable) {
            Log.d("hand","请求失败");
            Log.d("hand",throwable.getMessage());
        }
    });
  }
}
```

点击 MainActivity 中的“查询”按钮，调用 queryStudentRequest(newStuId)方法，在此方法中使用了 Retrofit 网络请求框架。

在 build.gradle 文件的 dependencies 节点中必须添加 retrofit 和 converter-gson 的依赖。

```
dependencies {
    implementation 'com.squareup.retrofit2:retrofit:2.1.0'
    implementation 'com.squareup.retrofit2:converter-gson:2.1.0'
}
```

com.example.studentmanager.service.StudentService.java 的主要代码如下：

```
public interface StudentService {
    @POST("studentQuery")
    @FormUrlEncoded
    Call<List<Student>> getStuById(@Field("stuId") String stuId);

    @GET("studentQuery")
    Call<List<Student>> getStuAll();
}
```

com.example.studentmanager.bean.Student.java 的主要代码如下：

```
public class Student {
    private String stuId;
    private String stuName;
    private float stuScore;
    …
}
```

3. 添加权限

注意在 EmployeeManager 工程的 AndroidManifest.xml 文件中必须添加访问网络的权限。

```
<uses-permission android:name="android.permission.INTERNET" />
```

10.5 相关阅读：Retrofit 注解

Retrofit 使用注解能够极大地简化网络请求数据的代码。在 Retrofit 2.x 中包含很多注解，可分为以下几类。

1. 网络请求 URL

Retrofit 把网络请求的 URL 分成了两部分：

网络请求的完整 URL = ① + ② 。

其中：①表示创建 Retrofit 实例时通过.baseUrl()设置。②表示网络请求接口的注解设置（path）。

小贴士： 加入接口的 URL(path)是一个完整的网址（URL），那么在创建 Retrofit 实例时可以不设置 URL，可视作 path=完整的 URL。如 path="http://host:port/NetServer/apath"，如果 baseUrl=不设置，则 URL=path。

2．网络请求方法注解

（1）@GET、@POST、@PUT、@DELETE、@PATCH、@HEAD 和 @OPTIONS：这些方法分别对应于 HTTP 中的网络请求方法，它们都接收一个网络地址 URL。当然也可以不指定，通过@HTTP 进行设置。

（2）@HTTP：用于替换以上这些注解的作用，并可以进行更多功能的扩展。使用时，可通过属性 method、path、hasBody 进行设置。这里 method 表示用以设置网络请求的方法（区分大小写），path 表示设置网络请求的地址路径，hasBody 则表示是否有请求体。

```
public interface GetRequest_Interface {
    @HTTP(method="GET", path="blog/{id}", hasbody=false)
    Call<ResponseBody>  getCall(@Path("id") int id);  // {id}表示是一个变量
}
```

@HTTP 可用于发送一个自定义的 HTTP 请求。

```
//自定义 HTTP 请求的标准样式
interface Service {
    @HTTP(method = "CUSTOM", path = "custom/endpoint/")
    Call<ResponseBody> customEndpoint();
}
//发送一个 DELETE 请求
interface Service {
    @HTTP(method = "DELETE", path = "remove/", hasBody = true)
    Call<ResponseBody> deleteObject(@Body RequestBody object);
}
```

3．标记注解

（1）@FormUrlEncoded：表示请求体是一个表单，可发送 form-encoded 的数据。用于修饰@Field 和@FieldMap，其实例代码如下：

```
public interface PostStudentService {
    @POST("studentQuery")
    @FormUrlEncoded
    Call<List<Student>> getCall(@Field("stuId") String stuId);
}
```

小贴士： 使用该@FormUrlEncoded 表示请求正文将使用表单网址编码，字段应该声明为参数。使用 FormUrlEncoded 注解的请求将具“application/x-www-form-urlencoded” MIME 类型。字段名称和值将先进行 UTF-8 进行编码，再根据 RFC-3986 进行 URI 编码。在代码中@POST 比起@GET 多了一个@FromUrlEncoded 的注解。如果去掉@FromUrlEncoded 在 POST 请求中使用@Field 和@FieldMap，那么程序会抛出 Java.lang.IllegalArgumentException: @Field parameters can only be used with form encoding. 的错误异常。所以使用 POST 请求千万别忘记了加上@FromUrlEncoded 注解。

使用 FormUrlEncoded 注解接口，有两种写法分别如下。

```
//第一种写法，使用@Field
@POST("/NewsServlet")
@FormUrlEncoded
```

```
    Call<ResponseBody> testFormUrlEncoded1(@Field("name") String name, @Field("age")
int age);
    //第二种写法，使用@FieldMap，Map的key作为表单的键
    @POST("/NewsServlet")
    @FormUrlEncoded
    Call<ResponseBody> testFormUrlEncoded2(@FieldMap Map<String, Object> map);
```

分别对应调用的两种方法。

第一种方法：

```
    NewsService newsService = createRetrofit().create(NewsService.class);
    Call<ResponseBody> answers = newsService.testFormUrlEncoded1("余小男", 28);
```

第二种方法：

```
    NewsService newsService = createRetrofit().create(NewsService.class);
    HashMap<String, Object> map = new HashMap<>();
    map.put("name", "余小男");
    map.put("age", 28);
    Call<ResponseBody> answers = newsService.testFormUrlEncoded2(map);
```

（2）@Multipart：表示请求体是一个支持文件上传的表单，类型为ContentType: multipart/form-data，带文件上传的网页就是用的这种请求方式。@Multipart注解一般在上传文件时使用，其接口的3种写法如下。

第一种写法

```
    //@Part后面支持3种类型：RequestBody、okhttp3.MultipartBody.Part、任意类型
    //除okhttp3.MultipartBody.Part外，其他类型都必须带上表单字段
    //okhttp3.MultipartBody.Part中已经包含了表单字段的信息
    @POST("NewsServlet")
    @Multipart
    Call<ResponseBody> testFileUpload1(@Part("name") RequestBody name, @Part("age")
RequestBody age,
            @Part MultipartBody.Part file);
```

第二种写法

```
    //@PartMap支持一个Map作为参数，支持RequestBody类型
    //如果有其他的类型，则会被retrofit2.Converter转换，如com.google.gson.Gson的
    //retrofit2.converter.gson.GsonRequestBodyConverter，所以MultipartBody.Part就
不适用了
    //文件只能用MultipartBody.Part
    @POST("NewsServlet")
    @Multipart
    Call<ResponseBody> testFileUpload2(@PartMap Map<String, RequestBody> args, @Part
MultipartBody.Part file);
```

第三种写法

```
    @POST("NewsServlet")
    @Multipart
    Call<ResponseBody> testFileUpload3(@PartMap Map<String, RequestBody> args);
```

其对应的3种调用方式如下。

第一种方式：

```
    NewsService newsService = createRetrofit().create(NewsService.class);
    MediaType textType = MediaType.parse("text/plain");
    RequestBody name = RequestBody.create(textType, "余小男");
    RequestBody age = RequestBody.create(textType, "28");
    //构建要上传的文件
    File file = new File(Environment.getExternalStorageDirectory(), "beauty.jpg");
//须申请权限
```

```
    RequestBody  requestFile  =  RequestBody.create(MediaType.parse("application/
otcet-stream"), file);
    MultipartBody.Part filePart = MultipartBody.Part.createFormData("fileUploader",
file.getName(), requestFile);
    Call<ResponseBody> answers = newsService.testFileUpload1(name, age, filePart);
```

第二种方式：

```
    NewsService newsService = createRetrofit().create(NewsService.class);
    MediaType textType = MediaType.parse("text/plain");
    RequestBody name = RequestBody.create(textType, "余小男");
    RequestBody age = RequestBody.create(textType, "28");
    Map<String, RequestBody> fileUpload2Args = new HashMap<>();
    fileUpload2Args.put("name", name);
    fileUpload2Args.put("age", age);
    //构建要上传的文件
    File file = new File(Environment.getExternalStorageDirectory(), "beauty.jpg");
//须申请权限
    RequestBody  requestFile  =  RequestBody.create(MediaType.parse("application/
otcet-stream"), file);
    MultipartBody.Part filePart = MultipartBody.Part.createFormData("fileUploader",
file.getName(), requestFile);
    Call<ResponseBody>  answers  =  newsService.testFileUpload2(fileUpload2Args,
filePart);
```

第三种方式：

```
    NewsService newsService = createRetrofit().create(NewsService.class);
    MediaType textType = MediaType.parse("text/plain");
    RequestBody name = RequestBody.create(textType, "余小男");
    RequestBody age = RequestBody.create(textType, "28");
    Map<String, RequestBody> fileUpload3Args = new HashMap<>();
    fileUpload3Args.put("name", name);
    fileUpload3Args.put("age", age);
    //构建要上传的文件
    File  file  =  new  File(Environment.getExternalStorageDirectory(),  "  beauty.
jpg");//须申请权限
    RequestBody  requestFile  =  RequestBody.create(MediaType.parse("application/
otcet-stream"), file);
    fileUpload3Args.put("fileUploader\"; filename=\"paoche3.jpg",requestFile);
    Call<ResponseBody> answers = newsService.testFileUpload3(fileUpload3Args);
```

（3）@Streaming：表示返回的数据以流的形式返回，适用于返回数据较大的场景。如果没有使用该注解，则默认把数据全部载入内存，之后获取数据也是从内存中读取。它作用于方法，并处理返回Response方法的响应体，即没有将body()转换为byte []。

4．参数注解

（1）@Query和 @QueryMap：@Query主要用于GET请求数据，拼接在URL问号之后的查询参数后。一个@Query相当于拼接一个参数，多个参数中间用“,”隔开。

```
public interface GetStudentService {
    @GET("studentQuery")
    Call<List<Student>>  getCall(@Query("stuName") String stuName);
}
```

@QueryMap作用等同于多个@Query参数拼接，主要用于GET请求网络数据。

```
@GET("http://ms.csdn.net/api/ask/all_questions")
Call<List<Repo>> getData(@QueryMap Map<String,String> params);
Map<String,String>params=newHashMap();
```

```
params.put("name","Mary");
params.put("age",18);
params.put("sex","man");
params.put("city","Shanghai");
```

这样等价于请求数据接口为：

```
http://ms.csdn.net/api/ask/all_questions?name=Mary&age=18&sex=man&city=Shanghai
```

（2）@Field 和 @FieldMap：@Field 的用法类似于@Query，就不再重复列举了，主要不同的是@Field 主要用于 POST 请求数据。@FieldMap 的用法类似于@QueryMap，两者主要区别是，如果请求为 POST 实现，则最好传递参数时使用@Field、@FieldMap 和@FormUrlEncoded。因为@Query 和 QueryMap 都是将参数拼接在 URL 后面的，而@Field 或@FieldMap 传递参数时则是放在请求体的。

（3）@Part 和 @PartMap：POST 请求时提交的表单字段，与@ Multipart 注解配合使用。@Part、@PartMap 与 @Field、@FieldMap 的区别是，功能相同，但携带的参数类型更加丰富，包括数据流，所以适用于有文件上传的场景。

（4）@Body：以 POST 方式传递自定义数据类型给服务器。使用@Body 注解可指定一个对象作为 RequestBody，为非表单请求体。

```
@POST("users/new")
Call<User> createUser(@Body User user);
```

5．其他注解

Retrofit 把网络请求的 URL 分成了两部分：

（1）@Path：@Path 主要用于 GET 请求，替换 URL 路径中的变量字符，其参数包含在 URL 路径中，具体使用代码如下：

```
public interface csdnService {
    @GET("users/{user}/question")
    Call<List<Repo>> getData(@Path("user") String user);
}
```

该接口定义了一个 getData()方法，该方法通过 GET 请求去访问服务器的 users/{user}/question 路径，其中通过@Path 注解会把路径中的{user}替换成参数 user 的具体值。如 user 的值如果是“zhangsan”，那么 URL 的路径就是“users/zhangsan/question”。

（2）@Url：直接传入一个请求的 URL 变量用于 URL 设置，其具体使用代码如下：

```
public interface GetRequest_Interface {
    @GET
    Call<ResponseBody> testUrlAndQuery(@Url String url, @Query("showAll") boolean
showAll);
    // 当有 URL 注解时，@GET 传入的 URL 就可以省略
    // 当 GET、POST、…、HTTP 等方法中没有设置 URL 时，则必须使用@Url 提供
}
```

（3）@Header 和 @Headers：添加不固定的请求头或者添加请求头。使用@Header 注解动态更新请求头，匹配的参数必须提供给@Header，若参数值为 null，这个请求头将会被省略。否则，会使用参数值 toString()方法的返回值，其具体使用代码如下：

```
// 使用@Header
@GET("user")
Call<User> getUser(@Header("Authorization") String authorization);
// 使用@Headers
@Headers("Authorization: authorization")
@GET("user")
Call<User> getUser();
// 以上的效果是一致的，区别在于使用场景和使用方式
```

```
// ①使用场景：@Header 用于添加不固定的请求头，@Headers 用于添加固定的请求头
// ②使用方式：@Header 作用于方法的参数；@Headers 作用于方法
```

10.6 本章小结

本章主要介绍Android网络通信的相关知识，包括HTTP和Socket的基础知识，HttpURLConnection、HttpClient等工具类进行Http通信，以及针对TCP和UDP两种协议进行Socket通信。另外，还介绍了WebView控件，阐述了如何利用WebView控件浏览网页、加载网页代码，以及与JavaScript进行交互。

HTTP属于OSI网络七层协议模型中的最上层"应用层"，由请求和响应构成，是一个标准的客户端服务器模型。在Internet中，所有的传输都是通过TCP/IP进行的。HTTP协议作为TCP/IP模型中应用层的协议也不例外。HTTP协议通常承载于TCP协议之上，有时也承载于TLS或SSL协议层之上，这个时候就成了HTTPS。

任何网络连接都需要经过Socket才能连接，HttpURLConnection不需要设置Socket，所以，HttpURLConnection并不是底层的连接，而是在底层连接上的一个请求。这就是为什么HttpURLConneciton只是一个抽象类，自身不能被实例化的原因。HttpURLConnection只能通过URL.openConnection()方法创建具体的实例。

虽然底层的网络连接可以被多个HttpURLConnection实例共享，但每一个HttpURL Connection实例只能发送一个请求。请求结束之后，应该调用HttpURLConnection实例的InputStream或OutputStream的close()方法以释放请求的网络资源，不过这种方式对于持久化连接没用。对于持久化连接，需要用disconnect()方法关闭底层连接的Socket。

OkHttp是一个优秀的网络请求框架，具有非常多的优势。

（1）能够高效地执行HTTP请求，数据加载速度快。

（2）支持GZIP压缩，提升速度，并节省流量。

（3）缓存响应数据，避免了重复的网络请求。

（4）使用简单，支持同步阻塞调用和带回的异步调用。

Retrofit是由Square公司出品的针对Android和Java类型安全的HTTP客户端，是一个高质量、高效率的开源HTTP库，它可以将开发的底层代码和细节都封装了起来。

Retrofit的本质过程：App应用程序通过Retrofit请求网络，实际上是使用Retrofit接口层封装请求参数、Header、URL等信息，之后由OkHttp完成后续的请求操作，在服务端返回数据之后，OkHttp将原始的结果交给Retrofit，Retrofit再根据用户的需求对结果进行解析。

Retrofit本质上是一个RESTful的HTTP网络请求框架的封装。网络请求的工作实际是由OkHttp完成的，而Retrofit仅负责网络请求接口的封装。

习 题 10

1．如何使用HttpURLConnection访问网络？试编程举例说明。

2．如何使用OkHttp访问网络？试编程举例说明。

3．如何使用Retrofit访问网络？试编程举例说明。

4．如何实现手机客户端与服务器端的Socket网络通信？试编程举例说明。

第三部分

开 发 篇

第 11 章　手机客户端解析 JSON

JSON 基于 JavaScript，是一种轻量级的数据交换格式，易于人阅读和编写，同时也易于机器解析和生成。JSON 采用完全独立于语言的文本格式，但也使用了类似于 C 语言家族的习惯（包括 C、C++、C#、Java、JavaScript、Perl、Python 等）。这些特性使 JSON 成为理想的数据交换语言，尤其是在智能手机客户端与服务器网络交互中。本章主要知识点有：（1）JSON 简介；（2）服务器端生成 JSON 数据；（3）在手机客户端中解析 JSON。

11.1　JSON 简介

1．JSON 数据格式的定义

JSON（JavaScript Object Notation）是一种轻量级的数据交换格式。它构建于两种结构：

（1）"名称/值" 对的集合（a collection of name/value pairs）。不同的语言中，它被理解为对象（Object）、记录（Record）、结构（Struct）、字典（Dictionary）、哈希表（HashTable）、有键列表（Keyed list），或者关联数组（Associative array）。

（2）值的有序列表（An ordered list of values）。在大部分语言中，它被理解为数组（Array）。

2．JSON 数据格式的特点

JSON 对象是一个无序的"'名称/值' 对"的集合，一个对象以"{" 开始，以"}" 结束，每个"名称" 后跟一个"："，"名称/值" 对之间使用"，" 分隔。

如{"name":"jackson", "age":100 }。

稍微复杂一点的情况，当数组是值的有序集合时，一个数组以"[" 开始，以"]" 结束，值之间使用"," 分隔。

如{"studengs":[{"name":"jackson","age":100},{"name": "michael", "age": 51}]}。

11.2　服务器端生成 JSON 数据

首先，进行服务器端的构建。

在 Eclipse（Java EE）中创建一个"Dynamic Web Project"即 jsonserver 工程。

代码和文件参见教学资源包中的第 11 章服务器端 jsonserver 工程。

需要说明的是，为了使得 Java 对象转换更为简单容易，这里使用了 json-lib 框架，需要用的相关 jar 包及其依赖包在 jsonserver 工程的 lib 文件夹中，如图 11-1 所示。

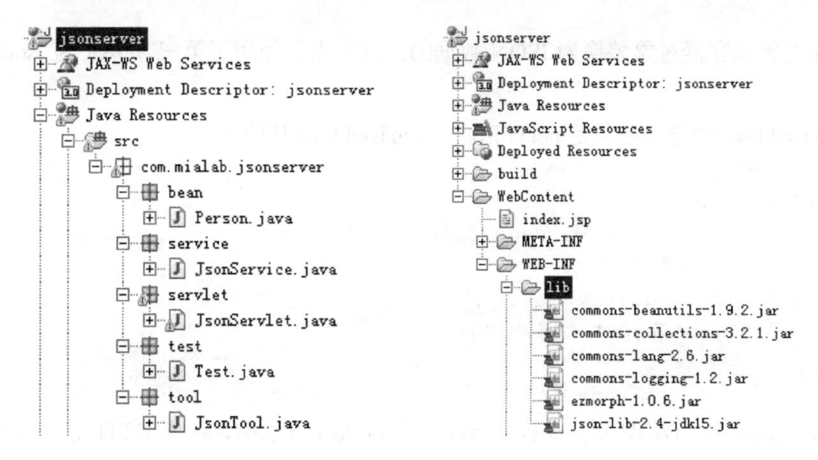

图 11-1　jsonserver 工程目录结构和 lib 文件夹

json-lib 是一个 Java 类库，提供将 Java 对象（包括 beans、maps、collections、java arrays 和 xml 等）转换成 JSON，或者反向转换的功能。

1．浏览器调用

将 jsonserver 工程编译发布到 Tomcat 上，并通过浏览器进行调用。测试时编者计算机的 IP 地址为：192.168.1.108。

访问地址一：

```
http://192.168.1.108:8080/jsonserver/servlet/JsonServlet?action_flag=person
```

输出以下结果：

```
{"person":{"id":1001,"address":"上海黄浦区","name":"Jack"}}
```

访问地址二：

```
http://192.168.1.108:8080/jsonserver/servlet/JsonServlet?action_flag=persons
```

输出以下结果：

```
{"persons":[{"id":1001,"address":"上海黄浦区","name":"Jack"},
{"id":1002,"address":"苏州吴中区","name":"李强"},
{"id":1003,"address":"苏州古城区","name":"Marry"}]}
```

访问地址三：

```
http://192.168.1.108:8080/jsonserver/servlet/JsonServlet?action_flag=listString
```

输出以下结果：

```
{"listString":["苏州","上海","南京"]}
```

访问地址四：

```
http://192.168.1.108:8080/jsonserver/servlet/JsonServlet?action_flag=listMap
```

输出以下结果：

```
{"listMap":[{"id":1001,"address":"上海","name":"Jack"},
{"id":1002,"address":"苏州","name":"李强"}]}
```

可以看到服务器端产生了四种类型的 JSON 数据。HTTP 请求传递参数为 action_flag，服务器端将根据 action_flag 返回不同的 JSON 数据给客户端。

2．JsonServlet 工程主要代码

客户端请求将提交给服务器端的 JsonServlet 进行处理。JsonServlet 中包含了 JsonService 的对象引用，JsonServlet 将调用 JsonService 和 JsonTool 的相关方法以得到 JSON 接口数据。

JsonService 提供了 getPerson()、getListPerson()等实例方法，用于产生包含有 JSON 数据的 Java 对象和容器对象。

而 JsonTool 则提供了类方法（静态方法）creatJsonString(String key, Object value)，creatJsonString()

方法是将 Java 对象或容器对象装换为 JSON 数据的。这里由于使用了第三方 JAR 包 json-lib，代码较为简单。

注意 JsonTool.java 中导入的包是 net.sf.json。JsonTool,java 代码如下：

```
import net.sf.json.JSONObject;
public class JsonTool {
 public static String creatJsonString(String key, Object value) {
    JSONObject jsonObject = new JSONObject();
       jsonObject.put(key, value);
       return jsonObject.toString();
    }
}
```

JsonServlet 的 doGet 方法调用了 doPost 方法，在 doPost 方法中，根据 HTTP 请求参数 action_flag 的不同，而调用 JsonTool 的不同实参的 createJsonString 方法，以在返回页面输出 JSON。

```
public class JsonServlet extends HttpServlet {
  private JsonService service;
  …
  public void  doPost(HttpServletRequest request, HttpServletResponse response)
         throws ServletException, IOException {
     response.setContentType("text/html;charset=utf-8");
     request.setCharacterEncoding("utf-8");
     response.setCharacterEncoding("utf-8");
     PrintWriter out = response.getWriter();
     String jsonString = "";
     String action_flag = request.getParameter("action_flag");
     if(action_flag.equals("person")){
        jsonString = JsonTool.creatJsonString("person", service.getPerson());
     }else if(action_flag.equals("persons")){
        jsonString = JsonTool.creatJsonString("persons", service. getListPerson());
     }else if(action_flag.equals("listString")){
        jsonString = JsonTool.creatJsonString("listString", service. getListString());
     }else if(action_flag.equals("listMap")){
        jsonString = JsonTool.creatJsonString("listMap", service.getListMap());
     }
     out.println(jsonString);
     out.flush();
     out.close();
  }

  public void init() throws ServletException {
     service = new JsonService();
  }
}
```

web.xml 中须配置 servlet 映射，其代码如下：

```
<servlet>
   <servlet-name>JsonServlet</servlet-name>
   <servlet-class>com.mialab.jsonserver.servlet.JsonServlet</servlet-class>
</servlet>
<servlet-mapping>
   <servlet-name>JsonServlet</servlet-name>
   <url-pattern>/servlet/JsonServlet</url-pattern>
</servlet-mapping>
```

11.3 在手机客户端中解析 JSON

启动 Android Studio，在 Ch11 工程中创建模块 ParseJsonDemo，来访问服务器的 JSON 数据。由 Ch11\ParseJsonDemo 中的 build.gradle 文件的第一行代码：apply plugin: 'com.android.application'，知道模块 ParseJsonDemo 为一个 Android 应用。

在模块 ParseJsonDemo 中创建文件 MainActivity.java、ResultActivity.java、OkHttpUtil.java（工具类，使用 OkHttpUtil 访问网络）、HttpUtil.java（工具类，使用 HttpURLConnection 访问网络）、JsonTool.java（工具类，把 JSON 字符串转换为容器对象及实体对象，也就是 JavaBean）、Person.java，以及用于布局界面的 activity_main.xml 和 activity_result.xml。相关代码和文件参见教学资源包中的 Ch11\ParseJsonDemo 模块。

Android 客户端用 GET 方式分别获取服务器端返回的 JSON 数据，并将四种不同的 JSON 数据解析成四种不同的结果类型（(Person、List<Person>、List<String>、List<Map<String、Object>>)，用 Intent 方式由 MainActivity 传递给 ResultActivity，并在 ResultActvity 中显示解析的结果。

1．ParseJsonDemo 工程目录结构和运行界面

（1）Ch11\ParseJsonDemo（源码）目录结构如图 11-2 所示。

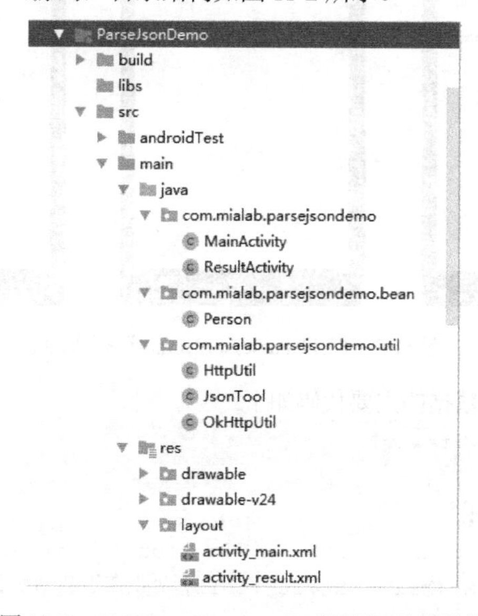

图 11-2 Ch11\ParseJsonDemo（源码）目录结构

（2）Ch11\ParseJsonDemo 运行界面如图 11-3 所示。

2．异步访问网络数据

当点击 MainActivity 主界面上的按钮时，便可启动新线程异步访问网络数据，根据访问网络数据的不同情况，发送不同标识的 Message 给 Handler 进行处理。

这里使用数字-2、1、2、3、4 作为发送 Message 的标识。

（1）-2：表示网络异常，访问网络数据失败。

（2）1：表示访问的是可转换成 Person 格式的 JSON 接口，且访问网络数据成功。

（3）2：表示访问的是可转换成 List<Person>格式的 JSON 接口，且访问网络数据成功。

（4）3：表示访问的是可转换成 List<String>格式的 JSON 接口，且访问网络数据成功。

（5）4：表示访问的是可转换成 List<Map<String,Object>>格式的 JSON 接口，且访问网络数据成功。

图 11-3　Ch11\ParseJsonDemo 运行界面

启动新线程异步访问网络数据的主要代码如下：

```
    public void onClick(View v) {
        switch (v.getId()) {
        case R.id.person:
            final String path1 = PATH + "?action_flag=person";
            Toast.makeText(MainActivity.this, "正在下载网络数据，请稍候…",
                    Toast.LENGTH_LONG).show();
            new Thread() {
                public void run() {
                    try {
                        //String jsonString1 = HttpUtil.getJsonContent(path1,"
utf-8");
                        String jsonString1 = OkHttpUtil.getJsonContent(path1);
                        demoperson = JsonTool.getPerson("person", jsonString1);
                    } catch (Exception e) {
                        e.printStackTrace();
                    }
                    if (demoperson != null) {
                        sendMessage(1, null);
```

```
                } else {
                    sendMessage(-2, getResources().getString(
                            R.string.network_exception));
                }
            }
        }.start();
        break;
    …
    }
}
```

3．Handler 的 handleMessage 方法

MainActivity 的 Handler 根据接收到 Message 的不同标识进行不同的处理，用 Intent 将数据封装后，设置不同的 click_flag 标识，分别为 person、persons、listString 和 listMap，再传递给 ResultActvity。handleMessage()方法的主要代码如下：

```
public void handleMessage(Message msg) {
 super.handleMessage(msg);
 switch (msg.what) {
    case -2:
        Toast.makeText(MainActivity.this, (String) msg.obj,
            Toast.LENGTH_LONG).show();
        break;
    case 1:
        Intent intent = new Intent(MainActivity.this, ResultActivity.class);
        intent.putExtra("click_flag", "person"); // 标记当前 intent 是按钮 person
        intent.putExtra("person", demoperson);   // 类 Person 是序列化的
        startActivity(intent);
        break;
    case 2:
        Intent intent2 = new Intent(MainActivity.this, ResultActivity.class);
        intent2.putExtra("click_flag","persons");// 标记当前intent按钮是persons
        intent2.putExtra("persons", (Serializable) list2);
        startActivity(intent2);
        break;
    case 3:
        …
        break;
    case 4:
        Intent intent4 = new Intent(MainActivity.this, ResultActivity. class);
        intent4.putExtra("click_flag","listMap");//标记当前 intent 按钮是 listMap
        intent4.putExtra("listMap", (Serializable) list4);
//用 Intent 传递 List<map<string,object>> 对象，其中 HashMap 实现了 Serializable
        startActivity(intent4);
        break;
    }
}
```

4．在 ResultActivity 中显示解析结果

ResultActivity 根据 Intent 对象中不同的 click_flag 标识，对封装数据进行相应的解析，并把解析结果显示到手机界面。

11.4 项目实战："移动商城"（三）

11.4.1 任务说明

在手机客户端异步访问网络数据的宽带列表和宽带详情的JSON接口，并在手机客户端解析JSON数据，显示到手机屏幕。

11.4.2 项目讲解

1．数据库安装和Tomcat配置

这里安装的数据库是PostgreSQL 9.3（x86），双击postgresql-9.3.1-1-windows.exe，（可在教学资源网站中的tools文件夹的postgresql-9.3.1-1-windows.exe下载，也可在PostgreSQL官网中下载），按默认安装即可，postgres的密码设置为1。

在pqAdmin III中创建数据库market，如图11-4所示。

选中market数据库，右键选择"恢复"选项，弹出"恢复数据库"对话框，选择备份文件newmarket.backup（在教学资源包第11章newmarket.backup中），点击"恢复"按钮，即可恢复market数据库的数据，如图11-5所示。

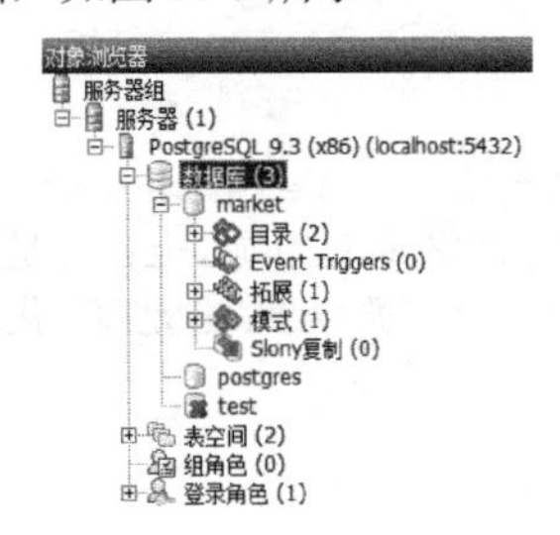

图11-4　创建数据库market

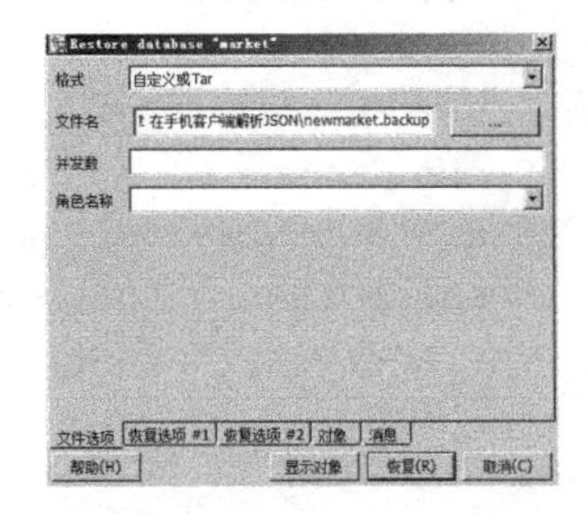

图11-5　恢复market数据库的数据

先把mialab-3gmarket.war文件复制到Tomcat的webapps文件夹下（详见教学资源包第11章mialab-3gmarket.war）；把postgresql-9.3-1100.jdbc41.jar复制到Tomcat的lib文件夹（详见教学资源包tools文件夹下postgresql-9.3-1100.jdbc41.jar）；在Tomcat的conf文件夹下的context.xml文件中加入代码如下：

```
    <Resource auth="Container" driverClassName="org.postgresql.Driver" maxActive=
"100"
    maxIdle="30" maxWait="10000" name="jdbc/MarketDB" password="1"
    type="javax.sql.DataSource"
    url="jdbc:postgresql://localhost:5432/market?characterEncoding=UTF8"
    username="postgres"/>
```

在保证PostgreSQL 9.3服务已启动的情况下，启动Tomcat即可。

2．服务器端JSON接口

宽带列表接口：

```
http://192.168.1.108:8080/mialab-3gmarket/3gmarket?cmd=market.list.product
&typeId=1000&page=1&pSize=20
```

宽带详情接口：

```
http://192.168.1.108:8080/mialab-3gmarket/3gmarket?cmd=market.product.info
&typeId=1000&id=10001003
```

3．Android 客户端总体框架和运行界面

启动Android Studio，在Ch11工程中创建模块MobileMall_C为一个Android应用。完成后的Android客户端Ch11\MobileMall_C（源码）目录结构和主要运行界面如图11-6和图11-7所示，其相关代码和文件参见教学资源包中的Ch11\MobileMall_C。

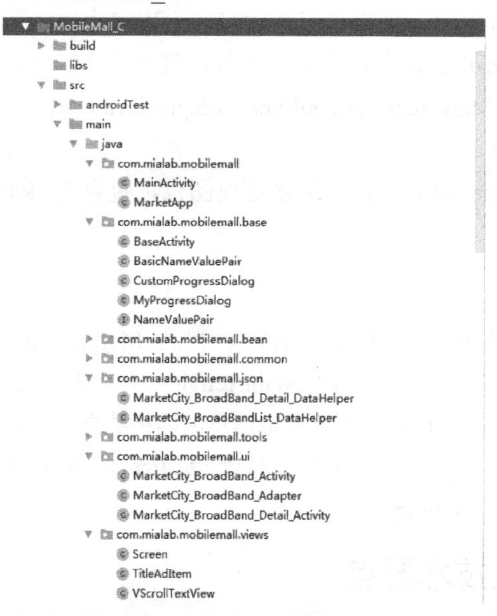

图 11-6　Ch11\MobileMall_C（源码）目录结构

图 11-7　Ch11\MobileMall_C 主要运行界面

Ch11\MobileMall_C 总体框架如下。

（1）com.mialab.mobilemarket 包：它是主应用包，包含有主界面 MainActivity.java 和一般用来放置全局变量的。android.app. Application 的子类为 MarketApp.java。

（2）com.mialab.mobilemarket.base 包：它是基础包，包含有 BaseActivity.java 等。

（3）com.mialab.mobilemarket.bean 包：它是 JavaBean 包，包含有 MarketCity_BroadBand_ Bean.java 和 MarketCity_BroadBand_Detail.java 等。

（4）com.mialab.mobilemarket.common 包：它是通用包，包含有 Constants.java 等。

（5）com.mialab.mobilemarket.json 包：它是 json 工具包，包含有 MarketCity_BroadBandList_DataHelper.java 和 MarketCity_BroadBand_Detail_DataHelper.java。

（6）com.mialab.mobilemarket.tools 包：它是核心工具包，包含有 OkHttpUtil.java、DiskCache.java、FileUtils.java、FunctionUtil.java、ImageCache.java 和 ImageWorker.java 等。

（7）com.mialab.mobilemarket.ui 包：它是界面包，包含有宽带列表界面 MarketCity_BroadBand_Activity.java 及其适配器 MarketCity_BroadBand_Adapter.java，还有宽带详情界面 MarketCity_BroadBand_Detail_Activity.java。

（8）com.mialab.mobilemarket.views 包：它是视图包，包含有 Screen.java、TitleAdItem.java 和 VScrollTextView.java。

4．程序主要流程

程序的入口是 MarketApp 的 onCreate 方法，得到 MarketApp 的单例，继而去执行主界面 MainActivity 的 onCreate 方法。在主界面 MainActivity 中点击宽带，便会跳转到宽带列表界面 MarketCity_BroadBand_Activity，再异步下载宽带信息的网络数据，并与宽带列表界面的 ListView 控件相绑定。点击 ListView 控件上具体的宽带产品 Item，便会跳转到相应的宽带详情界面 MarketCity_BroadBand_Detail_Activity。

11.4.3 典型代码及技术要点

1．宽带列表界面 MarketCity_BroadBand_Activity.java

（1）异步加载网络数据。

在 MarketCity_BroadBand_Activity 的 onCreate 方法中启动新线程 MyThread，通过加载并解析第 1 页数据，将每页具体宽带产品数据设为 6 条，其主要代码如下：

```
public class MarketCity_BroadBand_Activity extends BaseActivity {
    public static final int PAGECOUNT = 6;
    private ListView listView;
    private MarketCity_BroadBand_Adapter marketcity_BroadBand_Adapter;
    private String url = Constants.MarketCity_URL+ "?cmd=market.list. product&typeId
=1000&page=";
    public Vector<MarketCity_BroadBand_Bean> datas;
    private int page = 1;
    …
    @Override
    protected void onCreate(Bundle savedInstanceState) {
        …
        showWaitDialog("正在加载数据，请稍候…");
        new Thread(new MyThread()).start();
    }

    private class MyThread implements Runnable {
        @Override
        public void run() {
            String newurl = "";
            newurl = url + page + "&pSize=6";
            datas  =  MarketCity_BroadBandList_DataHelper.getBroadBand_Vector
(newurl);
            sendMessage(1, null);
        }
```

```
    }
    public void sendMessage(int index, String message) {
        Message msg = new Message();
        msg.obj = message;
        msg.what = index;
        handler.sendMessage(msg);
    }
}
```

通过 MarketCity_BroadBandList_DataHelper 的 getBroadBand_Vector(String url)方法获得服务器端的 JSON 数据，并转换成 Vector<MarketCity_BroadBand_Bean>类型。

数据获取成功后，给 Handler 发送标识为 1 的 Message。

（2）Handler 进行相应的处理。

Handler 在 handleMessage 方法中，通过适配器将第 1 页的宽带数据与显示控件 ListView 绑定，并通知界面更新，其主要代码如下：

```
marketcity_BroadBand_Adapter.setDatas_Search(datas);
listView.setAdapter(marketcity_BroadBand_Adapter);
marketcity_BroadBand_Adapter.notifyDataSetChanged();
```

（3）分页的解决。

如果具体宽带产品 Item 多于 6 条，就会存在第 2 页宽带产品 Item 数据，第 2 页的数据将放到另一个 Vector<MarketCity_BroadBand_Bean>容器中。这时页面底端会存在“点击加载更多”按钮，其背景色设置为透明的。点击此按钮，将启动另一线程 MyThread2，加载并解析第 2 页数据，每页具体宽带产品数据仍设为 6 条，如图 11-8 所示。

图 11-8　宽带列表界面分页的解决

数据获取成功后，给 Handler 发送标识为 2 的 Message。Handler 将存储第 2 页宽带数据的容器 datas_MarketCity_BroadBand_Info_Next 的每一个 MarketCity_BroadBand_Bean 添加到容器 datas 中，仍通过适配器与显示控件 ListView 绑定，并通知界面更新。

其主要代码如下：

```
public class MarketCity_BroadBand_Activity extends BaseActivity {
    …
    public Vector<MarketCity_BroadBand_Bean> datas;
```

```
    private Vector<MarketCity_BroadBand_Bean>  datas_MarketCity_BroadBand_ Info_
Next;
    public Button button;
    private Handler handler = new Handler() {
      public void handleMessage(Message msg) {
          super.handleMessage(msg);
          listView.addFooterView(loadingLayout);
          dismissDlg();
          switch (msg.what) {
          case 1:
              listView.addFooterView(loadingLayout);
              dismissDlg();
              if (datas == null || datas.size() == 0) {
                  …
              } else {
                  if (datas.size() < 6)
                      button.setVisibility(View.GONE);
                  button.setText("点击加载更多…");
                  button.setOnClickListener(new OnClickListener() {
                      @Override
                      public void onClick(View v) {
                          showWaitDialog("正在加载数据，请稍候…");
                          new Thread(new MyThread2()).start();
                      }
                  });
              }
              marketcity_BroadBand_Adapter.setDatas_Search(datas);
              listView.setAdapter(marketcity_BroadBand_Adapter);
              marketcity_BroadBand_Adapter.notifyDataSetChanged();
              break;
          case 2:
              if (datas_MarketCity_BroadBand_Info_Next.size() == 0) {
                  Toast toast = Toast.makeText(getApplicationContext(),
                          "没有更多数据", Toast.LENGTH_LONG);
                  toast.show();
                  button.setText("没有更多…");
                  button.setEnabled(false);
              } else {
                  if (datas.size() < 6)
                      button.setVisibility(View.GONE);
                  for (int i = 0; i < datas_MarketCity_BroadBand_Info_ Next.size();
 i++)
                      datas.add(datas_MarketCity_BroadBand_Info_Next. get(i));
                  marketcity_BroadBand_Adapter.notifyDataSetChanged();
              }
              break;
          }
      }
  };
    …
     private class MyThread2 implements Runnable {
        @Override
        public void run() {
           page++;
```

```
            String newurl = "";
            newurl = url + page + "&pSize=6";
            datas_MarketCity_BroadBand_Info_Next =
             MarketCity_BroadBandList_DataHelper.getBroadBand_Vector(newurl);
            sendMessage(2, null);
        }
    }
}
```

2．宽带详情界面 MarketCity_BroadBand_Detail_Activity.java

点击宽带列表界面 MarketCity_BroadBand_Activity 中 ListView 控件上具体宽带产品 Item 时，会将产品类型 type 和具体产品的 goodsId 封装到 Intent 对象中，传给 MarketCity_ BroadBand_Detail_Activity，其主要代码如下：

```
listView.setOnItemClickListener(new OnItemClickListener() {
     @Override
     public void onItemClick(AdapterView<?> arg0, View arg1, int arg2, long arg3){
        Intent intent = new Intent(MarketCity_BroadBand_Activity.this,
                          MarketCity_BroadBand_Detail_Activity.class);
         intent.putExtra("type", "1000");
         intent.putExtra("goodsId", String.valueOf(datas.get(arg2). getId()));
         startActivity(intent);
     }
});
```

MarketCity_BroadBand_Detail_Activity 从 Intent 对象中取出 type 和 goodsId，作为参数去异步访问相应的服务器端 JSON 数据，然后将其转换成 MarketCity_BroadBand_Detail 对象，再发送 Message 给 MarketCity_BroadBand_Detail_Activity 中的 Handler，通知宽带详情界面更新，其主要代码如下：

```
public class MarketCity_BroadBand_Detail_Activity extends BaseActivity {
  private MarketCity_BroadBand_Detail broadBandDetail;
  private String type;
  private String goodsId;
  …
  private Handler handler = new Handler() {
     public void handleMessage(Message msg) {
         switch (msg.what) {
         case -1:
             …
             break;
         case 1:
             dismissDlg();
             updateView();
             break;
         }
     }
  };
  @Override
 protected void onCreate(Bundle savedInstanceState) {
     …
     type = getIntent().getStringExtra("type");
     goodsId = getIntent().getStringExtra("goodsId");
     showWaitDialog("正在加载数据，请稍候…");
     new Thread(){
         public void run() {
```

```
                getDetail();
            };
        }.start();
    }
    private void getDetail() {
        broadBandDetail = MarketCity_BroadBand_Detail_DataHelper. getBroadBandDetail
(type, goodsId);
        if(broadBandDetail!= null){
            handler.sendEmptyMessage(1);
        }else{
            handler.sendEmptyMessage(-1);
        }
    }
}
```

11.5 本章小结

JSON 作为一种轻量级的数据交换格式，在前/后台数据交换中占据着非常重要的地位。JSON 的语法非常简单，采用键值对表示形式。JSON 可以将 JavaScript 对象中表示的一组数据转换为字符串，然后就可以在函数之间轻松地传递，或者在异步应用程序中将字符串从 Web 客户机传递给服务器端程序，也可以从服务器端传递 JSON 格式的字符串给前端并由前端解释。

JSON 的作用并不仅仅在于作为字符串在前/后台进行传递，在采用 JSON 传递数据时更主要考虑到的是它的传输效率。当两个系统需要进行数据交换时，如果传递的是经过序列化的对象，效率会非常低，如果传递的是存储大量对象的数组时效率就更不敢想象了，这时如果通过将对象或数据转换成 JSON 字符串进行传递，效率就会提高很多。

习 题 11

1. 什么是 JSON？JSON 数据格式是怎样的？
2. 怎样在服务器端生成 JSON 数据？试编程加以说明。
3. 在手机客户端中如何解析 JSON？试编程加以说明。

第 12 章　Maven

本章重点介绍 Maven 这一跨平台的项目管理工具。作为 Apache 组织中的一个颇为成功的开源项目，Maven 主要服务于基于 Java 平台的项目构建、依赖管理和项目信息管理。无论是小型的开源类库项目，还是大型的企业级应用；无论是传统的瀑布式开发，还是流行的敏捷模式，Maven 都能大显身手。本章主要知识点有：（1）Maven 简介；（2）Maven 安装和配置；（3）Maven 使用；（4）坐标和依赖；（5）构建支持 Servlet 3.0 的 MavenWeb 应用。

12.1　Maven 简介

1. 什么是 Maven

Maven 是一个项目管理工具，它包含了一个项目对象模型、一组标准集合、一个项目生命周期、一个依赖管理系统和用来运行定义在生命周期阶段中插件目标的逻辑。

当使用 Maven 的时候，可以用一个明确定义的项目对象模型来描述项目，然后 Maven 可以应用横切的逻辑，这些逻辑来自一组共享的（或者自定义的）插件。

Maven 有一个生命周期，当运行 mvn install 时被调用。这条命令告诉 Maven 执行一系列的有序步骤，直至指定的生命周期。

遍历生命周期中的一个影响就是，Maven 运行了许多默认的插件目标，这些目标完成了像编译和创建一个 JAR 文件这样的工作。

此外，Maven 能够方便于管理项目报告、生成站点、管理 JAR 文件等。

2. Maven 与 Ant

（1）Maven 和 Ant 针对构建问题的两个不同方面。Ant 为 Java 技术开发项目提供跨平台的构建任务。Maven 本身描述项目的高级方面，它从 Ant 借用了绝大多数构建任务。Ant 脚本是可以直接运行在 Maven 中的。Maven 和 Ant 最大差别在于 Maven 的编译及所有的脚本都有一个基础，就是 POM。这个模型定义了项目的方方面面，然后各式各样的脚本在这个模型上工作，而 Ant 则完全是自己定义。

（2）Maven 对所依赖的包有明确的定义，如使用哪个包，版本是多少，一目了然。而 Ant 则通常是简单的 inclde 所有的 jar，导致的最终结果就是，无法确定 JBoss 中 lib 的 common→logging 是哪个版本的，唯一方法就是通过打开 META→INF 目录的 MANIFEST.MF 查看。

（3）Maven 是基于中央仓库的编译，即把编译所需要的资源放在一个中央仓库里。当编译的时候，Maven 会自动在仓库中找到相应的包，如果本地仓库没有，则从设定好的远程仓库中下载到本地。这一切都是自动的，而 Ant 则需要自己定义了。这个好处导致的结果就是，用 Maven 编译的项目在发布的时候只需要发布源码，容量很小，反之，Ant 的发布则要把所有的包一起发布，显然 Maven 又胜了一筹。

（4）Ant 没有正式的约定（如一个一般项目的目录结构），必须明确告诉 Ant 去哪里找源代码，哪里放置输出内容。Maven 则拥有约定，因为遵循了约定，它就会知道源代码放在哪里了。它把字节码放到 target/classes，然后在 target 生成一个 jar 文件。

（5）Ant 是程序化的，必须要明确告诉 Ant 做什么，什么时候做。如必须要告诉它去编译，然后复制，再压缩。Maven 是声明式的，只需要创建一个 pom.xml 文件，然后将源代码放到默认的目录中，Maven 就会自动处理其他的事情。

（6）Ant 没有生命周期，必须要定义目标和目标之间的依赖，即手工为每个目标附上一个任务序列。Maven 有一个生命周期，当运行 mvn install 时被调用，这条命令会告诉 Maven 执行一系列的有序步骤，直至指定的生命周期。遍历生命周期中的一个影响就是，Maven 运行了许多默认的插件目标，这些目标完成了像编译和创建一个 jar 文件这样的工作。Maven 以插件的形式为一些项目任务提供了内置的智能。

（7）Maven 有大量的重用脚本可以利用，如生成网站，生成 javadoc、sourcecode reference 等。Ant 则需要自己去写。对于 Maven，有了 pom.xml，只要从命令行运行 mvn install，就会处理资源文件、编译源代码、运行单元测试、创建一个 jar，然后把这个 jar 安装到本地仓库以为其他项目提供重用性。不用做任何修改，就可以运行 mvn site，然后在 target/site 目录中找到一个 index.html 文件，这个文件链接了 JavaDoc 和一些关于源代码的报告。

（8）使用 Maven 还是 Ant 的决定不是非此即彼的，Ant 在复杂的构建中还是有其位置的。如果目前的构建包含一些高度自定义的过程，或者已经写了一些 Ant 脚本通过一种明确的方法完成一个明确的过程，而这种过程并不适合 Maven 标准时，仍然可以在 Maven 中用这些脚本。作为一个 Maven 的核心插件，Ant 还是可以用的。自定义的插件可以用 Ant 来实现，Maven 项目可以配置成在生命周期中运行 Ant 的脚本。

（9）Maven 不足的地方就是还没有像 Ant 那样成熟的 GUI 界面。目前使用 Maven 最好的方法就是命令行，实现起来又快又方便。

3．约定优于配置

约定优于配置（Convention Over Configuration）是一个简单的概念。系统、类库、框架应该假定合理的默认值，而非要求提供不必要的配置。Maven 通过给项目提供明智的默认行为来融合这个概念。

在没有自定义的情况下，源代码假定是在${basedir}/src/main/java 中，资源文件假定是在${basedir}/src/main/resources 中，测试代码假定是在${basedir}/src/test 中。

假定项目会产生一个 jar 文件。Maven 假定用户想要把编译好的字节码放到${basedir}/target/ classes 中，并且在${basedir}/target 中创建一个可分发的 jar 文件。

虽然这看起来无关紧要，但是大部分基于 Ant 的构建必须为每个子项目定义这些目录。Maven 对约定优于配置的应用不仅仅是简单的目录位置，它的核心插件还使用了一组通用的约定，以用来编译源代码、打包可分发的构件、生成 Web 站点，还有许多其他的过程。

Maven 有一个定义好的生命周期和一组知道如何构建和装配软件的通用插件。如果遵循这些约定，用户要做的仅仅是将源代码放到正确的目录下，Maven 就会帮用户处理剩下的事情。

使用“遵循约定优于配置”系统的一个副作用是，用户可能会觉得被强迫使用一种特殊的方法。当然 Maven 有一些核心观点不应该被怀疑，但是很多默认行为还是可配置的。如项目源码的资源文件的位置、jar 文件的名字，以及在开发自定义插件时，几乎任何行为都可以被裁剪以满足用户特定的环境需求。如果用户不想遵循约定，Maven 也会允许用户自定义默认值来适应自身的需求。

12.2　Maven 的安装和配置

1. 安装 JDK

在安装 Maven 之前，先确保已经安装 JDK1.6 及以上版本，并且已配置好环境变量。

2. 下载 Maven

在 http://maven.apache.org/download.cgi 下载 apache-maven-3.3.3-bin.zip 并解压，这里是解压到 D:\apache-maven-3.3.3。

3. 配置环境变量

配置 Maven3 的环境变量。

先配置 M2_HOME 的环境变量，新建一个系统变量 M2_HOME，路径是 D:\apache-maven-3.3.3，如图 12-1 所示。

再配置 path 环境变量，在 path 值的末尾添加“%M2_HOME%\bin”。

打开“cmd”窗口，输入“mvn -version”，出现如图 12-2 所示的内容表示安装成功。

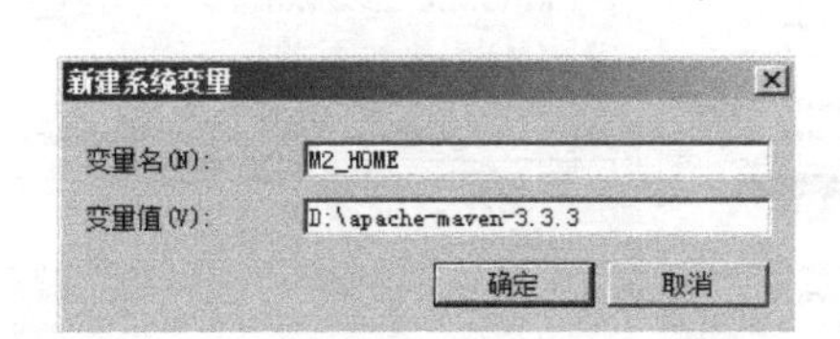

图 12-1　配置 M2_HOME 的环境变量

图 12-2　输入“mvn-version”

4. 给 Maven 添加本地仓库

默认情况下，不管是在 Windows 上还是在 Linux 上，每个用户在自己的用户目录下都有一个路径名为.m2/repository/的仓库目录。有时候，因为某些原因（如 C 盘空间不够），用户想要自定义本地仓库目录地址。这时，可以编辑文件~/.m2/settings.xml，设置 localRepository 元素的值为想要的仓库地址即可。这里“~”指代用户目录，如 C:\Users\Administrator。

此处本地仓库配置为：

```
<settings>
   <localRepository> D:\my_maven_repository\work\maven </localRepository>
</settings>
```

这样，该用户的本地仓库地址就被设置成 D:\my_maven_repository\work\maven。需要注意的是，默认情况下，~/.m2/settings.xml 文件是不存在的，用户需要从 Maven 安装目录复制$M2_HOME/conf/settings.xml 文件后，再进行编辑。

一个构件只有放在本地仓库中，才能由其他 Maven 项目使用，那么如何将构件放入到本地仓库中呢？最常见的是依赖 Maven 从远程仓库下载到本地仓库中。

还有一种常见的情况是，将本地项目的构件安装到 Maven 仓库中。如本地有两个项目 A 和 B，两者都无法从远程仓库获得，同时项目 A 又依赖于项目 B，为了能构建项目 A，项目 B 就必须首先得以构建并安装到本地仓库中。为此可以在项目中执行 mvn clean install 命令。Install 插件的 install 目标就会将项目的构建输出文件安装到本地仓库。

安装好 Maven 后，如果不执行任何 Maven 命令，本地仓库目录是不存在的。只有当用户输入第一条 Maven 命令之后，Maven 才会创建本地仓库，然后再根据配置和需要，从远程仓库下载构件至本地仓库。

由于最原始的本地仓库是空的，Maven 必须知道至少一个可用的远程仓库，才能在执行 Maven 命令时下载到需要的构件。中央仓库就是这样一个默认的远程仓库，Maven 的安装文件自带了中央仓库的配置。

私服是一种特殊的远程仓库，它是架设在局域网内的仓库服务，私服可代理广域网上的远程仓库，供局域网内的 Maven 用户使用。当 Maven 需要下载构件的时候，它向私服请求，如果私服不存在该构件，则从外部的远程仓库下载，缓存在私服之后，再为 Maven 的下载请求提供服务。此外，一些无法从外部仓库下载的构件也能从本地上传到私服供大家使用。

5. 配置用户范围 settings.xml

Maven 用户可以选择配置$M2_HOME/conf/settings.xml 或者~/.m2/settings.xml。前者是全局范围的，整台机器上的所有用户都会直接受到该配置的影响，而后者是用户范围的，只有当前用户才会受到该配置的影响，其具体配置如图 12-3 所示。

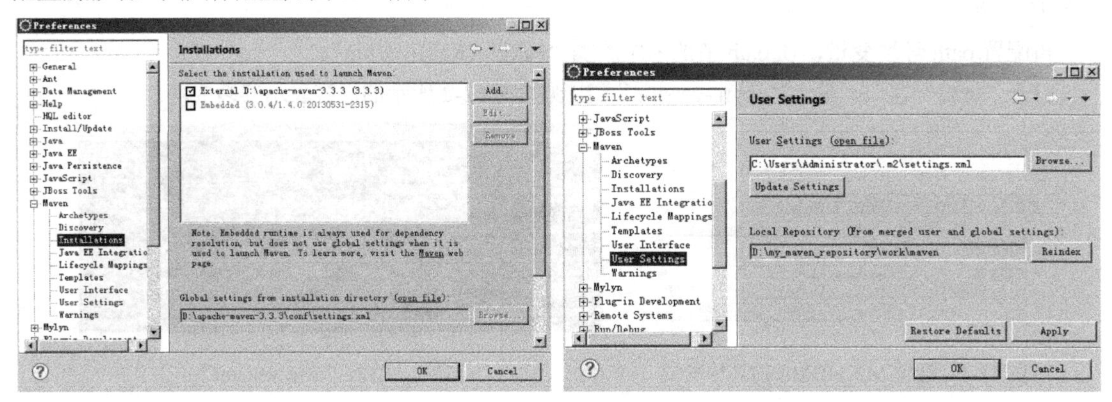

图 12-3　配置 settings.xml

6. 错误处理

在 Eclipse 使用 Maven 插件时，遇到运行 Run as→Maven clean 报错，其示例代码如下：

```
-Dmaven.multiModuleProjectDirectory system propery is not set.
Check $M2_HOME environment variable and mvn script match.
```

解决方法：选择 Window→Preferences→Java→Installed JREs，如图 12-4 所示。

点击“Edit”按钮，如图 12-5 所示，在 Default VM arguments 中设置：

```
-Dmaven.multiModuleProjectDirectory=$M2_HOME
```

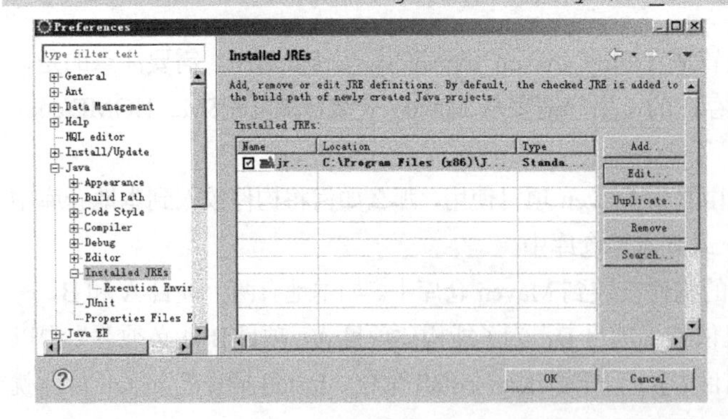

图 12-4　Installed JREs 对话框

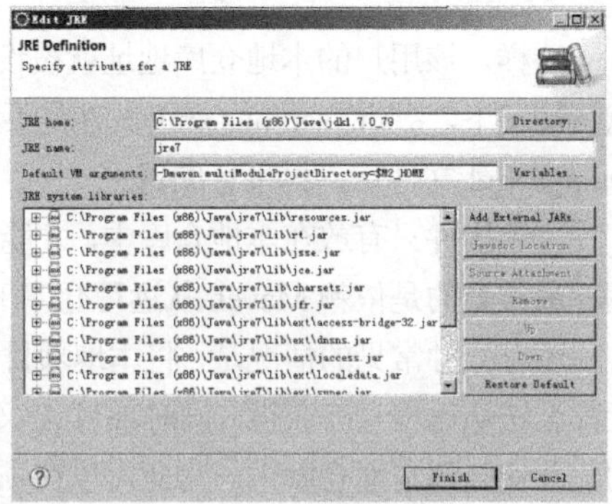

图 12-5　配置 JRE

7．设置 MAVEN_OPTS 环境变量

Maven 安装目录时，运行 mvn 命令实际上是执行了 Java 命令，那么运行 Java 命令可用的参数当然也应该在运行 mvn 命令时可用。这个时候，MAVEN_OPTS 环境变量就能派上用场了。

通常需要设置 MAVEN_OPTS 的值为-Xms128m -Xmx512m，因为 Java 默认的最大可用内存往往不能够满足 Maven 运行的需要，如在项目较大时，使用 Maven 生成项目站点就需要占用大量的内存，如果没有该配置，很容易得到 java.lang.OutOfMemeoryError 的提示。因此，一开始就进行配置该变量是推荐的做法。

关于如何设置环境变量，请参考相关设置 M2_HOME 环境变量的做法，尽量不要直接修改 mvn.bat 或者 mvn 这两个 Maven 执行脚本文件。因为如果修改了脚本文件，升级 Maven 时用户就不得不再次修改，一来麻烦，二来容易忘记。同理，应尽可能地不去修改任何 Maven 安装目录下的文件。

关于 Maven 中 JVM 参数的设置：

方法 1：在系统的环境变量中，设置 MAVEN_OPTS，用以存放 JVM 的参数，其具体设置的步骤，参数示例如下：

```
MAVEN_OPTS=-Xms256m -Xmx768m -XX:PermSize=128m -XX:MaxPermSize=256M
```

方法 2：找到 Maven 的安装目录，在 bin 目录下，编辑 mvn.bat（Linux 下的 mvn.sh），但这种方法不推荐。

12.3　Maven 的使用

1．生成项目

（1）新建一个 JAVA 项目：

```
mvn  archetype:generate  -DgroupId=com.mialab  -DartifactId=Exercise
-DarchetypeArtifactId=maven-archetype-quickstart  -Dversion=1.0
```

（2）新建一个 Web 项目：

```
mvn  archetype:generate  -DgroupId=com.mialab  -DartifactId=PalmSuda
-DarchetypeArtifactId=maven-archetype-webapp  -Dversion=1.0
```

其中相关参数如下。

- archetype:create：表示 archetype 插件的 create 目标。在 Maven 中一个插件可以拥有多个目标。
- archetype：表示一个内建插件，它的 create 任务将建立项目骨架。
- archetypeArtifactId：表示项目骨架的类型。
- groupId：表示项目的 Java 包结构，可修改。
- artifactId：表示项目的名称，生成的项目目录也是这个名字，可修改。
- Version：表示项目的版本。

2．Maven 常用命令

常用命令如下。

- mvn archetype:create：创建 Maven 项目。
- mvn compile：编译源代码。
- mvn test-compile：编译测试代码。
- mvn test：运行应用程序中的单元测试。
- mvn site：生成项目相关信息的网站。
- mvn clean：清除目标目录中的生成结果。

- mvn package：依据项目生成 jar 文件。
- mvn install：在本地 Repository 中安装 jar 文件。
- mvn deploy：将 jar 包发布到远程仓库。
- mvn eclipse:eclipse：生成 Eclipse 项目文件。

3．Maven 常用项目骨架

如果是 Maven3，可简单运行：

```
mvn archetype:generate
```

（1）internal→appfuse-basic-spring：创建一个基于 Hibernate、Spring 和 Spring MVC 的 Web 应用程序的原型。

（2）internal→maven-archetype-webapp：一个简单的 Java Web 应用程序。

（3）internal→maven-archetype-j2ee-simple：一个简单的 J2EE 的 Java 应用程序。

（4）internal→maven-archetype-mojo：一个 Maven 的 Java 插件开发项目。

（5）internal→maven-archetype-site-simple：一个简单的网站生成项目。

12.4 坐标和依赖

1．坐标

maven 的所有构件均通过坐标进行组织和管理。它的坐标通过 5 个元素进行定义，其中 groupId、artifactId、version 是必须的，packaging 是可选的（默认为 jar），classifier 是不能直接定义的。

- groupId：定义当前 Maven 项目所属的实际项目，跟 Java 包名类似，通常与域名反向一一对应。
- artifactId：定义当前 Maven 项目的一个模块。
- version：定义项目版本。
- packaging：定义项目打包方式，如 jar、war、pom、zip 等，默认为 jar。
- classifier：定义项目的附属构件，如 hibernate-core-3.6.6.Final-sources.jar、hibernate-core-3.6.6.Final-javadoc.jar，其中，sources 和 javadoc 就是这两个附属构件的 classifier。classifier 不能直接定义，通常由附加的插件帮助生成。

2．依赖

使用 Maven 可以方便地管理依赖，在 pom.xml 文件中声明依赖的示例代码如下：

```
<dependencies>
    <dependency>
        <groupId>org.springframework</groupId>
        <artifactId>spring-test</artifactId>
        <version>3.2.0.RELEASE</version>
        <type>jar</type>
        <scope>test</scope>
        <systemPath>${java.home}/lib/rt.jar</systemPath>
        <optional>false</optional>
        <exclusions>
            <exclusion></exclusion>
        </exclusions>
    </dependency>
</dependencies>
```

其中相关参数如下。

- type：指依赖类型，对应构件中定义的 packaging，可不声明，默认为 jar。
- scope：指依赖范围。
- optional：指依赖是否可选。
- exclusions：指排除传递依赖。

执行不同的 Maven 命令（mvn package、mvn test、mvn install 等），会使用不同的 classpath，Maven 对应的有三套 classpath：编译 classpath、测试 classpath 和运行 classpath。scope 选项的值，决定了该依赖构件会被引入到哪一个 classpath 中。

- compile：指编译依赖范围，默认值。此选项对编译、测试、运行三种 classpath 都有效，如 hibernate-core-3.6.5.Final.jar，表明在编译、测试、运行时都需要该依赖。
- test：指测试依赖范围。只对测试有效，在编译和运行时将无法使用该类依赖，如 junit。
- provided：指已提供依赖范围。编译和测试有效，运行无效。如 servlet-api，在项目运行时，tomcat 等容器已经提供，无须 Maven 重复引入。
- runtime：指运行时依赖范围。测试和运行有效，编译无效。如 jdbc 驱动实现，编译时只需接口，测试或运行时才需要具体的 jdbc 驱动实现。
- system：指系统依赖范围。同 provided 依赖范围一致，需要通过<systemPath>显示指定，且可以引用环境变量。
- import：指导入依赖范围。使用该选项，通常需要<type>pom</type>，将目标 pom 的 dependencyManagement 配置导入合并到当前 pom 的 dependencyManagement 元素中。

12.5 构建支持 Servlet 3.0 的 Maven Web 应用

1．创建工程 mavenweb

在 Eclipse 中创建 Maven Project，如图 12-6 所示，选择“maven-archetype-webapp”选项，在 Group ID 中输入“com.mialab.app”，在 Artifact ID 中输入“mavenweb”。

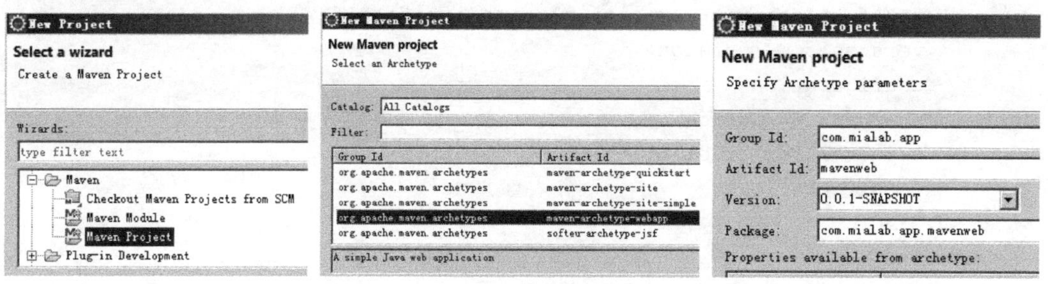

图 12-6　在 Eclipse 中创建 maven-archetype-webapp

此时，mavenweb 工程默认生成的 web.xml 内容如下。

```
<!DOCTYPE web-app PUBLIC "-//Sun Microsystems, Inc.//DTD Web Application 2.3//EN"
        "http://java.sun.com/dtd/web-app_2_3.dtd" >
<web-app>
        <display-name>Archetype Created Web Application</display-name>
</web-app>
```

2．完善 src 目录结构

mavenweb 工程默认是没有 java 目录和 test 目录的，如图 12-7 所示，必须补全。

因此在 src/main 目录下创建 Java 文件夹，再在 src 目录下创建 test 文件夹，src/test 文件夹下创建 java 和 resources 文件夹，如图 12-8 所示。

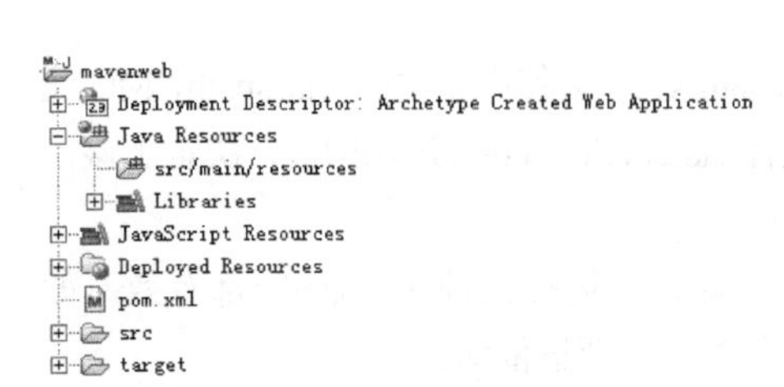

图 12-7　mavenweb 工程默认目录

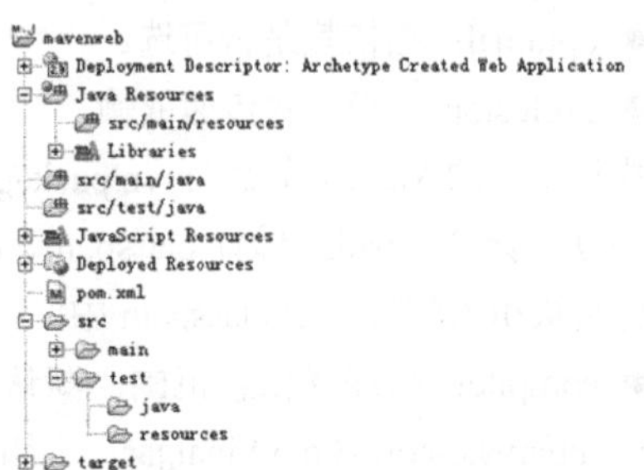

图 12-8　mavenweb 工程添加 java 目录和 test 目录

3．修改 WebApp 的版本

mavenweb 工程默认创建的 WebApp 的版本为 2.3，这里需要修改成支持 Servlet3.0 的 WebApp3.0。在工作区文件夹中找到 mavenweb 工程，编辑/mavenweb/.settings 文件夹下的 org.eclipse.wst.common.project.facet.core.xml，如图 12-9 所示。

```
1  <?xml version="1.0" encoding="UTF-8"?>
2  <faceted-project>
3    <fixed facet="wst.jsdt.web"/>
4    <installed facet="java" version="1.5"/>
5    <installed facet="jst.web" version="2.3"/>
6    <installed facet="wst.jsdt.web" version="1.0"/>
7  </faceted-project>
```

图 12-9　编辑 org.eclipse.wst.common.project.facet.core.xml 文件

（1）将<installed facet="java" version="1.5"/> 和<installed facet="jst.web" version="2.3"/>分别改为<installed facet="java" version="1.6"/>和<installed facet="jst.web" version="3.0"/>。

再重新启动 Eclipse，打开项目的 pom.xml 文件，在<build>节点中添加配置如下。

```
<plugins>
    <plugin>
      <artifactId>maven-compiler-plugin</artifactId>
      <version>3.0</version>
      <configuration>
      <source>1.6</source>
      <target>1.6</target>
      </configuration>
    </plugin>
</plugins>
```

保存更改后，在 Problems 标签中会出现错误提示：

```
Project configuration is not up-to-date with pom.xml.
Run Maven->Update Project or use Quick Fix.
```

（2）选中 mavenweb 工程，右击 mavenweb 工程的快捷菜单，选择“Maven”→“Update Project”，更新项目。

然后在 mavenweb 工程 Properties 对话框左侧选择“Project Facets”选项，将对话框右侧的 Dynamic Web Module 改为 3.0，Java 改为 1.6，最右边 Runtimes 选择 Tomcat7.0（Servlet3.0 需要改为 Tomcat7.0+），如图 12-10 所示。

最后删除 src/main/webapp/WEB-INF/web.xml，再次右击“mavenweb 工程”→“Maven”→“Update Project”，更新项目。这时错误就会全消失了。

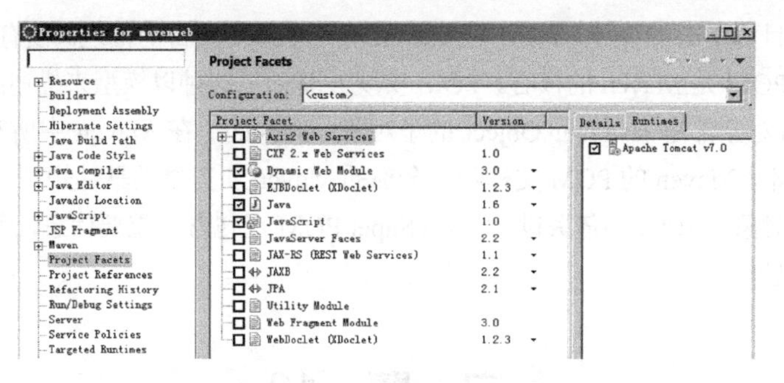

图 12-10　配置 Project Facets

（3）这时再在 Eclipse 中建立一个 Dynamic Web Project，如名称为 Hello，选择 Dynamic web module version 为 3.0。Hello 工程创建成功后把所得到的 web.xml 文件复制到 WEB-INF 目录下即可。这时就可顺利运行 index.jsp 了，如图 12-11 所示。

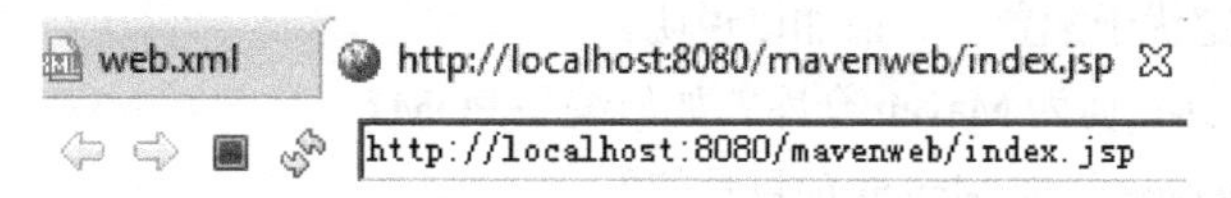

Hello World!

图 12-11　运行 index.jsp

4．创建 Servlet 并测试

如图 12-12 和图 12-13 所示，创建 HelloMavenServlet，并进行测试。

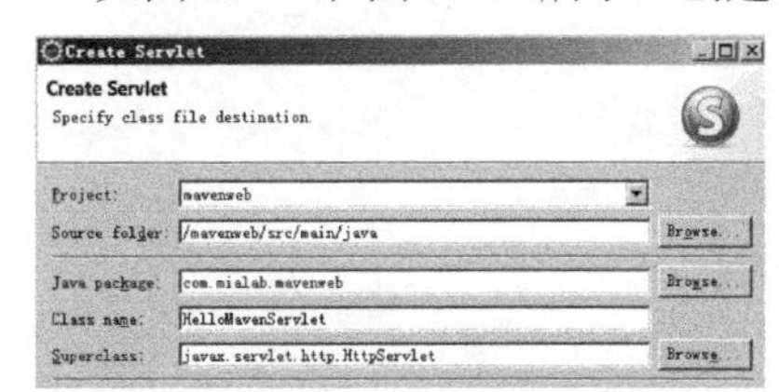

图 12-12　创建 HelloMavenServlet

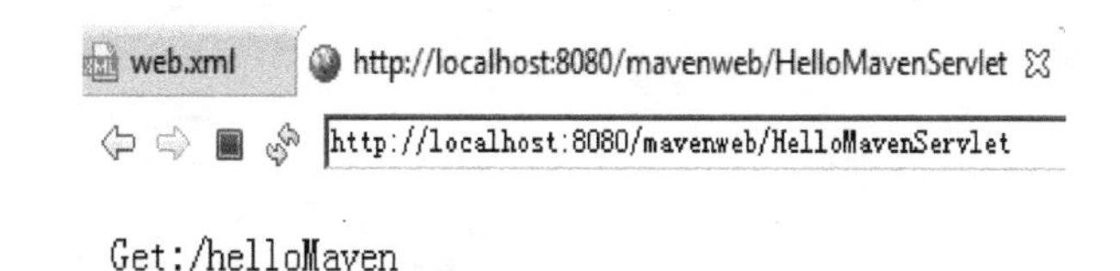

图 12-13　在浏览器中调用 HelloMavenServlet

HelloMavenServlet.java 主要代码如下：

```
public class HelloMavenServlet extends HttpServlet {
    protected void doGet(HttpServletRequest request, HttpServletResponse response)
            throws ServletException, IOException {
        response.getWriter().println("Get:/helloMaven");
    }
}
```

12.6　本章小结

Maven 提供了一套软件项目管理的综合性方案。无论是编译、发布、文档还是团队协作，Maven 都提供了必要的抽象，它鼓励重用，并做了除软件构建外的许多工作。

Maven 为构建、测试、部署项目定义了一个标准的生命周期。它提供了一个框架，允许遵循 Maven 标准的所有项目，可方便地重用公用的构建逻辑。Maven 项目存在的 Apache 软件基金会，是一个开源社区。它开发的软件工具，基于一个通用的软件对象模型（Project Object Model），也就是 POM。

Maven 是以项目为中心的设计。POM 是 Maven 对一个单一项目的描述。没有 POM 的话，Maven 是毫无用处的——POM 是 Maven 的核心。POM 实现并驱动了这种以模型来描述的构建方式。

在 Java 里所有对象都是 java.lang.Object 的子对象，同样的，在 Maven 中所有 POM 也都是一个 Super POM 的子脚本。Maven 的 POM 文档带有全部的 Maven 所鼓励的默认规则，这类似于 Java 语言的 java.lang.Object 对象。要记住的最关键一点是，Super POM 已包含了重要信息，因此在创建一个 POM 时不必重复这些信息。

习 题 12

1．Maven 与 Ant 的区别是什么？

2．如何安装配置 Maven？请简要加以说明。

3．如何理解“约定优于配置”？请加以说明。

4．何为 Maven 坐标？何为 Maven 仓库？如何编写 POM？

5．如何使用 Archetype 生成 Maven 项目骨架？

6．如何使用 Maven 构建支持 Servlet 3.0 的 Web 应用？

第 13 章　服务器端接口编程

如果将系统功能按前/后台来划分，前台即客户端，主要实现 UI 部分，负责和用户交互；后台即服务器端，负责根据用户请求进行业务处理，并与底层数据库进行交互，然后将数据返回前台。前/后台之间通过统一标准的数据交互协作实现整个系统的功能。本章服务器端接口编程主要介绍如何生成前/台所需的数据，即如何获取请求，怎样与数据库交互，最后采取何种方式返回数据给前台。

本章讲述服务器端接口编程相关技术知识，主要知识点有：（1）JNDI 数据源的基本概念，如何配置、获取 JNDI 数据源；（2）记录日志的主流方法 Log4J、SLF4J；（3）利用 MyBatis 框架进行接口编程；（4）比较两个主流的 ORM 框架 Mybatis 和 Hibernate。

13.1　JNDI 数据源配置

在 Java 开发中，通过 JDBC 的 API 可建立数据库连接、执行 SQL 语句进行数据的存取操作。常用的 JDBC 连接数据库代码如下：

```
Connection conn=null;
try {
    Class.forName("com.mysql.jdbc.Driver"); //加载 MySQL 驱动类
} catch (ClassNotFoundException e) {
    System.out.println("加载失败，找不到驱动器类！");
    e.printStackTrace();
}
//连接数据库的地址、账号、密码
String url="jdbc:mysql://127.0.0.1:3306/market";
String username="root";
String password="123";
try {
    //请求获取数据库连接对象
    conn=DriverManager.getConnection(url, username, password);
} catch (SQLException e) {
    System.out.println("数据库连接失败！");
    e.printStackTrace();
}
if(conn!=null){
    System.out.println("数据库连接成功，可以操作数据库！");
}
else{
    System.out.println("数据库连接失败，无法操作数据库！");
}
```

上述代码显示了 JDBC 连接数据库的基本步骤：通过 JDBC 先获取数据库驱动程序，然后根据数据库的地址、账号、密码创建数据库连接。当一个程序需要访问数据库时，就需要按照上述步骤连接数据库，但其开发效率较低。另外，一旦数据库连接信息发生变化，则需修改所有连接该数据库的地方。因此，是否有一种方法可以把对数据库访问的信息封装在一个对象（数据源）中，一个程序可以建立多个数据库连接，保存在数据库连接池中，程序访问数据库时，只需要从连接池中获取数据库连接即可，访问结束，释放连接，这比用 JDBC 直接访问数据库要更节省时间和资源。JNDI 就可以实现上述功能。

1．JNDI 简介

JNDI（Java Naming and Directory Interface）是 Java 命名和目录接口。它是 SUN 公司提供的一种标准的 Java 命名系统接口，JNDI 提供统一的客户端 API，通过不同的访问提供者接口 JNDI 服务供应接口（SPI）的实现，由管理者将 JNDI API 映射为特定的命名服务和目录系统，使得 Java 应用程序可以和这些命名服务和目录服务之间进行交互。目录服务是命名服务的一种自然扩展。

2．利用 JNDI 访问数据库

利用 JNDI 访问数据库主要分两个步骤：配置数据源和访问配置信息获取数据库连接。

（1）配置数据源。

根据配置的作用域不同，JNDI 配置数据库有三种方式。

① 仅供单个应用。这种配置方式是在服务器 Tomcat 的 server.xml 文件中，该项目的<Context>节点下添加数据源，该方式配置的数据源为某个应用所独享，其重用性较差。

具体配置代码如下：

```
<Context docBase="JNDI" path="/JNDI" reloadable="true"
 source="org.eclipse.jst.jee.server:JNDI">
<Resource name="jdbc/MarketDB01"
        auth="Container"
        type="javax.sql.DataSource"
        factory="org.apache.tomcat.dbcp.dbcp.BasicDataSourceFactory"
        driverClassName ="com.mysql.jdbc.Driver"
        url="jdbc:mysql://127.0.0.1:3306/market"
        username="root"
        password="123"
 />
</Context>
```

其中，<Resource>节点基本属性含义如下。

- name：表示 JNDI 名称，供应用程序调用。
- auth：表示管理数据源的方式，它的两个值为 Container 和 Application。通常设置为 Container，表示由容器进行创建和管理数据源，而 Application 则表示由 Web 应用进行创建和管理。
- type：表示数据源所属的类型，使用标准的 javax.sql.DataSource。
- factory：表示生成数据源的工厂类名。
- driverClassName：表示 JDBC 数据库驱动器。
- url：表示数据库 URL 地址。
- username：表示访问数据库所需用户名。
- password：表示数据库用户的密码。

② 配置全局数据源供单个应用程序使用。

与第一种方式不同的是，这种配置方式的数据源信息是添加在服务器 Tomcat 的 server.xml 文件中< GlobalNamingResources>节点下，而不是在项目的<Context>节点下。项目如需使用该数据源，只需在项目的<Context>节点下添加引用即可，其代码如下：

```
<Context docBase="JNDI" path="/JNDI" reloadable="true" source="org.eclipse.jst.
jee.server:JNDI">
    <ResourceLink global="jdbc/MarketDB02" name="jdbc/MarketDB02" type="javax.
sql.DataSource" />
</Context>
```

③ 配置全局数据源供所有应用使用。这种配置方式是在服务器 Tomcat 的 context.xml 文件中<Context>节点下添加一个数据源，数据源的配置方式与前面两种方式相同。

（2）获取数据库连接。

获取数据库连接的步骤如下：①初始化 JNDI 的上下文环境；②获取数据源；③获取数据库连接对象，其具体实现代码如下：

```
Connection conn=null;
DataSource ds=null;
try {
   Context ctx= new InitialContext();
   ds=(DataSource)ctx.lookup("java:comp/env/jdbc/MarketDB01");;
} catch (NamingException e) {
   e.printStackTrace();
}
try {
   conn=ds.getConnection();
} catch (SQLException e) {
   e.printStackTrace();
}
```

上述代码中，利用 InitialContext()方法初始化 JNDI 上下文环境，调用 Context.lookup(String name)方法来根据 name 查找并获得数据源，最后调用 DataSource.getConnection()方法根据数据源信息获取数据库连接对象，从而获得了访问数据库的许可，接下来即可实现对数据库的访问操作。

13.2 Log4J 与 SLF4J

Log4J 是一个日志信息输出的开源项目，SLF4J 为简单日志门面，它并不直接控制日志信息的输出，而是通过为各种日志系统提供一个统一接口，使最终用户能够在具体应用中使用其所希望的日志系统。

1. Log4J

通过 Log4J 提供的接口，用户可以控制日志信息输出的目的地，如控制台、文件等。可以通过设置日志信息等级从而更好地控制日志生成过程，还可以控制日志信息的输出格式。

（1）Log4J 配置。

Log4J 的配置主要包含 3 个部分。

① 根 Logger。

Logger 负责记录日志的大部分操作，包括定义日志等级及输出目的地 Appender 名称。

Logger 为日志划分了 7 个等级，通常使用 5 个等级，按优先级从高到低依次如下。

- FATAL：会导致应用程序退出的严重错误事件。
- ERROR：程序中有错误事件，但不影响程序的继续运行。
- WARN：警告，表示程序存在潜在的错误风险。
- INFO：强调程序运行过程中的信息提示，类似于 System.out.println 的功能。
- DEBUG：与 INFO 类似，主要在调试程序时使用。

Logger 通过等级设置控制应用程序中相应级别的日志信息，具体控制方式是只有等于及高于设置的级别才执行日志信息的输出。如设置了日志等级为 WARN，则程序中只有 ERROR 级别的日志信息才会被打印。

Logger 通过 Appender 名称与对应输出源（Appender）相关联，一个 Logger 可以有多个输出源。

② 日志信息输出目的地 Appender。

Appender 主要负责控制日志信息的输出目的地，主要有以下 5 种情况。

- ConsoleAppender：控制台。
- FileAppender：文件。
- DailyRollingFileAppender：每天产生一个日志文件。
- RollingFileAppender：文件大小到达指定尺寸时产生一个新的文件。
- WriteAppender：将日志信息以流格式发送到任意指定的地方。

③ 日志信息的输出格式 Layout。

输出格式主要包含两个部分：信息的布局方式和信息的格式化显示。

信息的布局方式有如下几种。

- HTMLLayout：以 HTML 表格形式布局。
- PatternLayout：采用指定布局模式，灵活地格式化输出日志信息。
- SimpleLayout：采用简单的方式输出日志，包含日志信息的级别和信息字符串。
- TTCCLayout：包含日志产生的时间、线程、类别等信息。

信息的格式化显示类似于 C 语言中格式化输出函数 printf，通过格式字符描述待输出的文字样式，Log4J 中提供的格式符有如下几个。

- %m：输出代码中指定的消息，如 log.error(message)中的 message。
- %M：输出打印该条日志的方法名。
- %p：输出优先级，即 DEBUG、INFO、WARN、ERROR、FATAL。
- %r：输出自应用启动到输出该 log 信息耗费的毫秒数。
- %c：输出所属的类目，通常就是所在类的全名。
- %n：输出一个回车换行符，Windows 平台为“rn”，UNIX 平台为“n”。
- %d：输出日志时间点的日期或时间，默认格式为 ISO8601，也可以在其后指定格式，如 %d{yyyy-MM-dd HH:mm:ss,SSS}，输出类似于 2015-07-05 10:20:20。

记录 Log4J 配置信息的文件有两种，一种是 XML 格式文件，另一种是 Java 特性文件 log4j.properties，通常采用第二种配置文件，本文详细介绍了第二种配置文件，并给出基本的 XML 配置文件示例。

根据上面介绍 Log4J 的 3 个部分，log4j.properties 文件中对应的语法结构如下：

① Logger。

```
log4j.rootLogger = [ level ] ,[appenderName], [appenderName], …
```

其中，level 表示日志等级，appenderName 表示 Appender 名称，通过该名称对应输出目的地。可同时指定多个输出目的地。

② Appender。

```
log4j.appender. appenderName=[destination]
```

其中，appenderName 为 Logger 中定义的名称，destination 表示目的地，值为上文介绍的 5 类目的地，如向控制台打印日志，则值为 org.apache.log4j.ConsoleAppender。

③ Layout。

```
log4j.appender.appenderName.layout=[layoutmethod] | [format]
```

其中，layoutmethod 为布局方式，format 为格式化字符串。

【示例】 使用 Log4J 输出日志，相关内容详见教学资源包第 13 章 LogDemo 工程。

创建工程，命名为 LogDemo，默认包名为 com.mialab.log。Log4J 采用 log4j.properties 文件描述，其定义代码如下：

```
log4j.rootLogger=INFO,logConsole,logFile
log4j.appender.logConsole=org.apache.log4j.ConsoleAppender
log4j.appender.logConsole.layout=org.apache.log4j.PatternLayout
log4j.appender.logConsole.layout.ConversionPattern=%c %d{yyyy-MM-dd} --> %m%n
log4j.appender.logFile=org.apache.log4j.DailyRollingFileAppender
log4j.appender.logFile.File=c:/log.log
log4j.appender.logFile.layout=org.apache.log4j.SimpleLayout
```

上述代码，定义了两个日志输出目的地，分为命名为 logConsole 和 logFile。logConsole 采用向控制台打印日志的信息，并根据 PatternLayout 的布局方式布局日志信息，每一条日志信息采用%c %d{yyyy-MM-dd}→%m%n 的方式格式化输出。LogFile 采用将日志信息储存在日志文件（c:/log.log）的方式，日志信息采用 SimpleLayout 的布局方式。

（2）Java 利用 Log4J 接口输出日志。

在 Java 项目中如需只用 Log4J 控制日志，首先应该导入 Log4J 对应的 jar 包，Log4J 的 jar 包可以在官网“http://logging.apache.org/log4j/2.x/download.html”中下载，此处选用的是“log4j-1.2.17.jar”版本的 jar 包。

Log4J 打印日志的基本步骤如下：

① Logger 对象的创建。通过 getLogger()方法获取 Logger 对象，其代码如下：

```
public static Logger Logger.getLogger(Class clazz)
```

② 打印日志。根据日志等级，调用相应的方法打印日志，其代码如下：

```
(Logger)log.xxx(String message)
```

其中，xxx 表示输出的日志等级，如输出的是 DEBUG 等级的日志，则 xxx 为 debug。

具体实现代码如下：

```
public class log4j {
   private static Logger logger= Logger.getLogger(log4j.class.getName());
   public static void main(String[] args) {
        logger.fatal("fatal 级别的信息");
        logger.error("error 级别的信息");
        logger.warn("warn 级别的信息");
        logger.info("info 级别的信息");
        logger.debug("debug 级别的信息");
   }
}
```

根据【示例】设置的配置文件，logConsole 向控制台打印日志效果如图 13-1 所示，logFile 将日志信息保存在 C:/log.log 日志文件中，效果如图 13-2 所示。

```
<terminated> log4j [Java Application] C:\Program Files
org.suda.app.log4j 2015-07-03 --> fatal级别的信息
org.suda.app.log4j 2015-07-03 --> error级别的信息
org.suda.app.log4j 2015-07-03 --> warn级别的信息
org.suda.app.log4j 2015-07-03 --> info级别的信息
```

图 13-1　Console 控制台输出

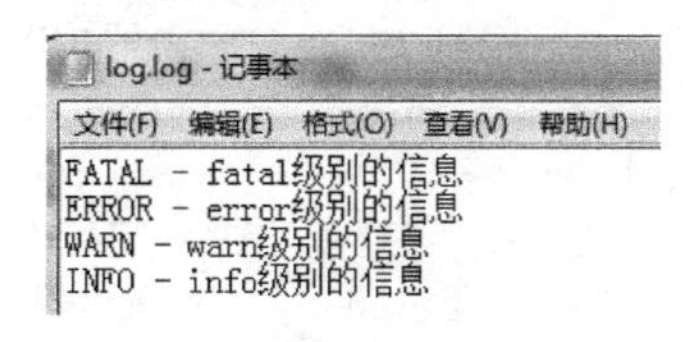

图 13-2　日志文件 log.log

从图 13-1 和图 13-2 中发现 Java 程序的 DEBUG 等级信息并没有被打印，这是由于在配置文件中定义的日志输出等级为INFO，因此，低于INFO 等级的DEBUG 日志信息无法显示，但该条打印DEBUG日志的语句仍被会执行，因此出于运行效率的考量，在该语句前添加了一个条件判断

logger.isDebugEnabled()，其代码可进行如下修改：

```
if(logger.isDebugEnabled()){
   logger.debug("debug 级别的信息");
}
```

2. SLF4J

SLF4J（Simple Logging Facade for Java，简单日志门面），它并不是具体的日志系统，而是一个日志系统的抽象，服务于其他日志系统。程序开发过程中，统一按照 SLF4J 标准进行开发，无须考虑用户采用的是哪种日志系统，只需要在部署的时候，选择对应日志系统的程序包，即可自动转换到不同的日志系统上。

与 Log4J 相比，除了上面提到的 SLF4J 独立于任意具体的日志系统，无须根据采用不同的日志系统而修改程序这一个优点。SLF4J 还有另一大特色，即可在程序中格式化日志。SLF4J 定义了一组格式占位符“{}”，该占位符的作用类似于 C 语言 printf 函数中的格式符，在运行时可被某个具体的字符串所替换。这不仅降低了代码中字符串拼接的次数，还提高了日志字符串的可读性，而且在运行时延迟字符串 String 的建立（只有需要 String 对象时建立）从而节省了新建的 String 对象，减少内存的消耗。如下面两条语句，分别采用 Log4J 和 SLF4J 打印日志信息：

```
logger.debug("时间: " + time + " 内容: " + content); //采用 Log4J 打印日志信息
logger.debug("时间:{} 内容:{} ", time, content);    //采用 SLF4J 打印日志信息
```

从上面代码可以直观地看出，采用 SLF4J 打印日志，代码更为简洁，日志信息输出字符串阅读也更为容易。根据不同的日志系统，在部署的时候，除了 SLF4J 的核心接口包，还需要具体日志系统相关的 jar 包，本文采用 Log4J 日志系统，因此，需要 log4j-1.2.17.jar、slf4j-api-1.7.12.jar、slf4j-log4j12-1.7.12.jar 三个程序包。SLF4J 相关的 jar 包可以到 SLF4J 官网 http://www.slf4j.org 中进行下载。

使用 SLF4J 打印日志的实现步骤与 Log4J 基本相似，但创建 Logger 对象方法不一样，SLF4J 创建 Logger 对象代码如下：

```
Logger LoggerFactory.getLogger(Class clazz)
```

【示例】 定义一个华氏温度转换为摄氏温度的方法，并使用 SLF4J 输出转换结果，相关内容参见教学资源包中的第 13 章 LogDemo 工程。

本例仍使用前面的 LogDemo 工程，将 SLF4J 所需的 jar 包：slf4j-api-1.7.12.jar 和 slf4j-log4j12-1.7.12.jar，导入到 LogDemo 工程的 lib 文件夹。创建 slf4j.java 的具体实现代码如下：

```
public class slf4j {
   final Logger logger=(Logger) LoggerFactory.getLogger(slf4j.class);
    public void TransferFC(Integer FTemp) {
        if(FTemp.intValue()>=0 && FTemp.intValue()<=100){
           Integer CTemp=(int)((FTemp-32)/1.8);
           logger.info( "华氏温度:{}度==>摄氏温度:{}",FTemp,CTemp);
         }else{
           logger.error("华氏温度:{}度超出范围! ",FTemp);
         }
    }
   public static void main(String[] args) {
      slf4j slj =  new slf4j();
      slj.TransferFC(50);
      slj.TransferFC(-10);
   }
}
```

在使用前面Log4J中定义的日志配置文件情况下，运行上述程序，控制台打印的日志信息如图13-3所示。

```
<terminated> slf4j [Java Application] C:\Program Files (x86)\Java\jre7\bin\javaw.exe (2015年7月4日 上午1:45:58)
org.suda.app.slf4j 2015-07-04 --> 华氏温度：50度==>摄氏温度：10
org.suda.app.slf4j 2015-07-04 --> 华氏温度：-10度超出范围！
```

图13-3 使用SLF4J向控制台打印日志

13.3 项目实战："移动商城"（四）

13.3.1 任务说明

本项目主要负责实现"移动商城"的后台接口编程。

商品数据表market_broadband_info结构如表13-1所示。

表13-1 商品数据表结构

字 段	类 型	说 明
id	integer	主键，商品ID
subject	character(50)	标题
body	character(1000)	内容
provider_id	integer	供应商ID
buy_price	real	购买价
sell_price	real	卖价
discount_price	real	折扣价
stock	integer	库存
img	character(255)	商品图片
package_category	integer	宽带产品分类（0-单产品包月，1-单产品包年，2-融合宽带）
owner	integer	收款方
type	integer	产品类型
start_time	timestamp	产品有效开始时间
end_time	timestamp	产品有效终止时间
create_time	timestamp	产品创建时间
last_modify_time	timestamp	产品最新修改时间
tag	smallint	备用
img_url	character (255)	列表小图标
city_id	integer	城市ID

13.3.2 项目讲解

1. 总体框架

MyBatis的功能架构可分为API接口层、数据处理层和基础支撑层，因此本项目在第12章建立的Maven Web项目基础上，建立如图13-3所示的项目结构，相关内容参见教学资源包中的第13章mialab-3gmarket-demo工程。

```
mialab-3gmarket-demo
├─ JAX-WS Web Services
├─ Deployment Descriptor: mialab-3gmarket-demo
└─ Java Resources
   └─ src/main/java
      ├─ org.suda.app
      ├─ org.suda.app.common
      ├─ org.suda.app.common.json
      ├─ org.suda.app.common.mgr
      ├─ org.suda.app.common.util
      ├─ org.suda.app.db
      ├─ org.suda.app.market
      ├─ org.suda.app.market.persistence
      ├─ org.suda.app.market.vo
      └─ org.suda.app.mgr
```

图 13-3　mialab-3gmarket-demo 工程（源码）目录

其各个包的作用如下。

（1）org.suda.app.common：存放通用的功能类文件。

（2）org.suda.app.db：基础支撑层，存放配置文件及数据库访问类。

（3）org.suda.app.market.persistence：数据持久层，放置数据访问接口及映射文件。

（4）org.suda.app.market.vo：POJO 实体类。

（5）org.suda.app. mgr：业务层。

2．程序主要流程

程序的入口是 MarketServlet 的 doGetPost()方法。首先请求是否包含 cmd 参数，若无则执行打印；若有则调用 RequestManager 的 getManager()方法获取 MarketMgr 对象。getManager()方法是调用 MarketMgr 的 getInstance()方法来获取 MarketMgr 对象实例。MarketMgr 继承了 AbstractMgr，而 AbstractMgr 实现了接口 IMgr，因此，MarketMgr 实现了 IMgr 中定义的接口方法 processRequest()。因此，，获取到 IMgr 接口对象后，执行 IMgr 接口对象的 processRequest()方法来处理请求。

processRequest()方法主要实现根据请求参数 cmd 的值执行相应的业务，本例中执行的是调用 writeResponse()方法在浏览器中打印 showProductInfo()方法返回的 JSON 字符串。writeResponse()方法是在 AbstractMgr 中定义的，用于利用服务器返回的响应对象向页面打印响应信息。showProductInfo()使用 MyBatis 获取查询得到的商品信息。

13.3.3　典型代码及技术要点

1．通用功能包的类实现

common 包主要有以下几个功能类。

（1）JSONObject.java：它是一个接口，定义了转换为 JSON 字符串的方法 toJSONString()。

（2）TextUtils.java：定义了判断字符串是否为空的方法 isNotEmpty()。

（3）IMgr.java 和 AbstractMgr.java：IMgr 为一个接口，定义了 processRequest()方法。AbstractMgr 是实现 IMgr 接口的抽象类，该类定义了 writeResponse()方法。

（4）CommonConstants.java：该类用于存储常量。

（5）ResponseMessage、ResponseSuccessMessage、ResponseErrorMessage：这三个类均是用于操作结果消息的类。其中，ResponseMessage 封装一个操作消息所需要的数据成员，而 ResponseSuccessMessage、ResponseErrorMessage 继承了 ResponseMessage，并分别定义了用于返回操作成功和操作失败的方法。

2．Action 层

MarketServlet.java：M 无论是采用 GET 请求还是 POST 请求，均调用 doGetPost()方法。该方法根

据 HTTP 请求，判断是否有请求参数 CommonConstants.PARAMS_CMD。若无则向页面执行打印。如果存在，则通过 RequestManager 类中 getManager()方法获取 MarketMgr 对象实例。若返回的是空对象，则打印错误日志；若返回的是非空对象，则调用 processRequest ()方法，将 JSON 字符串打印到返回页面，其主要代码如下：

```
    public class MarketServlet extends HttpServlet {
        …
        public void doGet(HttpServletRequest request, HttpServletResponse response)
             throws ServletException, IOException {
            doGetPost(request, response);
        }
        public void doPost(HttpServletRequest request, HttpServletResponse response)
                throws ServletException, IOException {
            doGetPost(request, response);
        }
        private void doGetPost(HttpServletRequest request, HttpServletResponse
response)
             throws ServletException,IOException {
            response.setContentType("text/html; charset=UTF-8");
            String command = request.getParameter(CommonConstants.PARAMS_CMD);
            PrintWriter out = response.getWriter();
            if (command == null) {
                out.println("ok");
                out.flush();
                out.close();
            } else {
                IMgr mgr = RequestManager.getManager(command);
                if (mgr == null) {…
                } else {
                    try {
                        mgr.processRequest(request, response);
                    } catch (Exception e) {…
                    }
                }
            }
        }
    }
```

3. 业务层

接口层负责接收业务请求，然后调用数据处理层来完成具体的数据处理。该层主要由两个类组成：RequestManager.java 和 MarketMgr.java。

RequestManager 获取并返回 MarketMgr 实例，实际上也是 IMgr 接口对象。

```
     public class RequestManager {
        public static IMgr getManager(String command) {
            if(command.startsWith(Command.CMD_MARKET_PRODUCT_INFO)) {
             return MarketMgr.getInstance();
            }
            return null;
        }
        …
     }
```

MarketMgr 继承了 AbstractMgr，而 AbstractMgr 实现了 IMgr 接口，因此在 MarketMgr 中需要实现

IMgr 接口定义的 processRequest()方法。

```
    public class MarketMgr extends AbstractMgr {
       private static MarketMgr mgr;
       private SqlSessionFactory sqlMapper;
       public MarketMgr() {
          sqlMapper = DBOperatorMgr.getInstance().getSqlMapper();
       }
       public static synchronized MarketMgr getInstance() {
          if (mgr == null) {
             mgr = new MarketMgr();
          }
          return mgr;
       }
       @Override
       public void processRequest(HttpServletRequest request, HttpServletResponse
response) {
          String command = request.getParameter(CommonConstants.PARAMS_CMD);
          if (command.equalsIgnoreCase(Command.CMD_MARKET_PRODUCT_INFO)) {
             writeResponse(response, showProductInfo(request, response));
          }
       }
       private String showProductInfo(HttpServletRequest request,
       HttpServletResponse response) {
          StringBuffer sb = new StringBuffer();
          SqlSession session = null;
          String strTypeId = request.getParameter(Params.PARAM_TYPEID);
          String strId = request.getParameter(Params.PARAM_ID);
          String strPhoneId = strId;
          try {
             session = sqlMapper.openSession();
             MarketMapper marketMapper = session.getMapper (MarketMapper. class);
             if (strTypeId.equals("1000")) {
                List<MarketProductInfo> list1 = marketMapper.getMarketBroadbandInfo
(Integer.parseInt(strId));
                sb.append("{\"BroadbandInfo\":[");
                if (list1 != null) {
                   for (int i = 0; i < list1.size(); i++) {
                      sb.append(list1.get(i).toJSONString());
                      if (i < list1.size() - 1) {
                         sb.append(",");
                      }
                   }
                }
                sb.append("]}");
             }
          } catch (Exception e) {
            e.printStackTrace();
          } finally {
            if (session != null) {
               session.close();
               session = null;
            }
          }
          return sb.toString();
```

```
        }
    }
```

MarketMgr 定义了 getInstance()方法用来实例化一个 MarketMgr 对象，从而获取 DBOperatorMgr 类定义的 SqlSessionFactory 实例。processRequest()方法主要实现根据请求进行数据处理，从而输出 JSON 字符串的功能。

processRequest()方法判断请求的业务与实际相应业务是否一致，只有一致，才能调用父类 AbstractMgr 中定义的 writeResponse ()方法，实现 JSON 字符串的输出。writeResponse()方法的第二个参数即为待输出的 JSON 字符串，该字符串通过调用 showProductInfo()返回得到。

showProductInfo()方法中创建 SqlSession，调用 SqlSession 的 getMapper()方法，该方法根据配置文件生成 Mapper 对象，通过这个 Mapper 对象可以访问 MyBatis。Mapper 对象访问 MyBatis 是通过接口来访问的，因此也可以说 getMapper()方法获取了接口对象。此处调用 MarketMapper 接口中定义的 getMarketBroadbandInfo()方法来实现商品详情信息。另外，在 showProductInfo()方法中还引用了 Params 类定义的两个静态变量 PARAM_TYPEID（值为 typeId）和 PARAM_ID（值为 ID）。

4．数据层

（1）根据商品数据表 market_broadband_info，建立对应的实体类文件 MarketProductInfo.java，存放于 vo 包中。该类根据数据表中的字段，建立了一一对应的对象属性及生成属性的 get 和 set 方法。该类实现了 org.suda.app.common.json 接口，并实现了接口中定义的方法 toJSONString()，将获取得到的 MarketProductInfo 对象转换为符合要求的 JSON 字符串。

（2）建立接口文件 MarketMapper.java，定义了显示宽带详情的操作。

```
    public interface MarketMapper {
        List<MarketProductInfo> getMarketBroadbandInfo(int par); //显示宽带详情
    }
```

（3）建立与 MarketMapper.java 对应的映射文件 MarketMapper.xml。

```
    <!-接口的实现，namespace 的路径必须和与其对应的接口类路径一致-->
     <mapper namespace="org.suda.app.market.persistence.MarketMapper">
    <!-定义查询映射时返回类型-->
        <resultMap id="MarketProductInfoMap" type="MarketProductInfo">
        <result property="id" column="id" />
        <result property="subject" column="subject" />
        …
        </resultMap>
         <select id="getMarketBroadbandInfo" parameterType="int" resultMap=
"MarketProductInfoMap">
            SELECT id,subject,body,provider_id,buy_price, sell_price,discount_
price, stock,img,
                package_category,owner,type,start_time,
end_time,create_time,last_modify_time,tag,img_url
            FROM market_broadband_info
            WHERE id=#{id}
            AND  tag = 1
       </select>
    </mapper>
```

（4）建立配置文件 Configuration.xml，配置文件包含对 MyBatis 系统的核心设置，以及获取数据库连接实例的数据源、决定事务范围和控制的事务管理器。

（5）数据库连接类 DBOperatorMgr.java，DBOperatorMgr 的作用是建立数据库的连接。getInstance()方法利用 Read 对象读取 MyBatis 配置文件，然后通过 SqlSessionFactoryBuilder 对象的 build()方法来获得 SqlSessionFactory 对象的实例，再调用 getSqlMapper()方法返回 SqlSessionFactory 实例。

（6）配置 Web.xml，其主要代码如下：

```
<servlet>
    <servlet-name>MarketServlet</servlet-name>
    <servlet-class>org.suda.app.MarketServlet</servlet-class>
</servlet>
<servlet-mapping>
    <servlet-name>MarketServlet</servlet-name>
    <url-pattern>/3gmarket</url-pattern>
</servlet-mapping>
```

（7）mialab-3gmarket-demo 工程打包后，将 mialab-3gmarket-demo.war 部署到 Tomcat 服务器的 webapps 目录下，并启动 Tomcat 服务器。在浏览器中输入 URL：

http://localhost:8080/mialab-3gmarket-demo/3gmarket?cmd=market.product.info&typeId=1000&id=10001001，页面显示效果如图 13-4 所示。

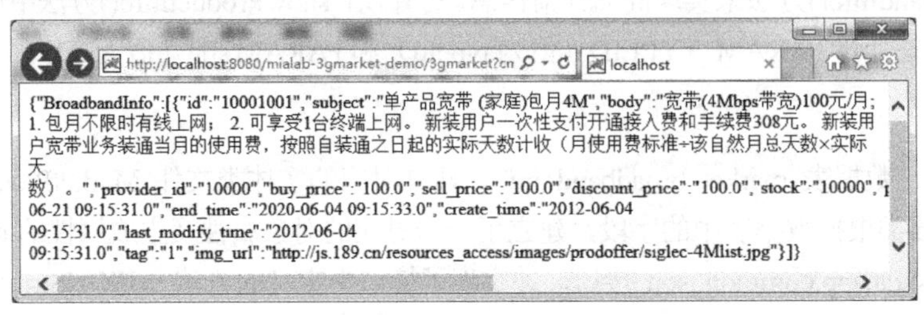

图 13-4　商品详情接口数据显示

13.4　MyBatis 与 Hibernate

MyBatis 和 Hibernate 都是目前主流的 ORM（Object/Relation Mapping，对象/关系数据库映射）框架。

1. Hibernate

Hibernate 是一个面向 Java 环境的对象/关系数据库映射工具，它将对象模型表示的 Java 对象 POJO 映射到基于 SQL 的关系数据结构中。POJO 作为普通的 JavaBean 对象，并不具备持久化操作能力，Hibernate 通过定义映射文件将 POJO 映射到相应的数据库，获得持久化对象 PO，再通过操作 PO，实现对数据库的增加、删除、修改、查询操作。要实现对数据库的操作，应首先建立数据库连接，Hibernate 采用类似于 MyBatis 的方式将数据库连接信息（数据源）封装在配置文件中，并通过 Hibernate 提供的 API 读取配置文件，获取数据库连接。

1）Hibernate 配置文件

Hibernate 的配置文件既可以使用.properties 文件，也可以使用 XML 文件配置。在实际开发过程中，通常使用后者。Hibernate 配置文件 Hibernate.cfg.xml 的主要代码如下：

```
<hibernate-configuration>
    <session-factory>
        <!--根据连接的数据库，设置数据库所用的驱动器-->
        <property name="hibernate.connection.driver_class"> </property>
        <!--设置连接数据库所需的密码-->
        <property name="hibernate.connection.password"></property>
        <!--设置连接的数据库 url-->
        <property name="hibernate.connection.url"></property>
        <!--设置连接数据库所需的用户名-->
        <property name="hibernate.connection.username"></property>
        <!--设置连接的数据库方言-->
```

```
        <property name="hibernate.dialect"></property>
        …
        <!--项目中所有的映射文件-->
        <mapping resource=""/>
    </session-factory>
</hibernate-configuration>
```

上述代码描述了如何配置数据源所需的信息，以及项目包含的所有映射文件。

2）Hibernate 映射文件

映射文件是一个描述 POJO 和关系数据库之间映射的 XML 文档。该文档通过 XML 节元素的设置，根据 POJO 的定义，将 POJO 和数据表建立起一一对应的关系。

Hibernate 的映射文件主要有四类节点：根节点<hibernate-mapping>、类节点<class>、主键节点<id>、普通属性节点<property>。Hibernate 为每个节点定义了丰富的属性，从而能够更好地完成映射工作。

（1）< hibernate-mapping>。

一个映射文件有且只有一个根节点，该节点可以包含多个<class>节点。根节点有一个属性 package，该属性指定了包的前缀，对于映射文件中若<class>没有指定全限定的类名，则默认使用该包前缀。例如：

```
<hibernate-mapping package="com.mialab.domain">
        <class name="Student">
</hibernate-mapping>
```

等价于：

```
<hibernate-mapping>
        <class name="com.mialab.domain .Student">
 </hibernate-mapping>
```

（2）<class>。

每个<class>节点代表一个 POJO 与数据表的映射。节点有两类子节点：主键<id>和普通属性<property>。<class>节点通过类名 name 属性指定映射的一方——POJO，类名为该 POJO 的 Java 全限定的类名，以及通过 table 属性指定映射的另一方——数据库表，table 值为映射的数据库表名。

（3）<id>。

<id>节点通过子节点<generator>指定主键的生成策略，与数据表的主键字段对应起来，其中的 name 属性为 POJO 属性名，class 属性则指定了 Hibernate 定义的主键生成器，常见的生成器有 assigned、hlo、seqhilo、increment、identity、sequence、native、uuid、foreign、select、guid。

（4）<property>。

<property>通过其子节点<column>，设置相关属性，实现 POJO 的属性与数据表中的普通字段之间的一一对应，<id>对应于主键字段，因此一个<class>有且仅有一个<id>，但可以有多个<property>。

<property>的 name 属性为 POJO 属性名，type 属性为 POJO 属性的数据类型，而<column>的 name 属性为数据表的字段名。

3）使用 Hibernate 进行持久化操作

在进行操作之前，需要将 Hibernate 提供的功能 jar 包导入项目中，Hibernate 的 jar 包可于 http://www.hibernate.org/downloads 中下载，这里使用的是 Hibernate3.6.0。对下载的压缩包进行解压，从解压的文件夹中获取所需的 jar 包，并将其导入项目。它主要包括：Hibernate 核心 jar 包 hibernate3.jar 和 lib 文件夹中存放运行 Hibernate 所需的第三方 jar 包。

除此以外，还需要导入数据库 JDBC 驱动包，用于访问数据库，此处使用 MySql 数据库，因此导入的 jar 包是 mysql-connector-java-5.1.22-bin.jar。

利用 Hibernate 实现数据的持久化操作，其具体步骤如下。

（1）获取 Configuration 对象。

Hibernate 的配置对应于一个 Configuration 对象，实例化 Configuration 并读取项目中的配置文件，从而获取了应用程序到数据库的映射配置，其实现代码如下：

```
Configuration config=new Configuration().configure();
```

实例化的 Configuration 调用 Configure()方法从 classpath 中寻找 Hibernate.cfg.xml 文件，若该文件不存在，则程序会抛出异常。

（2）创建 SessionFactory。

SessionFactory 采用工厂模式生成 Session 实例，一个项目通常只创建一个 SessionFactory，它既缓冲了 Hibernate 自动生成的 SQL 语句和一些其他的映射数据，还缓冲了一些将来有可能重复利用的数据。创建 SessionFactory 代码如下：

```
SessionFactory factory=config.buildSessionFactory();
```

（3）获取 Session。

Session 接口负责实现持久化对象的操作，其代码如下：

```
Session session=factory.openSession();
```

（4）操作数据库。

对于数据库的操作主要分为增加、删除、修改、查找，它分别有两个接口负责完成。这两个接口是 Transaction 和 Query。

Transaction 接口负责管理 Hibernate 事务，实现底层事务如 JDBC 事务的封装；Query 接口实现对数据库及持久对象进行查询，这里的查询可以使用 HQL 语言或本地数据库的 SQL 语句。Transaction 接口和 Query 接口提供了较多的方法来实现对数据库的操作，将在下面的实例中进行详细介绍。

（5）关闭 Session，关闭事务 SessionFactory。

【示例】 使用 MySql 建立数据库 market，该数据库中有一张数据表 market_goods_ provider，其结构如表 13-2 所示。利用 Hibernate 对数据库的操作，实现对 market_goods _provider 数据表中数据的增加、删除、修改、查询等操作。相关内容参见教学资源包第 13 章 HibernateDemo 工程。

表 13-2　数据表 market_goods_provider 结构

字 段 名	类　型	说　明
provider_id	int	供应商 ID，主键
provider_name	char(32)	供应商姓名
email	char(50)	邮箱地址
mobile	char(13)	联系电话

（1）创建数据库 market，新建数据表 market_goods_provider。

本例使用 MySql Server 5.1 来管理 market 数据库。

具体操作步骤如下：打开 MySql.exe，在命令窗口中输入密码，从而进入命令编辑窗口。输入创建数据库和数据表的命令，其代码如下：

```
create database market; //创建数据库
use market;             //选择market数据库
//创建数据表market_goods_provider
create table market_goods_provider(
   provider_id int(11) not null,
   provider_name char(32),
   email char(50),
   mobile char(13)
);
```

（2）创建一个 Java 工程，名为 HibernateDemo，按照上面介绍需求，导入相关的 jar 包，如图 13-5 所示，并建立项目的目录结构，如图 13-6 所示。

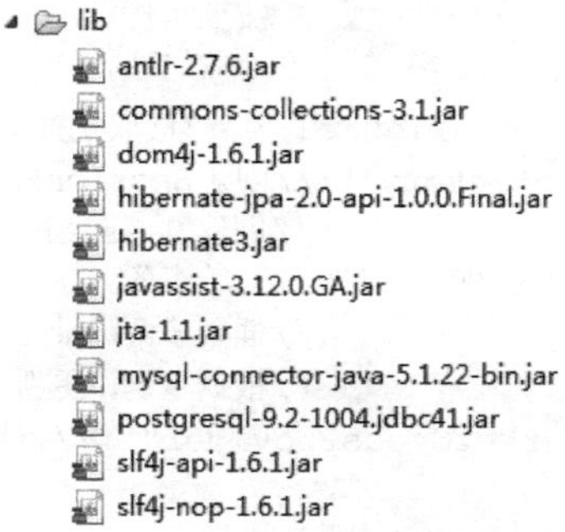

```
lib
    antlr-2.7.6.jar
    commons-collections-3.1.jar
    dom4j-1.6.1.jar
    hibernate-jpa-2.0-api-1.0.0.Final.jar
    hibernate3.jar
    javassist-3.12.0.GA.jar
    jta-1.1.jar
    mysql-connector-java-5.1.22-bin.jar
    postgresql-9.2-1004.jdbc41.jar
    slf4j-api-1.6.1.jar
    slf4j-nop-1.6.1.jar
```

图 13-5　项目所需的 jar 包

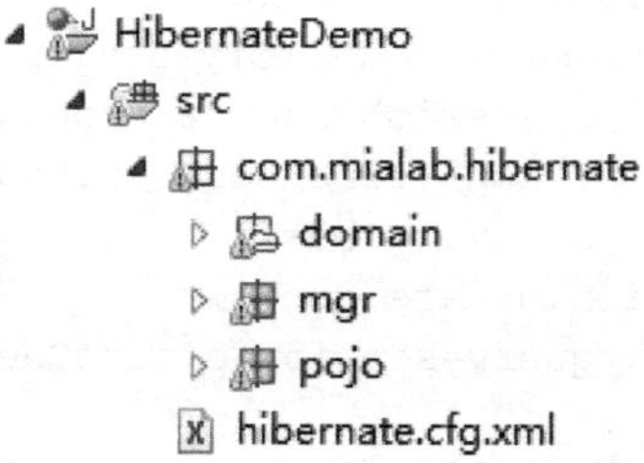

```
HibernateDemo
    src
        com.mialab.hibernate
            domain
            mgr
            pojo
        hibernate.cfg.xml
```

图 13-6　项目的目录结构

HibernateDemo 工程中 src 目录下各个包的作用如下。

- com.mialab.hibernate.pojo 包：放置 POJO 对象。
- com.mialab.hibernate.domain：放置映射文件。
- com.mialab.hibernate.mgr：放置 Java 应用程序。

（3）创建 POJO 实体类 MarketGoodsProvider.java，在 Eclipse 中自动生成其属性的 set 方法和 get 方法。（右键 MarketGoodsProvider.java，选择“Source”→“Generate Getters and Setters”）

```
public class MarketGoodsProvider{
    private int provider_id;
    private String provider_name;
    private String email;
    private String mobile;
    …
}
```

（4）根据 MarketGoodsProvider.java，定义映射文件 MarketGoodsProvider.hbm.xml，实现实体类与数据库的映射。

```
<hibernate-mapping>
   <class name="com.mialab.hibernate.pojo.MarketGoodsProvider"  table= "market_
goods_provider">
       <id name="provider_id" type="java.lang.Integer">
           <generator class="assigned"  />
       </id>
       <property name="provider_name" type="java.lang.String">
           <column name="provider_name" />
       </property>
       <property name="email" type="java.lang.String">
           <column name="email" />
       </property>
       <property name="mobile" type="java.lang.String">
           <column name="mobile" />
       </property>
   </class>
</hibernate-mapping>
```

（5）定义配置文件 hibernate.cfg.xml，相关内容参见教学资源网站的 HibernateDemo 工程。

（6）创建 Java 应用程序，实现对数据表中数据的增加、删除、修改、查询等操作。

```
public class MarketManager {
   public static void main(String[] args) {
```

```
        MarketGoodsProvider goods=new MarketGoodsProvider();  //实例化实体类
        goods.setProvider_id(10003);
        goods.setProvider_name("沃达丰 vodafone");
        goods.setEmail("vodafone@163.com");
        goods.setMobile("18812312345");
        Configuration config=new Configuration().configure(); //读取配置文件
        SessionFactory factory=config.buildSessionFactory();//创建 SessionFactory
        Session session=factory.openSession();              //创建 Session
        Transaction trans=session.beginTransaction();       //开始事务
        session.save(goods);                                //插入一条记录
        session.update(goods);                              //根据主键，更新一条记录
        Query query=session.createQuery("from MarketGoodsProvider");//根据 HQL 语
//句，查询记录
        List<MarketGoodsProvider> goodslist= query.list();
        for(MarketGoodsProvider mgp:goodslist){
           System.out.println(mgp.getProvider_id()+"-"+mgp.getProvider_ name()
   +"-"+mgp.getEmail()+"-"+mgp.getMobile());
        }
        session.delete(goods);     //根据主键，删除一条记录
        trans.commit();            //提交事务
        session.close();           //关闭 Session
        factory.close();           //关闭 SessionFactory
    }
  }
```

若只执行添加和显示功能，则运行程序，通过控制台打印数据，如图 13-7 所示。

```
<terminated> MarketManager [Java Application] C:\Program Files (x86)\Java\jre7\bin\javaw.exe
10000-电信-chinanet@163.com-10000
10001-移动-cmcc@163.com-10086
10002-联通-chinaunicom@163.com-10010
10003-沃达丰vodafone-vodafone@163.com-18812312345
```

图 13-7　控制台数据显示效果

2．MyBatis 与 Hibernate 比较

MyBatis 和 Hibernate 均属于 ORM 框架，实现持久化类与数据库之间相互映射，通过操作持久化类，从而实现对数据库的操作。它们的基本实现过程是相似的，首先它们都由配置文件获取数据源，然后采用工厂模式（MyBatis：SqlSessionFactory，Hibernate：SessionFactory），获取与之交互的接口（MyBatis：SqlSession，Hibernate：Session），进而获得操作持久化类的接口对象（MyBatis：Mapper，Hibernate：Transaction 和 Query）。但在具体实施细节上还略有不同，大致有以下几个方面。

（1）对象与数据库映射。

MyBatis 通过映射配置文件，接收 SQL 所需参数，并将执行 SQL 的结果映射到相应的 POJO 对象，这种映射方式，在使用 MyBatis 时仍需手动编写 SQL 语句来管理 POJO 对象，因此开发人员需要熟练掌握 SQL 语句。

Hibernate 通过 POJO 对象和映射文件组成的对象关系模型，可自动生成 SQL，从而实现对数据库的操作，基于这种方法，开发人员对于 SQL 语句的掌握要求并不高。

（2）SQL 优化。

一方面，Hibernate 的 SQL 大多是自动生成的，无法直接维护 SQL，虽然配备了 Hibernate 查询语言（HQL），但 HQL 在功能上还是不如 SQL 丰富。另一方面，Hibernate 查询数据表会直接将所有字段查询出来，这样就会影响查询效率。如果 Hibernate 采用自己编写 SQL 来查询的话，又破坏了 Hibernate 开发自动生成 SQL 的特点，因此在这一点上，手动编写 SQL 的 MyBatis 更为灵活，可以按需指定查询字段，在查询效率上更具优势。

（3）数据库移植性。

Hibernate 通过对象与数据库的映射机制及 HQL 语言，使得与具体数据库的关联只需在 XML 文件中配置即可，与数据库无关，这大大降低了对象与数据库（Oracle、Mysql 等）之间的耦合性，而 MyBatis 由于需要手写 SQL，当 SQL 不具通用性而使用了很多为某数据库所特有的 SQL 语句时，可移植性就会随之降低很多，移植成本也会很高。

（4）日志系统。

Hibernate 日志系统非常健全，涉及很广泛，包括 SQL 记录、关系异常、优化警告、缓存提示、脏数据警告等；MyBati 则除了基本记录功能，其他功能要薄弱很多。

13.5 本章小结

本章讲述服务器端接口编程的相关技术知识。为了更好地统一管理数据源，引入了 JNDI 配置数据源，介绍了 JNDI 的基本概念、JNDI 数据源的三种配置方式，以及 Java 程序使用 JNDI 的基本步骤和具体实现。同时掌握程序运行过程及维护程序，介绍了两种记录日志的主流方法 Log4J、SLF4J。通过综合运用前面介绍的相关知识，详细阐述了本章的重点，利用 MyBatis 框架进行接口编程。最后，还对主流的 ORM 框架 Mybatis 和 Hibernate 进行了比较。

习　题　13

1．比较 JNDI 三种数据源的配置方法，并分别实现三种配置。配置的数据源信息如下：

数据库：PostgreSQL；地址：192.158.2.45；端口：3306；

用户名：postgres；密码：postgres_pwd。

2．根据如表 13-3 所示数据，建立数据库及数据表。

表 13-3　course 表

课程编号	课程名称	学　分	备　注
001	C 语言	3	暂无
002	Java 程序设计	4	暂无
003	Android 开发	2	暂无
004	iOS 开发	2	暂无

（1）利用 MyBatis 框架实现对课程信息的增、删、改、查接口编程。

增：添加一条课程记录；删：根据课程编号，删除一条课程记录；

改：根据课程编号，修改课程记录；查：查询学分大于某个数值的课程记录。

（2）查询课程记录时，利用 SLF4J 将查询结果以 INFO 等级日志打印至 D 盘的 error.log 文件中，并采用 PatternLayout 方式进行布局，其字符串输出格式如：001-C 语言-3。

第四部分

拓 展 篇

第 14 章 MVP 模式

本章导读

MVP 模式从 MVC 模式演变而来，它们的基本思想有相通的地方，Controller/Presenter 负责逻辑的处理，Model 负责提供数据，View 负责显示。MVP 模式作为一种新的模式，与 MVC 模式的区别：在 MVP 中 View 并不直接使用 Model，它们之间的通信是通过 Presenter 来进行的，所有交互都发生在 Presenter 内部。MVP 模式的核心思想是：将 Activity 中的 UI 逻辑抽象成 View 接口，把业务逻辑抽象成 Presenter 接口，Model 类还是原来的 Model。本章主要内容有：（1）MVP 模式简介；（2）MVP 模式与 MVC 模式；（3）MVP 模式的优点和缺点；（4）MVP 模式的使用示例。

14.1 MVP 模式简介

MVP 模式（Model View Presenter）可以说是 MVC 模式（Model View Controller）在 Android 开发上的一种变种、进化模式。

在 Android 项目中，Activity 和 Fragment 占据了大部分的开发工作。

按照 MVC 模式的分层，Activity 和 Fragment（后面只讲 Activity）属于 View 层，用于展示 UI 界面，以及接收用户的输入，此外还要承担一些生命周期的工作。

Activity 在 Android 开发中充当非常重要的角色，可以把一些业务逻辑直接写在 Activity 里面，这样虽然非常直观方便，但 Activity 会变得越来越臃肿，经常会出现超过 1000 行的代码，而且如果是一些可以通用的业务逻辑（如用户登录），写在具体的 Activity 里就意味着这个逻辑不能复用了。

如果有进行代码重构经验的人，看到 1000 多行代码的类肯定会有所顾虑。

Activity 不仅承担了 View 的角色，还承担了一部分的 Controller 角色，这样一来，View 和 Controller 就耦合在一起了。虽然这样写方便，但是如果进行业务调整，维护代码就会变得困难，而且在一个臃肿的 Activity 类查找业务逻辑的代码也会非常麻烦。所以，有必要在 Activity 中，把 View 和 Controller 抽离出来，这就是 MVP 模式的工作。

MVP 模式把 Activity 中的 UI 逻辑抽象成 View 接口，把业务逻辑抽象成 Presenter 接口，Model 类还是原来的 Model。这就是 MVP 模式的核心思想。

（1）Model：依然是业务逻辑和实体模型，负责业务逻辑和数据处理（数据库存储操作、网络数据请求、复杂算法等）。

（2）View：对应于 Activity，负责 View 的绘制和与用户交互。

（3）Presenter：负责完成 View 与 Model 间的交互。

MVP 模式与其说是设计模式，还不如说是一种程序架构范式。MVP 模式的架构如图 14-1 所示。

图 14-1　MVP 模式的架构

MVP 模式表示层的优势体现在以下三个方面：

（1）View 与 Model 完全隔离。

Model 和 View 之间具有良好的松耦合设计。这意味着，如果 Model 或 View 中的一方发生变化，只要交互接口不变，另一方就没必要对上述变化做出改变。这使得 Model 层的业务逻辑具有很好的灵活性和可重用性。

（2）Presenter 与 View 的具体实现技术无关。

应用程序可以用同一个 Model 层适配多种技术构建的 View 层。

（3）可以进行 View 的模拟测试。

过去，由于 View 和 Model 之间的紧耦合，在开发完成之前对其中一方进行测试是不可能的。出于同样的原因，对 View 或 Model 进行单元测试很困难。

现在，MVP 模式解决了所有的问题。在 MVP 模式中，View 和 Model 之间没有直接依赖，开发者能够借助模拟对象注入测试两者中的任一方。

14.2 MVP 模式与 MVC 模式

MVP 模式和 MVC 模式的区别如图 14-2 所示。MVC 模式中是允许 Model 和 View 进行交互的，而 MVP 模式中 Model 与 View 之间的交互用 Presenter 完成。Presenter 与 View 之间的交互是通过接口完成的。MVC 模式中 View 对应的是布局文件，MVP 模式中 View 对应的是 Activity。

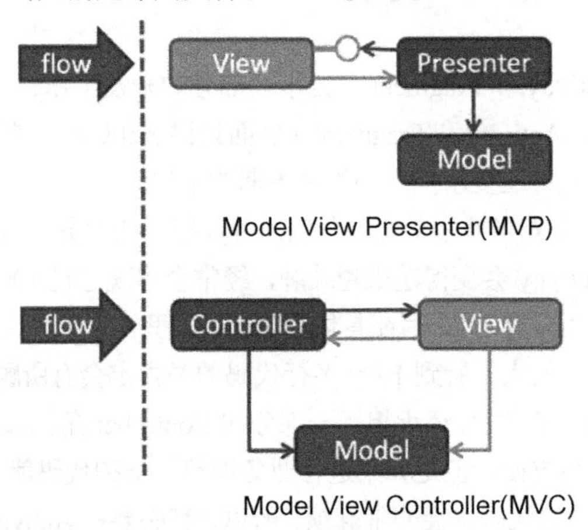

图 14-2　MVP 和 MVC 的区别

在 MVC 模式里，View 是可以直接访问 Model 的。从而，View 里会包含 Model 信息和一些业务逻辑。在 MVC 模式里，虽然 Model 不依赖于 View，但是 View 是依赖于 Model 的。不仅如此，因为有一些业务逻辑在 View 里实现了，导致要更改 View 也是比较困难的，至少那些业务逻辑是无法重用的。

在 MVP 模式里，Presenter 完全把 Model 和 View 进行了分离，将主要的程序逻辑在 Presenter 里实现。Presenter 与具体的 View 是没有直接关联的，而是通过定义好的接口进行交互，从而在变更 View 时可以保持 Presenter 不变，即重用。

不仅如此，还可以编写测试用的 View，模拟用户的各种操作，从而实现对 Presenter 的测试，而不需要使用自动化的测试工具，甚至可以在 Model 和 View 都没有完成时，就可以通过编写 MockObject（实现了 Model 和 View 的接口，但没有具体的内容）来测试 Presenter 的逻辑。

在 MVP 模式里，应用程序的逻辑主要在 Presenter 里实现，其中的 View 是很薄的一层。因此就有人提出了 PresenterFirst 的设计模式，即根据 UserStory 来设计和开发 Presenter。在这个过程中，View

是很简单的，能够把信息显示清楚就可以了。以后可根据需要再随时更改 View，而对 Presenter 没有任何影响。

如果要实现的 UI 比较复杂，而且相关的显示逻辑还跟 Model 有关系，就可以在 View 和 Presenter 之间放置一个 Adapter。由这个 Adapter 来访问 Model 和 View，可避免两者之间的关联。同时，因为 Adapter 实现了 View 的接口，从而可以保证与 Presenter 之间的接口不变。这样就可以使 View 和 Presenter 之间的接口既简洁又不失去 UI 的灵活性。

在 MVP 模式里，View 只应该有简单的 Set/Get 方法、用户输入和设置界面显示的内容，除此就不应该再有更多的内容。绝不容许直接访问 Model，这就是 MVP 模式与 MVC 模式的最大不同之处。

14.3 MVP 模式的优点和缺点

1. MVP 模式的优点

MVP 模式与 MVC 模式的主要区别是 View 与 Model 不直接交互，而是通过 Presenter 来完成交互，这样可以修改视图而不影响模型，达到解耦的目的，实现了 Model 和 View 真正的完全分离。视图的变化总是比较频繁的，将业务逻辑抽取出来，放在表示器中实现，可使模块职责划分明显，层次清晰。一个表示器能复用于多个视图，而不需要更改表示器的逻辑（当然是在该视图的改动不影响业务逻辑的前提下），这就增加了程序的复用性。

数据的处理由模型层完成，隐藏了数据。在数据显示时，表示器可以对数据进行访问控制，提高数据的安全性。以前的 Android 开发是难以进行单元测试的，但是随着项目变得复杂，测试是保证应用质量的关键。MVP 模式中，表示器对视图是通过接口进行的，可以利用测试驱动，模拟出视图对象，实现视图相对于表示器的接口，就可以对表示层进行不依赖于 UI 环境的单元测试了，这大大降低了 Android 应用开发过程中业务逻辑测试的难度和复杂度。

引入 MVP 模式后，视图层可完全不依赖于模型层，相当于将视图从特定的业务场景中脱离出来，做到了对业务完全不可知的状态，因此可以将视图层组件化，提供一系列接口供表示层操作，这样就可以制作出高度可复用的视图组件了。

2. MVP 模式的缺点

MVP 模式中每个 View 都有对应的 Presenter，类相对比较多。这就增加了代码的复杂度，特别是针对小型 Android 应用的开发，可使程序冗余。

由于 MVP 模式中视图的渲染过程会放在 Presenter 中，从而导致视图与 Presenter 交互过于频繁。如果某特定视图的渲染很多，就会造成 Presenter 与该视图联系过于紧密，一旦该视图需要变更，那么 Presenter 也要进行变更，就不能如预期的那样降低耦合度和增加复用性了。

14.4 MVP 模式的使用示例

MVP 模式的使用交互流程如图 14-3 所示。

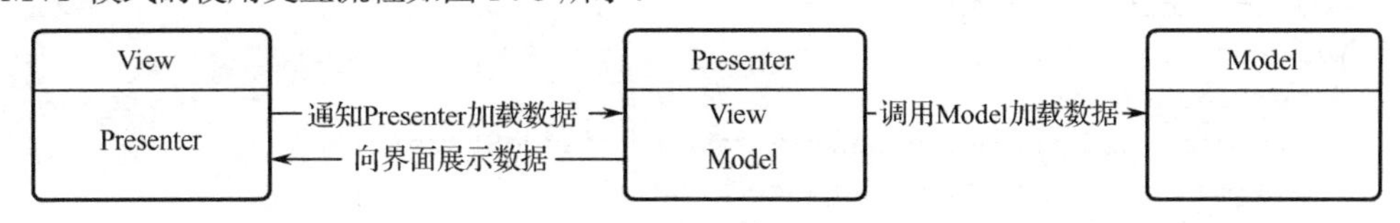

图 14-3　MVP 模式的使用交互流程

Presenter 是 View 与 Model 交互的中间人。Presenter 要持有 View 对象，而 View 对象往往是 Activity、Fragment。如果 Activity 退出时 Presenter 正在执行一个耗时的网络请求，那么将导致 Activity 的内存无法被释放而造成内存泄漏。因此，需要定义一个含有关联、取消关联 View 角色的 Presenter。

【示例】 MVP 模式的基本使用。

启动 Android Studio，在 Ch14 工程中创建 MVPDemo 模块，并在该模块中创建文件，如图 14-4 所示。com.mialab.mvpdemo.login 包中的文件有 LoginView.java、LoginActivity.java、LoginPresenter.java 和 LoginModel.java。com.mialab.mvpdemo.info 包中的文件有 InfoView.java、InfoActivity.java、InfoPresenter. java、InfoModel.java、Employee.java 和 EmployeeAdapter.java。res\layout 目录下的文件有 activity_login.xml、activity_info.xml 和 view_main_item.xml。

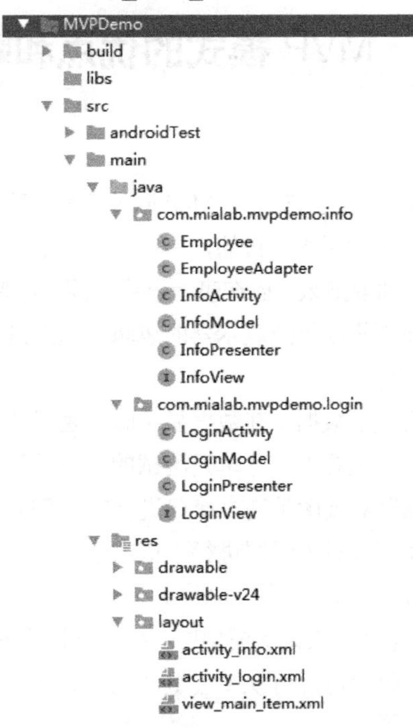

图 14-4 MVPDemo（源码）目录结构

MVPDemo 模块的 LoginView.java 主要代码如下：

```
public interface LoginView {
   void showProgress();
   void hideProgress();
   void setUsernameError();
   void setPasswordError();
   void navigateToHome();
}
```

MVPDemo 模块的 LoginActivity.java 主要代码如下：

```
public class LoginActivity extends AppCompatActivity implements LoginView{
   private ProgressBar progressBar;
   private EditText username;
   private EditText password;
   private LoginPresenter presenter;
   @Override
   protected void onCreate(Bundle savedInstanceState) {
      super.onCreate(savedInstanceState);
      setContentView(R.layout.activity_login);
```

```
        progressBar = findViewById(R.id.progress);
        username = findViewById(R.id.username);
        password = findViewById(R.id.password);
        findViewById(R.id.button).setOnClickListener(v -> validateCredentials());
        presenter = new LoginPresenter(this, new LoginModel());
    }
    @Override
    protected void onDestroy() {
        presenter.onDestroy();
        super.onDestroy();
    }
    @Override
    public void showProgress() {
        progressBar.setVisibility(View.VISIBLE);
    }
    @Override
    public void hideProgress() {
        progressBar.setVisibility(View.GONE);
    }
    @Override
    public void setUsernameError() {
        username.setError(getString(R.string.username_error));
    }
    @Override
    public void setPasswordError() {
        password.setError(getString(R.string.password_error));
    }
    @Override
    public void navigateToHome() {
        startActivity(new Intent(this, InfoActivity.class));
        finish();
    }
    private void validateCredentials() {
        presenter.validateCredentials(username.getText().toString(), password.
getText().toString());
    }
}
```

注意：LoginActivity 的 onDestroy()方法调用了 LoginPresenter 实例的 onDestroy()方法。

MVPDemo 模块的 LoginPresenter.java 主要代码如下：

```
public class LoginPresenter implements LoginModel.OnLoginFinishedListener {
    private LoginView loginView;
    private LoginModel loginModel;
    LoginPresenter(LoginView loginView, LoginModel loginModel) {
        this.loginView = loginView;
        this.loginModel = loginModel;
    }
    public void validateCredentials(String username, String password) {
        if (loginView != null) {
            loginView.showProgress();
        }
        loginModel.login(username, password, this);
    }
    public void onDestroy() {
```

```
            loginView = null;
        }
        @Override
        public void onUsernameError() {
            if (loginView != null) {
                loginView.setUsernameError();
                loginView.hideProgress();
            }
        }
        @Override
        public void onPasswordError() {
            if (loginView != null) {
                loginView.setPasswordError();
                loginView.hideProgress();
            }
        }
        @Override
        public void onSuccess() {
            if (loginView != null) {
                loginView.navigateToHome();
            }
        }
    }
```

LoginPresenter 实例的 onDestroy()方法中释放了持有的 LoginView 对象的引用（loginView = null;），以避免 LoginActivity 退出时内存无法释放而造成内存泄漏。

MVPDemo 模块的 LoginModel.java 主要代码如下：

```
public class LoginModel {
    interface OnLoginFinishedListener {
        void onUsernameError();
        void onPasswordError();
        void onSuccess();
    }
    public void login(final String username, final String password, final
OnLoginFinishedListener listener) {
        // Mock login. I'm creating a handler to delay the answer a couple of seconds
        new Handler().postDelayed(() -> {
            if (TextUtils.isEmpty(username)) {
                listener.onUsernameError();
                return;
            }
            if (TextUtils.isEmpty(password)) {
                listener.onPasswordError();
                return;
            }
            listener.onSuccess();
        }, 2000);
    }
}
```

限于篇幅，其余相关文件及代码参见教学资源包中的 Ch14\MVPDemo 模块。

运行 Ch14\MVPDemo，其结果如图 14-5 所示。

图 14-5　运行 MVPDemo 模块

14.5　本章小结

MVP 模式通过 Presenter 实现数据和视图之间的交互，减轻了 Activity 的职责。MVP 模式既避免了 View 和 Model 的直接联系，又通过 Presenter 实现两者之间的沟通。

MVP 模式简化了 Activity 中的代码，将复杂的逻辑代码提取到了 Presenter 中进行处理，模块职责划分明显，层次清晰。与之对应的好处就是，耦合度变低，可以方便地进行测试。

习　题　14

1．什么是 MVP 模式？MVP 模式和 MVC 模式有什么不同？
2．MVP 模式的优点和缺点是什么？
3．如何遵循 MVP 模式进行 Android 客户端编程？试举例加以说明。

第 15 章 Java 设计模式

本章导读

设计模式（Design pattern）是一套被反复使用、多数人知晓的、经过分类编目的、代码设计经验的总结。使用设计模式是为了可重用代码，让代码更容易为他人理解，以及保证代码的可靠性。设计模式使代码“编制”真正工程化，设计模式是软件工程的基石。项目中合理运用设计模式可以完美地解决很多问题，每一个模式都描述了一个不断重复发生的问题，以及该问题的核心解决方案，这也是它能被广泛应用的原因。本章主要内容有：（1）设计模式的分类和设计原则；（2）创建模式；（3）结构模式；（4）行为模式。

15.1 设计模式的分类和设计原则

15.1.1 设计模式的分类

总体来说，设计模式可分为 3 大类：创建型模式、结构型模式和行为型模式。

创建型模式共 5 种，包括工厂方法模式、抽象工厂模式、单例模式、建造者模式和原型模式。

结构型模式共 7 种，包括适配器模式、装饰器模式、代理模式、外观模式、桥接模式、组合模式和享元模式。

行为型模式共 11 种，包括策略模式、模板方法模式、观察者模式、迭代子模式、责任链模式、命令模式、备忘录模式、状态模式、访问者模式、中介者模式和解释器模式。

15.1.2 设计模式的设计原则

每个设计模式都是很重要的一种思想，在 Java 本身的设计之中处处都有体现，像 AWT、JDBC、集合类、IO 管道，或者是 Web 框架。

在使用软件架构工作时，需要遵循面向对象的设计原则。这些原则同样体现在各类设计模式、架构模式之中，在学习过程中可以通过类图、时序图、示例代码等形式，不断体会这些原则在解决“依赖”和“变化”中的效果。

在使用面向对象的思想进行系统设计时，前人共总结出了 7 条原则，它们分别是：开闭原则、里氏代换原则、依赖注入原则、接口分离原则、迪米特原则、合成复用原则和单一职责原则。

1. 开闭原则

开闭原则（Open Close Principle）是指对扩展开放，对修改关闭。在程序需要进行拓展的时候，不能去修改原有的代码，实现一个热插拔的效果。

因此为了使程序的扩展性好，易于维护和升级，就需要使用接口和抽象类，在后面的具体设计中还会提到相关内容。

2. 里氏代换原则

里氏代换原则（Liskov Substitution Principle，LSP）是面向对象设计的基本原则之一。里氏代换原则中说，任何基类可以出现的地方，子类也一定可以出现。LSP 是继承复用的基石，只有当衍生类可以替换掉基类。当软件单位的功能不受到影响时，基类才能真正被复用，而衍生类也能够在基类的基础上增加新的行为。

里氏代换原则是对开闭原则的补充。实现开闭原则的关键步骤就是抽象化。基类与子类的继承关系就是抽象化的具体实现，所以里氏代换原则是对实现抽象化具体步骤的规范。

3. 依赖注入原则

依赖注入原则（Dependence Inversion Principle）是开闭原则的基础，具体内容为针对接口编程，依赖于抽象化而不依赖于具体。

4. 接口分离原则

接口分离原则（Interface Segregation Principle）是指使用多个隔离的接口，比使用单个接口更好，同时还能降低类之间的耦合度。因此设计模式就是一个软件的设计思想，从大型软件架构出发，且为了升级和维护方便。所以上文中多次出现：降低依赖和耦合的内容。

5. 迪米特法则

迪米特法则（Demeter Principle）是较少知道的原则，是指一个实体应当尽量少地与其他实体之间发生相互作用，使得系统功能模块相对独立。

6. 合成复用原则

合成复用原则（Composite Reuse Principle）是指尽量使用合成/聚合的方式，而不是使用继承。

7. 单一职责原则

单一职责原则（Single Responsibility Principle）的核心思想是，系统中的每个对象都应该只有一个单独的职责，而所有对象所关注的就是自身职责的完成。

15.2 创建模式

15.2.1 工厂方法模式和抽象工厂方法模式

1. 工厂方法模式

工厂方法模式（Factory Method）分为以下 3 种：普通工厂方法模式、多个工厂方法模式、静态工厂方法模式。

（1）普通工厂方法模式：建立一个工厂类，对实现了同一接口的一些类进行实例的创建。

（2）多个工厂方法模式：对普通工厂方法模式的改进。在普通工厂方法模式中，如果传递的字符串出错，则不能正确地创建对象；多个工厂方法模式是提供多个工厂方法，分别创建对象。

（3）静态工厂方法模式：将多个工厂方法模式里的方法置为静态的，不需要创建实例，直接调用即可。

总体来说，工厂方法模式适合的场景是：有大量的产品需要创建，并且具有共同的接口时，可以通过工厂方法模式进行创建。

在以上 3 种模式中，如果第 1 种传入的字符串有误，则不能正确创建对象。当第 3 种相对于第 2

种，则不需要实例化工厂类，所以大多数情况下会选用第 3 种静态工厂方法模式。

2. 抽象工厂方法模式

抽象工厂方法模式（Abstract Factory）有一个问题就是，类的创建依赖工厂类，也就是说，如果想要拓展程序，必须对工厂类进行修改，这就违背了开闭原则。所以从设计角度考虑，存在一定的问题，如何解决呢？这就用到抽象工厂模式，通过创建多个工厂类，可以在需要增加新的功能时，直接增加新的工厂类，而不需要再修改之前的代码了。

15.2.2 单例模式

单例模式（Singleton）是一种常用的设计模式。在 Java 应用中，单例对象能保证在一个 JVM 中，该对象只有一个实例存在。这样的模式有以下好处：

（1）某些类创建比较频繁，对于一些大型的对象是一笔很大的系统开销。

（2）省去了 new 操作符，降低了系统内存的使用频率，可减轻 GC 的压力。

（3）有些类犹如交易所的核心交易引擎，控制着交易流程，如果该类可以创建多个的话，系统就会完全乱了。所以只有使用单例模式，才能保证核心交易服务器独立控制整个流程。

15.2.3 建造者模式

工厂方法模式提供的是创建单个类的模式，而建造者模式（Builder）则是将各种产品集中起来进行管理，用来创建复合对象，所谓复合对象就是指某个类具有不同的属性。

从这点可以看出，建造者模式将很多功能集成到一个类里，这个类可以创造出比较复杂的东西。所以它与工厂方法模式的区别在于：工厂方法模式关注产品整体，而建造者模式则关注构建过程。因此，是选择工厂方法模式还是建造者模式，应依实际情况而定。

15.2.4 原型模式

原型模式（Prototype）也是创建型的模式，该模式的思想就是将一个对象作为原型，对其进行复制，产生一个和原对象类似的新对象。

在 Java 中，复制对象是通过 clone()实现的，先创建一个原型类：

```
public class Prototype implements Cloneable {
    public Object clone() throws CloneNotSupportedException {
            Prototype proto = (Prototype) super.clone();
            return proto;
        }
}
```

很简单，一个原型类只需要实现 Cloneable 接口和覆写 clone 方法，此处 clone 方法可以改成任意的名称，因为 Cloneable 接口是个空接口，可以任意定义实现类的方法名，如 cloneA 或者 cloneB。其中，super.clone()调用的是 Object 的 clone()方法。在这里将结合对象的浅复制和深复制来进行说明。

（1）浅复制：指将一个对象复制后，其基本数据类型的变量都会重新创建，而引用类型指向的则还是原对象所指向的。

（2）深复制：指将一个对象复制后，不论是基本数据类型还是引用类型都是重新创建的。简单来说，就是深复制进行了完全彻底的复制，而浅复制则不彻底。

要实现深复制，需要采用流的形式读入当前对象的二进制输入，再写出二进制数据对应的对象。

15.3 结构模式

本节将介绍以下 7 种结构型模式：适配器模式、装饰模式、代理模式、外观模式、桥接模式、组合模式和享元模式，其中对象的适配器模式是各种模式的起源，如图 15-1 所示。

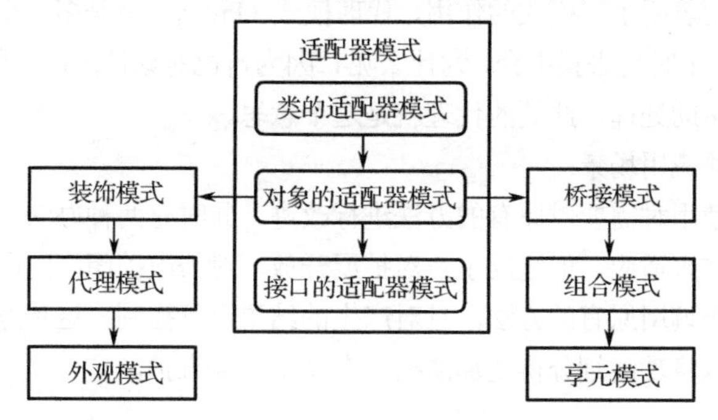

图 15-1　7 种结构型模式

15.3.1　适配器模式和装饰模式

1．适配器模式

适配器模式（Adapter）指将某个类的接口转换成客户端期望的另一个接口，目的是消除由于接口不匹配所造成的类的兼容性问题。它主要分为三类：类的适配器模式、对象的适配器模式和接口的适配器模式。

（1）类的适配器模式：它的核心思想是有一个 Source 类拥有一个方法待适配，目标接口是 Targetable，通过 Adapter 类，将 Source 的功能扩展到 Targetable 里。这样 Targetable 接口的实现类就具有了 Source 类的功能。

（2）对象的适配器模式：它的基本思路和类的适配器模式相同，只是将 Adapter 类进行修改，使其不继承 Source 类，而是持有 Source 类的实例，以达到解决兼容性的问题。

（3）接口的适配器模式：在写的一个接口中有多个抽象方法时，必须实现该接口的所有方法，这会造成浪费，因为并不是所有的方法都是需要的，有时只需要某一些，为了解决这个问题，引入了接口的适配器模式。借助于一个抽象类，该抽象类实现了该接口的所有方法，因此只要写一个类，继承该抽象类，重写需要的方法就行了。

总结 3 种适配器模式的应用场景：

（1）类的适配器模式：当希望将一个类转换成满足另一个新接口的类时，可以使用类的适配器模式，创建一个新类，继承原有的类，实现新的接口即可。

（2）对象的适配器模式：当希望将一个对象转换成满足另一个新接口的对象时，可以创建一个 Wrapper 类，持有原类的一个实例。在 Wrapper 类的方法中，调用实例的方法就行。

（3）接口的适配器模式：当不希望实现一个接口中所有的方法时，可以创建一个抽象类 Wrapper，实现所有方法，在写别的类的时候，继承抽象类即可。

2．装饰模式

装饰模式（Decorator）就是给一个对象增加一些新的功能，而且是动态的。它要求装饰对象和被装饰对象实现同一个接口，装饰对象持有被装饰对象的实例。

装饰器模式的应用场景：（1）需要扩展一个类的功能；（2）动态的为一个对象增加功能，而且还能动态撤销。继承不能做到这一点，继承的功能是静态的，不能动态进行增加和删除。

缺点：会产生过多相似的对象，且不易排错。

15.3.2 代理模式和外观模式

1. 代理模式

其实每个模式名称就表明了该模式的作用，代理模式（Proxy）就是多一个代理类出来，替原对象进行一些操作，如在租房子时会去找中介，为什么呢？因为自己对该地区房屋的信息掌握的不够全面，希望找一个更熟悉的人帮助处理，此处的代理就是这个意思。

下面介绍代理模式的应用场景：

如果已有的方法在使用时需要对原有的方法进行改进，此时有两种办法：

（1）修改原有的方法来适应，但违反了“对扩展开放，对修改关闭”的原则。

（2）采用一个代理类调用原有的方法，且对产生的结果进行控制。这种方法就是代理模式。

使用代理模式，可以将功能划分得更加清晰，有助于后期维护。

2. 外观模式

外观模式（Facade）要求一个子系统的外部与其内部的通信必须通过一个统一的对象。外观模式提供了一个高层次的接口，使得子系统更易于使用，也就是说除了这个接口不允许有任何访问子系统的行为发生。

外观模式可以像 Spring 一样，将类和类之间的关系配置到配置文件中，而外观模式就是将这种关系放在一个 Facade 类中，降低了类与类之间的耦合度。

15.3.3 桥接模式和组合模式

1. 桥接模式

桥接模式（Bridge）就是把事物和其具体实现分开，使它们可以各自进行独立的变化。桥接的用意是，将抽象化与实现化解耦，使得二者可以独立变化，如 JDBC 进行连接数据库时，可以在各个数据库之间进行切换，基本不需要动太多的代码，甚至丝毫不用动，原因就是 JDBC 提供了统一接口，每个数据库提供各自的实现，然后用一个数据库驱动程序进行桥接就行了。

2. 组合模式

组合模式（Composite）有时又叫部分-整体模式，在处理类似树形结构的问题时比较方便。

使用场景：将多个对象组合在一起进行操作，常用于表示树形结构，如二叉树、树等。

15.3.4 享元模式

享元模式（Flyweight）的主要目的是实现对象的共享，即共享池。当系统中对象多的时候，可以减少内存的开销，通常与工厂方法模式一起使用。

享元工厂模式负责创建和管理享元单元，当一个客户端请求时，工厂需要检查当前对象池中是否有符合条件的对象，如果有则返回已经存在的对象，如果没有则创建一个新对象。享元模式是超类。一提到共享池很容易联想到 Java 里面的 JDBC 连接池，根据每个连接的特点，不难总结出：适用于共享的一些对象，它们有一些共有的属性，就拿数据库连接池来说，url、driverClassName、username、password 及 dbname 等属性对于每个连接来说都是一样的，所以就适合用享元模式来处理。建一个工厂类，将上述类似属性作为内部数据，其他的作为外部数据。在方法调用时，可作为参数传进来，这样就节省了空间，减少了实例的数量。

15.4 行为模式

行为模式（Behavioral Pattern）共有 11 种：策略模式、模板方法模式、观察者模式、迭代器模式、责任链模式、命令模式、备忘录模式、状态模式、访问者模式、中介者模式、解释器模式，这 11 种模式的关系如图 15-2 所示。

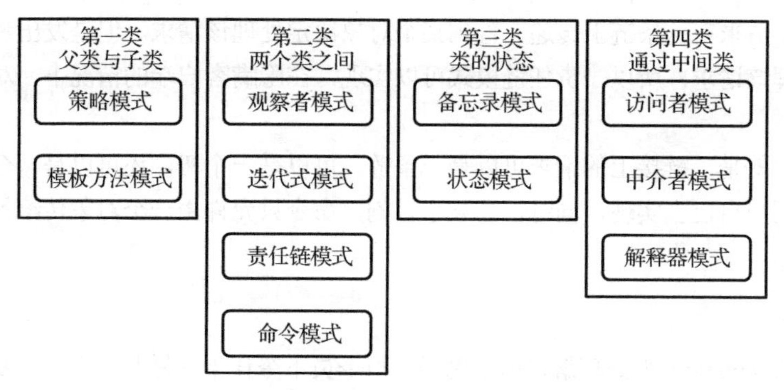

图 15-2　11 种行为模式

15.4.1　策略模式和模板方法模式

1．策略模式

策略模式（Strategy）定义了一系列算法，并将每个算法封装起来，使它们可以相互替换，且算法的变化不会影响到使用算法的客户。

需要设计一个接口，为一系列实现类提供统一的方法，多个实现类实现该接口，可设计一个抽象类（可有可无，属于辅助类），提供辅助函数。

策略模式的决定权在用户，系统本身提供不同算法的实现，如新增或者删除，可对各种算法做封装。因此，策略模式多用在算法决策系统中，外部用户只需要决定用哪个算法即可。

2．模板方法模式

模板方法模式（Template Method）定义了一个操作中的算法框架，而将一些步骤延迟到子类中。使得子类可以不改变一个算法的结构，即可重定义该算法的某些特定步骤。

模板方法模式是指一个抽象类中，有基本方法和模板方法。基本方法是由子类实现的方法，模板方法是实现对基本方法的调度，可完成固定的逻辑。一般模板方法都要加上 final 关键字，不允许被覆写。因此，定义一个类，继承该抽象类，并重写抽象方法，通过调用抽象类，即可实现对子类的调用。

15.4.2　观察者模式、迭代器模式、责任链模式和命令模式

接下来的四个模式，都是类和类之间的关系，不涉及继承。

1．观察者模式

观察者模式（Observer）很好理解，类似于邮件订阅和 RSS 订阅，当浏览一些博客或 Wiki 时，经常会看到 RSS 图标，它的意思是，当你订阅了该文章后，如果后续有更新，就会及时通知你。简单来说，就一句话：是指当一个对象发生变化时，其他依赖该对象的对象都会收到通知，并且会随着变化。对象之间是一种一对多的关系。

2. 迭代器模式

迭代器模式（Iterator）提供一种既能访问一个容器对象中各个元素，而又不暴露该对象内部细节的方法。

顾名思义，迭代器模式就是顺序访问聚集中的对象。它在集合中非常常见。

3. 责任链模式

责任链模式（Chain of Responsibility）是指有多个对象，每个对象持有对下一个对象的引用，这样就会形成一条链，请求在这条链上传递，直到某个对象决定处理该请求。但是发出者并不清楚到底最终哪个对象会处理该请求，所以，责任链模式可以实现，在隐瞒客户端的情况下，对系统进行动态的调整。

此处强调一点就是，链接上的请求可以是一条链，可以是一个树，还可以是一个环，模式本身不约束这个，需要用户自己去实现，同时，在某个时刻，命令只允许由一个对象传给另一个对象，而不允许传给多个对象。

4. 命令模式

命令模式（Command）很好理解，举个例子，司令员下令让士兵做某件事情，从整个事情的角度来考虑，司令员的作用是，发出口令，口令传递到士兵，士兵去执行。这个过程的优点是，三者相互解耦，任何一方都不用去依赖其他人，只需要做好自己的事儿就行，司令员要的是结果，不会去关注士兵到底是怎么实现的。

命令模式的目的就是达到命令的发出者和执行者之间解耦，实现请求和执行分开，熟悉 Struts 的学生应该知道，Struts 其实就是一种将请求和呈现分离的技术，其中必然涉及命令模式的思想。

15.4.3 备忘录模式和状态模式

1. 备忘录模式

备忘录模式（Memento）主要目的是保存一个对象的某个状态，以便在适当的时候恢复对象。例如，假设有原始类 A，A 中有各种属性，A 可以决定需要备份的属性，备忘录类 B 是用来存储 A 的一些内部状态，类 C 呢，就是一个用来存储备忘录的，且只能存储，不能有修改等操作。

2. 状态模式

状态模式（State）的核心思想就是，当对象的状态改变时，同时改变其行为。如 QQ 在线、隐身、忙碌等不同状态，每个状态都对应不同的操作，而且你的好友也能看到你的状态，所以，状态模式就是：（1）可以通过改变状态来获得不同的行为。（2）你的好友能同时看到你的状态变化。

15.4.4 访问者模式、中介者模式和解释器模式

1. 访问者模式

访问者模式（Visitor）指把数据结构和作用于结构上的操作解耦合，使得操作集合可相对自由地演化。访问者模式适用于数据结构相对稳定且算法又易变化的系统，因为访问者模式使算法操作增加变得容易。若系统数据结构对象易于变化，经常有新的数据对象增加进来，则不适合使用访问者模式。访问者模式的优点是增加操作很容易，因为增加操作意味着增加新的访问者。访问者模式将有关行为集中到一个访问者对象中，其改变不影响系统数据结构。访问者模式的缺点是要增加新的数据结构很困难。

简单来说，访问者模式就是一种分离对象数据结构与行为的方法，通过这种分离，可达到为一个

被访问者动态添加新操作，且无须进行其他修改的效果。

该模式适用场景：如果想为一个现有的类增加新功能，需要考虑以下问题：（1）新功能会不会与现有功能出现兼容性的问题？（2）以后会不会再需要添加？（3）如果类不允许修改代码怎么办？面对这些问题，最好的解决方法就是使用访问者模式，因为访问者模式适用于数据结构相对稳定的系统，可以把数据结构和算法解耦。

2．中介者模式

中介者模式（Mediator）用来降低类与类之间的耦合，因为如果类与类之间有依赖关系的话，不利于功能的拓展和维护，只要修改一个对象，其他关联的对象都要进行修改。如果使用中介者模式，只需关心和 Mediator 类的关系，具体类与类之间的关系及调度交给 Mediator 就行了，这有点像 Spring 容器的作用。

User 类统一接口，User1 和 User2 分别是不同的对象，二者之间有关联，如果不采用中介者模式，则需要二者相互持有引用，这样二者的耦合度会很高。为了解耦引入了 Mediator 类，可提供统一接口。MyMediator 为其实现类，里面持有 User1 和 User2 的实例，用来实现对 User1 和 User2 的控制。这样 User1 和 User2 两个对象相互独立，只要保持好和 Mediator 之间的关系即可。

3．解释器模式

解释器模式（Interpreter）主要应用在 OOP 的编译器的开发中。

15.5　本章小结

（1）设计模式的核心是对象及类之间的关系，所以动机部分是最重要的。工厂方法模式的关键是零件的二维分类：kind 维和 family 维中的 Abstract Factory。建造者模式的关键是零件的依赖关系：一个零件的创建和另一个零件的创建有关，其代码如下：

```
builer.part1();
builer.part2();
builer.retreiveProduct();
```

产品 product 的最终结果可能与调用 part1();part2();的顺序有关，所以在不改变程序的情况下，不同的建造过程（调用 part1();part2();...的顺序）生成不同的产品。也就是说，同样的构建过程可以创建不同的表示。

（2）装饰模式是指在不改变接口的前提下，动态扩展对象的功能。代理模式是指在不改变接口的前提下，控制对象的访问。装饰模式强调功能扩展，如 A 对象的 B 方法，运用装饰模式后，在调用 B 方法前后，实现新的功能，此时 B 方法的效果就会与原来的不同。装饰模式的重点在于装饰方法，增加方法的功能，添加装饰对于用户是主动的。

（3）代理模式是指为其他对象提供一种代理以控制对这个对象的访问，又称为委托模式。代理模式是把当前的行为或功能委托给其他对象执行，代理类负责接口限定：是否可以调用真实角色，以及是否对发送到真实角色的消息进行变形处理。它不对被代理角色（被代理类）的功能做任何处理，保证原汁原味的调用。代理模式重在控制，（指代理类控制真实角色的使用），可以在 A 对象（被代理角色）的 B 方法调用前、调用后增加行为，甚至彻底改变方法的行为。代理模式着重点在于对方法的控制，添加行为对于用户是被动的。

（4）适配器模式是指将一个类的接口，转换成客户期望的另外一个接口。适配器让原本接口不兼容的类可以很好地合作。装饰者模式可动态地将责任附加到对象上（因为利用组合而不是继承来实现，而组合是可以在运行时进行随机组合的）。若要扩展功能，装饰者提供了比继承更富有弹性的替代方案（同样地，通过组合可以很好地避免类暴涨，也规避了继承中的子类必须无条件继承父类所有属性的弊端）。

（5）策略模式是指把易于变化的行为分别封装起来，让它们之间可以互相替换，使这些行为的变化独立于拥有这些行为的客户。Command 命令模式是一种对象行为型模式，它主要解决的问题是：将一个请求封装为一个对象，从而可使不同的请求对客户进行参数化；对请求排队或记录请求日志，以及支持可撤销的操作。

从这点来看，策略模式是通过不同的算法做同一件事情，如排序；命令模式是通过不同的命令做不同的事情，常含有（关联）接收者，且隐藏接收者的执行细节，多见于菜单事件。策略模式做相同的事情常含有不同的算法。两者区别在于是否含有接收者，命令模式含有，而策略模式则不含。

习　题　15

1. 什么是单例模式？试编程加以说明。

2. 什么是工厂方法模式？什么是抽象工厂方法模式？试编程加以说明。

3. 什么是代理模式？什么是原型模式？试编程加以说明。

4. 什么是责任链模式？责任链模式的使用场景是怎样的？责任链模式的优点和缺点是什么？试编程加以说明。

5. 什么是观察者模式？观察者模式的使用场景是怎样的？观察者模式的优点和缺点是什么？试编程加以说明。

6. 什么是适配器模式？适配器模式的使用场景是怎样的？适配器模式的优点和缺点是什么？试编程加以说明。

7. 什么是门面模式？门面模式（Facade Design，外观模式）的使用场景是怎样的？门面模式的优点和缺点是什么？试编程加以说明。

8. 什么是装饰模式？装饰模式的使用场景是怎样的？装饰模式的优点和缺点是什么？试编程加以说明。

9. 什么是享元模式？享元模式的使用场景是怎样的？享元模式的优点和缺点是什么？试编程加以说明。

附录 A　Eclipse 的编码问题（包括 ADT）

在 Windows 环境下，ADT 编译器默认编码方式 GBK。一般设置默认为 UTF-8，这样能更好地解决乱码问题，其设置方式如下：

（1）改变整个工作空间的编码格式。

点击“Eclipse”→“Windows”→“Preferences”→“General”→“WorkspaceTypes”→“Other”→“UTF-8”→“OK”，如图 A-1 所示。

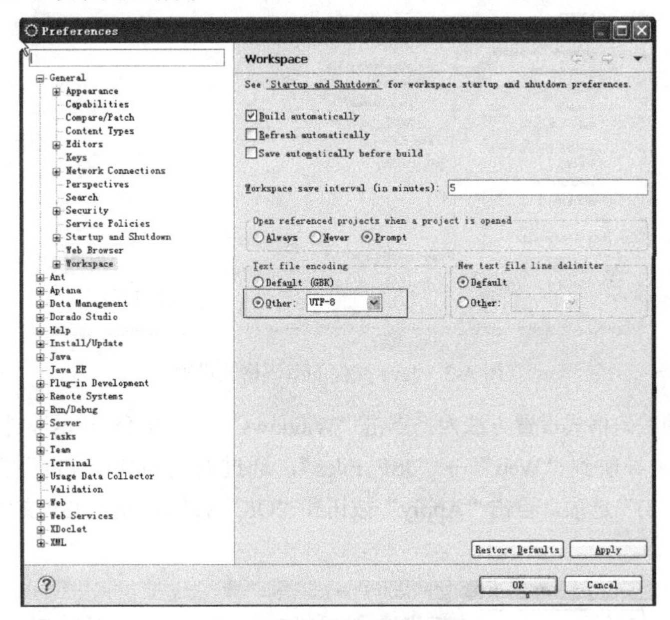

图 A-1　工作空间的编码格式设置

（2）项目范围的编码格式设置。

点击“Project”→“Properties”→“General”→“Resource”→“Other”→“UTF-8”→“OK”，如图 A-2 所示。

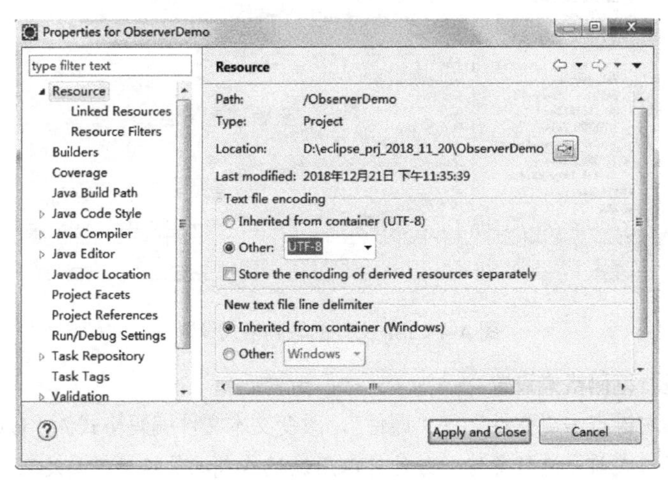

图 A-2　项目范围的编码格式设置

（3）某类型文件的编码格式设置。

点击“Eclipse”→“Windows”→“Preferences”→“General”→“Content Types”→在右侧找到要修改的文件的类型（JAVA 等），在下方的 Default encoding 输入“UTF-8”，再选择“Update”→“OK”。

如 Java 文件的编码格式设置方法为：选择“Windows”→“Preferences”，打开“首选项”对话框，通过左侧导航树导航到“General”→“Content Types”，通过右侧 Context Types 树，打开“Text”→“Java Source File”，在下方的 Default encoding 输入框中输入“UTF-8”，点击“Update”按钮，设置 Java 文件编码为 UTF-8，如图 A-3 所示。

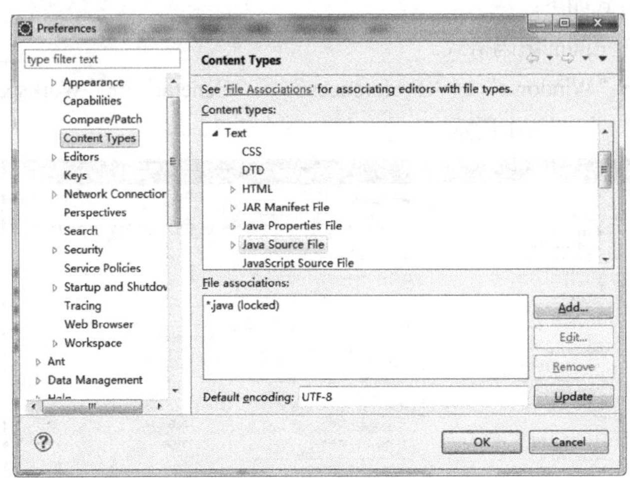

图 A-3　Java 文件的编码格式设置

又如 JSP 文件的编码格式设置方法为：点击“Windows”→“Preferences”，打开“首选项”对话框，通过左侧导航树导航到“Web”→“JSP Files”，通过右侧的 Encoding 的下拉框选择“ISO 10646/Unicode (UTF-8)”选项，点击“Apply”按钮和“OK”按钮，设置 JSP 文件编码为 UTF-8，如图 A-4 所示。

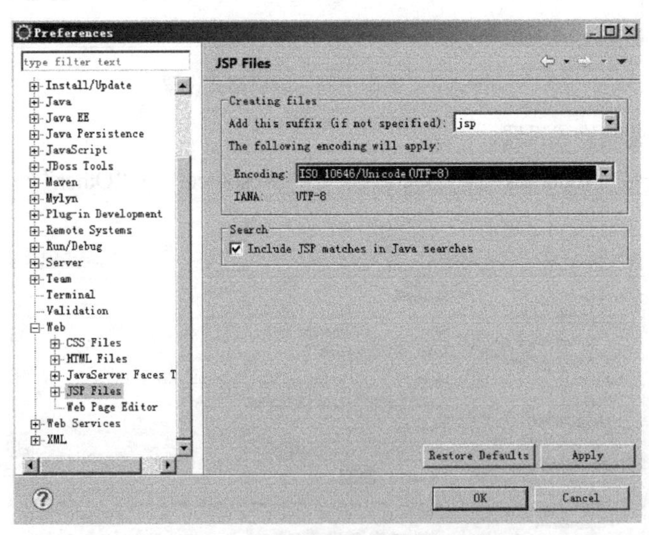

图 A-4　JSP 文件的编码格式设置

（4）单个文件进行编码格式设置。

在包资源管理器右键点击“文件”→“属性”，改变文本文件编码格式为“UTF-8”。

注意：改变编码格式前应进行备份，另外可以用编码格式批量转换工具处理。

附录 B　Eclipse 自动部署项目到 Tomcat 的 webapps 目录

（1）配置 Server 的 Runtime Enviroments，添加 Apache Tomcat v8.0。

在 Eclipse 中选择“Windows”→“Preferences”，再选择左侧 Server 下的“Runtime Enviroments”选项，如图 B-1 所示。

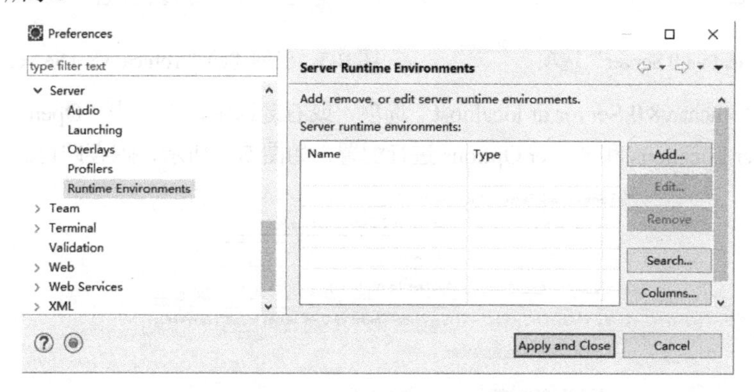

图 B-1　Server Runtime Enviroments

点击“Add”按钮，添加 Apache Tomcat v8.0，再配置其安装路径，如图 B-2 所示。

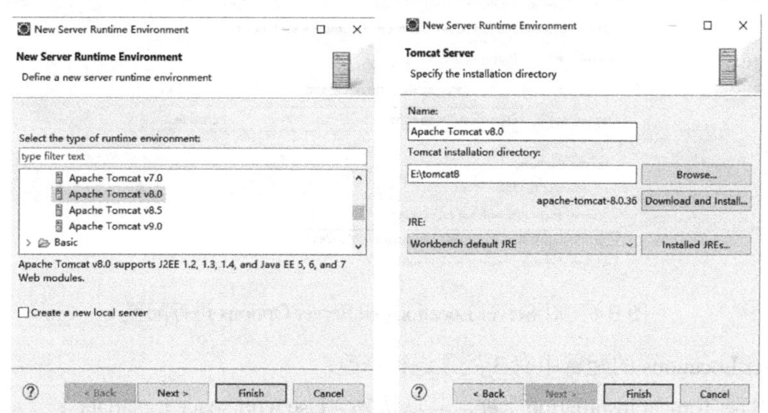

图 B-2　添加 Tomcat 8 并配置

（2）创建运行的 Server，并对 Server Locations 和 Server Options 进行配置。

在 Eclipse 中，选择“Windows”→“Show View”→“Servers”，打开如图 B-3 所示的界面。

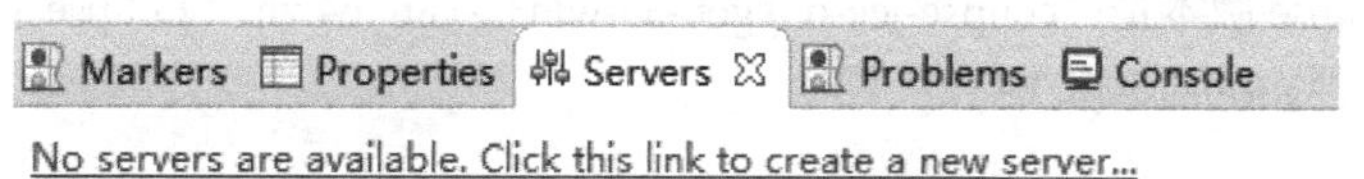

图 B-3　Servers 标签

点击链接，将弹出“创建 Server”对话框，如图 B-4 所示，选择“Tomcat v8.0 Server”选项；点击“Next”按钮，接着再点击“Finish”按钮即可。这时可得到如图 B-5 所示的界面，即已创建“Tomcat v8.0 Server at localhost”。

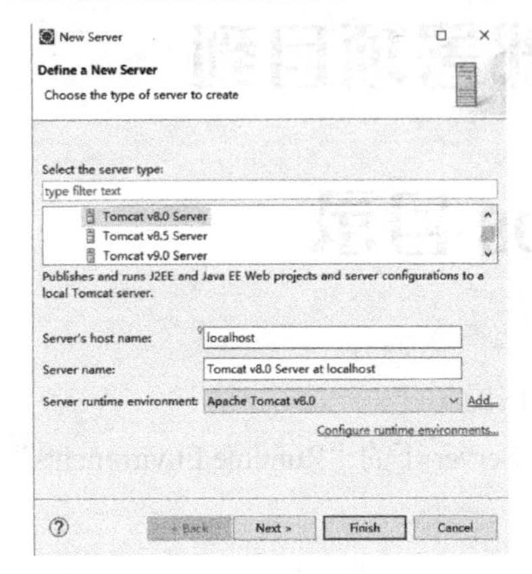

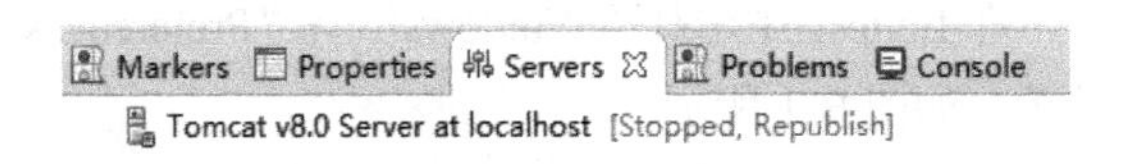

图 B-4　选择“Tomcat v8.0 Server”选项

图 B-5　已创建的“Tomcat v8.0 Server at localhost”选项

鼠标双击“Tomcat v8.0 Server at localhost”选项，或右键选中，再点击“Open”按钮。在弹出的界面中，对 Server Locations 和 Server Options 进行配置，如图 B-6 所示，保存即可。

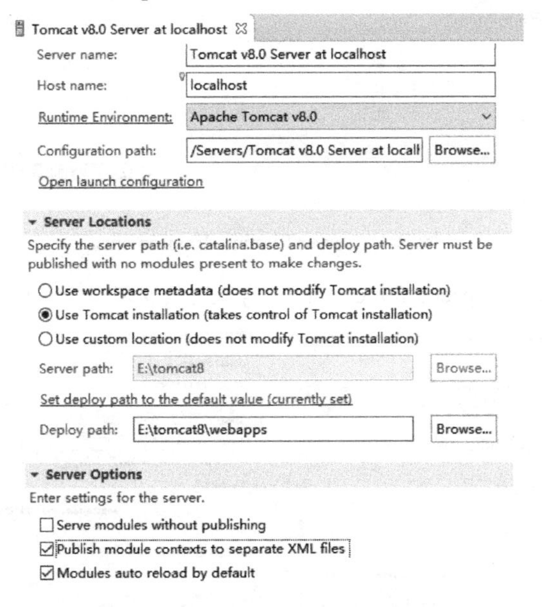

图 B-6　对 Server Locations 和 Server Options 进行配置

其中，Server Locations 的配置中有 3 处须进行修改：

（1）选中“Use Tomcat installation”项，其默认为“Use workspace metadata”；

（2）选中“Use Tomcat insallation”项，Server Path 会自动修改为 Tomcat8 实际的安装位置；

（3）Deploy Path 处须修改为“Tomcat 根目录\webapps”，其默认为“wtpwebapps”。

在 Server Options 的配置中，勾选其复选框的第 2 项和第 3 项即可。

这里使用的 Eclipse 版本是：eclipse-jee-oxygen-3a-win32-x86_64.zip、Eclipse Java EE IDE for Web Developers、Oxygen.3a Release (4.7.3a)。

附录C ADB命令

ADB（Android Debug Bridge）起到调试桥的作用。通过ADB可以在Eclipse的DDMS中调试Android程序。借助ADB工具除了可以管理设备或手机模拟器的状态，还可以进行很多操作，如安装软件、系统升级、运行shell命令等。ADB就是连接Android手机与PC端的桥梁，可以让用户在计算机上对手机进行全面的操作。

（1）显示系统中全部Android平台：android list targets。

（2）显示系统中全部AVD（模拟器）：android list avd。

（3）创建AVD（模拟器）：android create avd -n <avd名称> -t <Android版本>。

（4）启动模拟器：emulator -avd <avd名称>。

（5）删除AVD（模拟器）：android delete avd -n <avd名称>。

（6）创建SDCard：mksdcard <size> <file>。

（7）显示当前运行的全部模拟器：adb devices。

（8）对某个模拟器执行命令：abd -s <模拟器编号> <命令>。

（9）安装APK程序：adb install [-r] [-s] <file>，其中，-r表示重新安装该APK包，-s表示将APK包安装到SD卡上（默认是安装到内部存储器上），<file>表示要安装的APK包。

（10）获取模拟器中的文件：adb pull <remote> <local>。

（11）向模拟器中写文件：adb push <local> <remote>。

（12）进入模拟器的shell模式：adb shell。

（13）启动Android SDK Manager（SDK、文档、实例等下载管理器）：Android。

（14）卸载apk包：adb uninstall [-k] <package>，其中，-k表示只删除该应用程序，但保留该程序所用的数据和缓存记录。或者：

```
adb shell
cd data/app
rm apk包
exit
```

（15）查看ADB命令的帮助信息：adb help。

（16）在命令行中查看LOG信息：adb logcat -s <标签名>。

（17）删除系统应用：

```
adb remount （重新挂载系统分区，使系统分区重新可写）
adb shell
cd system/app
rm *.apk
```

（18）访问数据库SQLite3：

```
adb shell
sqlite3
```

参考文献

[1] [美]Brian Goetz，等. Java 并发编程实战[M]. 童云兰，等译. 北京：机械工业出版社，2012.

[2] [美]Adam，等. Android Studio 实战：快速、高效地构建 Android 应用[M]. 靳晓辉，等译. 北京：清华大学出版社，2016.

[3] 郭霖. 第一行代码——Android（第 2 版）[M]. 北京：人民邮电出版社，2016.

[4] 罗升阳. Android 系统源代码情景分析（第 3 版）[M]. 北京：电子工业出版社，2017.

[5] [美]Bill Phillips，等. Android 编程权威指南（第 3 版）[M]. 王明发译. 北京：人民邮电出版社，2017.

[6] 秦小波. 设计模式之禅[M]（第 2 版）[M]. 北京：机械工业出版社，2015.

[7] 何红辉. Android 开发进阶：从小工到专家[M]. 北京：人民邮电出版社，2016.

[8] 明日学院. Android 开发从入门到精通（项目案例版）[M]. 北京：中国水利水电出版社，2017.

[9] 徐宜生. Android 群英传神兵利器. 北京：电子工业出版社，2016.

[10] 许晓斌. Maven 实战[M]. 北京：机械工业出版社，2011.

[11] 任玉刚. Android 开发艺术探索[M]. 北京：电子工业出版社，2015.

[12] 何红辉，关爱民. Android 源码设计模式解析与实战（第 2 版）[M]. 北京：人民邮电出版社，2017.

[13] [美]lan G. Clifton. 基于 Material Design 的 Android 用户界面设计[M]. 郑磊译. 北京：电子工业出版社，2016.

[14] 李刚. 疯狂 Android 讲义（第 4 版）[M]. 北京：电子工业出版社，2019.

[15] 欧阳燊. Android Studio 开发实战：从零基础到 App 上线（第 2 版）[M]. 北京：清华大学出版社，2018.

[16] 李宁宁，等. 基于 Android Studio 的应用程序开发教程[M]. 北京：电子工业出版社，2016.

[17] [美]Ken Kousen. 巧用 Gradle 构建 Android 应用[M]. 李建译. 北京：电子工业出版社，2017.

[18] [美] Erich Gamma，等. 设计模式：可复用面向对象软件的基础[M]. 李英军，等译. 北京：机械工业出版社，2013.